U0496421

人工智能
本科专业知识体系
与课程设置 第2版

郑南宁 ◎ 主编

清华大学出版社
北京

内 容 简 介

本书针对高等学校人工智能本科专业人才培养的专业内涵、定位和知识体系，设置了数学与统计、科学与工程、计算机科学与技术、人工智能核心、认知与神经科学、先进机器人技术、人工智能与社会、人工智能工具与平台等课程群，重点介绍了八大课程群中各门课程的概况和知识点，为培养具有科学家素养的工程师奠定知识和能力的基础。

本书可为各类学校人工智能专业构建宽口径和学科交叉的课程体系提供参考和引导示范，也可为研究生相关课程体系建设和专业学习提供指引。

本书封面贴有清华大学出版社防伪标签，无标签者不得销售。

版权所有，侵权必究。举报：010-62782989，beiqinquan@tup.tsinghua.edu.cn。

图书在版编目(CIP)数据

人工智能本科专业知识体系与课程设置/郑南宁主编.—2版.—北京：清华大学出版社，2023.10（2025.6重印）

ISBN 978-7-302-63814-8

Ⅰ.①人… Ⅱ.①郑… Ⅲ.①人工智能-课程设置-高等学校 Ⅳ.①TP18

中国国家版本馆 CIP 数据核字(2023)第 106012 号

责任编辑：王　芳
封面设计：刘　键
责任校对：韩天竹
责任印制：曹婉颖

出版发行：清华大学出版社
　　　网　　址：https://www.tup.com.cn，https://www.wqxuetang.com
　　　地　　址：北京清华大学学研大厦 A 座　邮　编：100084
　　　社 总 机：010-83470000　　邮　购：010-62786544
　　　投稿与读者服务：010-62776969，c-service@tup.tsinghua.edu.cn
　　　质量反馈：010-62772015，zhiliang@tup.tsinghua.edu.cn
　　　课件下载：https://www.tup.com.cn，010-83470236
印 装 者：三河市铭诚印务有限公司
经　　销：全国新华书店
开　　本：185mm×230mm　　印　张：28.25　　字　数：623 千字
版　　次：2019 年 9 月第 1 版　2023 年 10 月第 2 版　　印　次：2025 年 6 月第 4 次印刷
印　　数：3001～3500
定　　价：139.00 元

产品编号：100556-01

知识的变与不变
（第二版代序）

时间过得真快，仿佛一转身，五年就已过去。本书内容的实践始于 2018 年，成书于 2019 年夏，如今这本书构建的人工智能本科专业知识体系和课程设置，经历了完整两届人工智能本科专业学生培养的教学实践，它伴随着我们人工智能本科专业人才培养的每一步足迹，见证了我们从最初的探索，到如今的再版修订。

近年来，人工智能领域正经历着深刻而又令人激动的变化，犹如璀璨的繁星，闪耀出一道道引人瞩目的创新之光。自动驾驶技术从实验室走向了真实场景，语言大模型为我们解答了无数复杂的问题，生成式 AI 创造了一幅又一幅令人惊叹的画面，现代物理学和神经科学与人工智能的交叉融合正在开创一种新的研究范式，将为人工智能领域带来更多新的发现和创新。面对这些惊艳的进展，我们需要对该书的内容进行必要的调整、更新和修订，以保持人工智能本科专业知识体系与时俱进，使其课程设置跟上时代的进步。将人工智能领域的最新知识和前沿技术融入教学内容，我们可以有更全面、更深入、更具时代性的课程设置，这无疑会为学生打开一扇通向未来的大门。

人工智能的发展正在触动着人类社会的每一个角落，教育也由此迎来新的机遇和变革。适应"变"，并秉承"不变"，是人工智能教育所特有的探索和挑战。事实上，我们每一位教师和学生都是在"知识的变与不变"的学习过程中而成长起来。知识的"变"寓示着知识的不断更新，而"不变"则凝聚着深厚的学科基础和科学精神。

人才培养的根本使命促使我们深刻思考：如何在"知识的变与不变"中，实现研究性的教、探究性的学、创新性的做。只有这样，我们才能超越单纯的教与学的角色，向着更高、更多元的创新价值链迈进，从而成为真正的"创新者"。不积跬步，无以至千里；不积小流，无以成江海。不断地学习吸收新知识，不断地巩固已学到的知识，并运用到实践中，才能在未来创新之路上找到我们自身的位置。

没有永恒不变的学科，也没有永恒不变的教育，教育即成长，成长即变化。真正的教育不仅关乎知识的传授，更在于激发学生的思维活力，培养学生持续学习和热情探索的精神。教学过程不仅仅是知识传授的过程，更是教师和学生共同探索、发现、创新的过程。在这个过程中，教师和学生都是学习的主体，都在相互影响和相互促进中实现自我成长。这就要求教师自身保持科学研究的热情，终身学习，并引导学生洞悉事物变化的本质，将"知识的学习"与"动手的实践"相结合培养学生的创新思维能力，将科学精神与科学素养的培养潜移默化地融入到教学过程中，进而点燃学生创造的热情。只有这样，教育才能更加丰富、更有意义地传递知识，更深刻地引导学生成长，使他们在不断变化的世界中实现真正的发展。

我们再次启航，驶向充满变化与无限可能的人工智能广阔领域。

郑南宁教授

2023 年 7 月 23 日于西安

教育也是一种创造
（第一版代序）

在本书付梓之际，窗外细雨蒙蒙，梧桐滴翠，沁人心脾。自本书酝酿至今，迁思回虑、推敲琢磨，历时两年有余，可谓"千呼万唤始出来"，这本书的背后浸润着我们三十余年来在人工智能领域的研究探索和人才培养的实践。编写过程中一次次思想碰撞，一遍遍伏案执笔，反复"自我教育和学习"的洗礼，更深感教育是一个缓慢而优雅的过程；正是这样的"缓慢"诠释了我们对构建全新一流的人工智能专业知识体系与课程设置的认识。

问渠那得清如许？为有源头活水来。教育的基本问题是培养什么样的人，怎样培养人，而专业知识体系构建与课程设置是为人才培养提供保障的重要载体之一。专业知识体系构建与课程设置是一个不断完善的过程，只有坚持实践、与时俱进，使其知识体系的建设和教育质量的提高成为一种创造的追求，才能使我们通过教育的实践培养出优秀的人才。潜心教育与课程建设不仅是对教师科学生命的一种延续，也是不断焕发教师科学青春的一剂良方。

人工智能是一门新兴的技术科学，具有多学科综合、高度复杂的特征以及渗透力和支撑性强等特点，它涉及信息科学、认知科学、脑科学、神经科学、数学、心理学、人文社科与哲学等多学科的深度交叉融合。这是我们构建人工智能专业知识体系和课程设置的基本指导思想，同时我们在专业知识体系的构建中注重"脑"（Mind）与"手"（Hand）相结合，即"知识的学习"与"动手的实践"相融相长，为学生今后能成为"大科学家"、成为具有"科学家素养"的工程师和人工智能相关领域的领军人才奠定知识和能力的基础。

教育者，非为已往，非为现在，而专为将来。人工智能的人才培养应重视思考未来的人工智能需要从哪些学科获得灵感。在集体智慧的交互、精益求精的斟酌与研讨的基础上，形成了人工智能专业知识体系中的数学与统计、科学与工程、计算机科学与技术、人工智能核心、认知与神经科学、先进机器人技术、人工智能与社会、人工智能工具与平台等八大课程群。通过这些理论、方法、工具和系统等课程设置以及高水平的跨学科师资团队，培养人工智能专业学生具有创新、创业、跨学科交叉、全球化和伦理道德等思维能力，有助于学生在毕业以后拥有更强的可塑性和更广阔的发展空间，在各行各业担当起创新、创业的重要角色，成为人工智能领域的高层次人才。

  以学生为本,以创造为源。教育不是注满一桶水,而是点燃一把火,打开一扇门。期望在这本书指导下的教学工作能点燃学生内心探索人工智能奥秘的火种,帮助学生走进未来,并将在未来某个时刻,他们能放射出更加灿烂的光芒。

郑南宁教授

2019 年 7 月 22 日于西安

PREFACE
第二版前言

自本书第一版面世以来,被国内诸多高校作为人工智能本科专业和相关学科研究生培养的主要参考书籍之一。在深感欣慰之余,我们深知,本书中仍存在一些有待提高和完善之处。近年来,不少热心读者、尤其是高等学校的老师们对本书提出了许多很好的建议。为此,我们从 2022 年开始,着手本书的第二版修订工作。我们尤为关注来自学生和在教学一线老师们对课程设置和教学内容的意见和建议,其中不乏真知灼见,给我们留下深刻印象。我们多次召开本书第二版修订的专门会议,就知识体系、课程设置、教学大纲和知识点与各课程主讲老师交流,并倾听学生在学习过程中的体会和感想。

正是广大读者充满热忱的支持和期待,激励着我们倾心投入此次修订工作。第二版修订和课程调整的主要依据来自人工智能学科最新的学术研究和技术发展、教师的课堂教学实践和学生的反馈等。在本书的第二版中,为让学生有足够的探索时间和空间,促进学生探究性地学习,避免低效的课程教学,我们减少了专业课程学习的总学分数,对课程的设置和内容进行了优化,以使人工智能本科专业的知识体系和课程设置跟上时代的进步。

当前,人工智能(AI)与物理学的交叉融合正在开创一种全新的研究范式,我们由此对大学物理课程的内容设置进行了改革,为人工智能本科专业学生专门开设"现代物理与人工智能(Physics for AI)"课程,这也是对大学物理教学改革的探索。与现在的普通大学物理课程不同,该课程在注重经典物理知识衔接的基础上,学习和探究先进物理理论在 AI 领域的实际应用和研究,理解物理学与人工智能之间存在的深刻交叉和互动,培养学生运用物理学知识解决 AI 实际问题的能力,为探究人工智能未来发展的新方向奠定现代物理知识基础。同时,还增设了其他新的课程,如"先进自动驾驶技术与系统""生成式 AI 与大语言模型""创新设计思维"等课程;此外,对其他 21 门课程的内容进行了细致的修订和调整,以保持我们的知识体系与时俱进,课程设置更加合理,让学生能更好地了解并掌握人工智能领域的学科基础理论、最新知识和技术。

在本书第二版的修订过程中,我们心中时时铭记第一版序"教育也是一种创造"中所言"教育者,非为以往,非为现在,而专为将来"。我们学院的老师们亲身践行这一教育的核心内涵,尽力使本书的内容能够传承经典、跨越时代,并与时俱进、求新思变,期待能为我国人工智能创新型人才培养贡献应有的力量！

　　本书不足之处在所难免，殷切希望广大教师、学者同仁和读者不吝赐教与批评指正！

《人工智能本科专业知识体系与课程设置》
第二版修订工作组
2023 年 8 月

主要修订内容

　　对本书第一版的主要修订如下：

1. 新增 4 门课程

　　(1)"现代物理与人工智能"课程与普通"大学物理"课程不同，在注重与经典物理知识衔接的基础上，学习和探究先进物理理论在 AI 领域的实际应用和研究，主要包括经典物理概述、量子力学与量子计算、相对论与电磁学、统计物理和机器学习、物理驱动的 AI、非线性动力学与复杂系统、虚拟现实中的物理模拟等内容，培养学生运用物理学知识解决 AI 实际问题的能力，为探究人工智能未来发展的新方向奠定现代物理知识基础，为学生未来在人工智能领域的前沿研究和工作做好准备。

　　(2)"生成式 AI 与大语言模型"课程着重介绍生成式 AI 和大语言模型的核心概念、原理和应用领域以及大语言模型的优势和局限性，并通过实践完成基于大语言模型的文本生成、对话系统设计等任务，以及思考和讨论生成式 AI 对社会和伦理的影响。

　　(3)"先进自动驾驶技术与系统"课程全面、系统地讨论了自动驾驶的传感器系统、交通场景感知与理解、地图构建与定位、行为决策与规划、控制系统、仿真测试与验证系统、自动驾驶的操作系统等核心算法、关键技术及其系统实现。

　　(4)"创新设计思维"课程通过对一个具体的移动智能体完成从设计到构建的全过程，帮助学生理解、应用和初步掌握创新设计思维的五个阶段——"共情、定义、构想、原型、验证"，在实践中提升学生解决复杂问题和工程创新的素养和能力。

2. 取消 5 门课程

　　(1)"信息论"（其相关内容并入"现代物理与人工智能"课程）。
　　(2)"大学物理"（其相关内容整合至"现代物理与人工智能"课程）。

（3）"虚拟现实与增强现实"（其相关内容整合至"现代物理与人工智能"课程中的物理模拟与虚拟现实章节）。

（4）"神经生物学与脑科学"（其相关内容整合至"计算神经工程"课程）。

（5）"仿生机器人"。

3. 更改名称的3门课程

（1）原"机器学习工具与平台"更名为"深度学习工具与平台"。

（2）原"三维深度感知"更名为"3D深度感知"。

（3）原"强化学习与自然计算"更名为"强化学习"。

上述3门课程的内容也做了更新。

4. 内容重组与整合的3门课程

（1）原"人工智能的现代方法Ⅰ：问题表达与求解"的内容经修订，独立设置为"人工智能概论"课程。

（2）原"人工智能的现代方法Ⅱ：机器学习"的内容经修订，独立设置为"机器学习"课程。

（3）原"无人驾驶平台"相关内容整合至"先进自动驾驶技术与系统"课程。

5. 专业综合实验课的调整

（1）原"游戏AI设计与开发"课程调整为"游戏AI设计与开发"实验课。

（2）取消原"自主无人系统实验"课（其相关内容整合至"机器人导航技术"实验课）。

（3）取消原"虚拟现实与仿真实验"课（其相关内容整合至"现代物理与人工智能"课程中的实验）。

6. 本书第二版22门课程的主要修订

（1）"计算机科学与人工智能的数学基础"课程删除了原有"点集的勒贝格测度"、"线性规划"等章节，增加了"三维空间刚体运动"、"多元相关性与条件不相关性"等章节，微调了数值计算的部分内容，并将优化理论调整为"优化基础"、"无约束优化"、"有约束优化"等3章。

（2）"电子技术与系统"课程将原有第二部分拆分为"数字电路与系统"和"计算机体系结构基础"两部分，增加了"绪论"以及数字电路、计算机体系架构部分的课内实验。

（3）"现代控制工程"课程大幅更新调整了原有内容，如第二至八章分别修改为"连续控制系统的数学模型"、"连续控制系统的时域分析与设计"、"连续控制系统的根轨迹法"、"连续控制系统的频率法"、"连续控制系统的状态空间法"、"离散控制系统分析与综合"和"自动控制理论的应用举例"等。

（4）"计算机程序设计"课程增加了"Python 编程拓展"以提升学生 C++ 和 Python 两种计算机编程语言的实践能力。

（5）"计算机体系结构"课程增加了"通用图形处理器"以替换原有"仓库级计算机"。

（6）"3D 计算机图形学"课程调整了知识点，重新设计实验环节，并更新了参考文献。

（7）"智能感知与移动计算"课程简化了知识点的表述方式，更新了基本内容和参考文献。

（8）"人工智能概论"课程对原"人工智能的现代方法Ⅰ：问题表达与求解"进行了全面修订，内容更新调整为"知识表示"、"知识图谱"、"搜索求解"、"博弈搜索"、"确定性推理"、"不确定性推理"和"多智能体系统"等章节。

（9）"机器学习"课程对原"人工智能的现代方法Ⅱ：机器学习"进行了调整，围绕机器学习全面修改了内容，增加了"线性模型"、"注意力模型"等内容。

（10）"自然语言处理"课程增加了"基于神经网络的语言模型与大语言模型"和"语义分析"等内容。

（11）"计算机视觉与模式识别"课程删除了原有"视觉生理学与视觉认知基础"和"图像合成方法"，并在"视觉认知与识别方法"中增加了"Transformer"。

（12）"强化学习"课程内容源自原"强化学习与自然计算"中的强化学习部分，修正了部分概念名词，删除了原有的"基于证据推理的多智能体分布式决策"和"多智能体系统的协同控制"，而且原自然计算的内容调整至"机器学习课程"和"多智能体与人机混合智能"课程。

（13）"人工智能的科学理解"课程在"主题 1 控制论与人工智能"中增加了"智能体的自组织与分布式认知"；在"主题 2 智能系统的信念"中删除了原有"科学方法"，增加了"可信的通用人工智能系统"。

（14）"认知心理学基础"课程删减了原有认知神经科学相关章节和实验以及感觉适应、运动知觉、意识状态、前摄及后摄干扰、错误记忆、概念形成等知识点，更新调整了相关知识点的内容编排。

（15）"计算神经工程"课程围绕峰电位、脑电、功能磁共振等 3 种脑信号调整了相关处理与分析方法的内容编排，并融入了原有"神经生物学与脑科学"的相关内容，增加了"神经形态计算"。

（16）"机器人学基础"课程删除了原有"并联机器人"、"串联机器人"等章节，调整了"机械臂的机构设计"、"并联机器人运动学实验"等内容，强化了通用的机器人分析建模等内容，更新了参考文献，并增加了"逆向运动学的数值解"等内容。

（17）"多智能体与人机混合智能"课程对原课程内容做了更新和精简，分"受脑认识启发的混合增强智能"和"人机协同的混合增强智能"两个主题，增加了"蚁群算法"、"生物智能与脑认知过程的信息处理"等内容。

（18）"认知机器人"课程从认知机器人架构、视觉、导航、人机交互、知识表示与推理等方面重新调整内容，补充了认知机器人领域的最新研究进展。

（19）"深度学习工具与平台"课程删减了原"机器学习工具与平台"课程中的 MATLAB、C++、Python 等编程语言相关内容以及 TensorFlow、keras、MXNet 和 CNTK 等开源框架内容，新增了深度学习第一、二代框架的演进及特点，以及 AutoML、模型压缩等相关内容。

（20）"人工智能芯片设计导论"课程缩减了有关 RISC-V 的知识点，更加聚焦于数据驱动的 AI 计算架构。

（21）"脑信号处理"实验课将原"基于 EEG 的脑机接口"和"基于 fMRI 的视觉编解码"调整为"基于一般线性模型的大脑激活区分析"、"基于 fMRI 数据的图像分类"、"fMRI 功能连接网络的构建"、"脑电视觉刺激设计及呈现"、"脑电的采集及伪迹"、"情绪脑电识别"、"运动想象解码"和"基于脉冲神经网络的数字分类"等 8 个具体实验。

（22）"游戏 AI 设计与开发"实验课替代了原"游戏 AI 设计与开发"课程，以丰富的游戏平台为载体提升人工智能算法开发和动手实践能力。

PREFACE
第一版前言

40亿年以来,地球上的生命遵循着最基本的自然进化法则缓慢地演化。然而,随着人工智能等科学技术的发展,人类历史上将会出现按照有机化学规则演变的生命和无机的智慧生命并存的形态,或者说人类有可能利用计算机和人工智能去设计生命。目前,尽管我们无法描述人工智能技术在未来几十年后会形成什么样的具体形态,但可以确定的是,人工智能技术的发展一定会给人类带来革命性的变化,并且这个变化一定会远超人类过去千年所发生的变化。人工智能将成为未来30年影响最大的技术革命。

近年来,布局发展人工智能已经成为世界许多国家的共识与行动。中国高度重视人工智能的发展,习近平总书记多次重要讲话强调加快人工智能发展的重要性和紧迫性,强调"人工智能是新一轮科技革命和产业变革的重要驱动力量,加快发展新一代人工智能是事关我国能否抓住新一轮科技革命和产业变革机遇的战略问题""要加强人才队伍建设,以更大的决心、更有力的措施,打造多种形式的高层次人才培养平台,加强后备人才培养力度,为科技和产业发展提供更加充分的人才支撑"。

自2017年起,人工智能已连续三年写入《政府工作报告》,加快新一代人工智能发展已成为国家重大战略。2017年7月,国务院正式发布《新一代人工智能发展规划》,将我国人工智能技术与产业的发展上升为国家重大发展战略,提出要"完善人工智能教育体系"。2018年4月,为贯彻落实国家《新一代人工智能发展规划》,教育部印发了《高等学校人工智能创新行动计划》,明确提出了设立人工智能专业、推动人工智能领域一级学科建设、建立人工智能学院以及完善人工智能领域人才培养体系等重要任务。今年3月,教育部批准全国35所大学建设首批"人工智能"新本科专业。

早在1986年,西安交通大学就成立了"人工智能与机器人研究所"(简称人机所),该研究所依托"模式识别与智能系统"国家重点二级学科开展教学、科研和人才培养工作,并逐步形成了独特的育人文化,培养了一批学术界和产业界的领军人才,成为高水平创新人才培养的重要基地。30余年来,人机所始终坚持人工智能,特别是计算机视觉与模式识别的应用基础理论研究,并积极与国家重大需求相结合,培养了以人工智能领域世界一流科学家孙剑为代表的一大批优秀人才,取得了一系列重大科研成果,获国家自然科学基金委员会"创新研究群体科学基金"首批资助，成为"视觉信息处理与应用"国家工程实验室、教育部和国家外国专家局"机器视觉与认知计算"高等学校学科创新引智基地、教育部混合增强智能示范中心、"认知科学与工程"国际研究中心的支撑单位，为西安交通大学在人工智能领域奠定了坚实的基础，有力支撑了我国人工智能发展。

为满足国家重大战略需求，服务国家和地方社会经济发展，紧抓人工智能发展的重大历史机遇，充分发挥西安交通大学在人工智能领域的科学研究和人才培养的优势，加快人工智能创新人才培养，2017年由中国工程院院士郑南宁教授领衔，在学校的大力支持下，创办了"人工智能拔尖人才培养试验班"，探索培养人工智能方向本科生，并于2018年招收第一批本科生。2018年11月在人机所的基础上成立了人工智能学院，2019年3月获教育部首批"人工智能"本科新专业建设资格。

人工智能具有多学科综合、高度复杂的特征以及渗透力和支撑性强等特点，其课程体系必须以学科交叉为重要指导思想。因此，人工智能人才培养具有高度的挑战性。人工智能专业的建设目标是培养扎实掌握人工智能基础理论、基本方法、应用工程与技术，熟悉人工智能相关交叉学科知识，具备科学素养、实践能力、创新能力、系统思维能力、产业视角与国际视野，未来能在我国人工智能学科与产业技术发展中发挥重要作用，并有潜力成长为一流的人工智能领域人才。

自2017年初开始，历经两年多的深入研讨和推敲，借鉴国际一流大学人工智能方向的课程设置和培养理念，最终形成了包括"数学与统计""科学与工程""计算机科学与技术""人工智能核心""认知与神经科学""人工智能与社会""先进机器人技术""人工智能工具与平台"等八大课程群计，共39门课程，其中必修27门、选修12门（完成至少9学分）。在实践方面，特设了"专业综合性实验"课程群，培养学生综合运用所学知识动手解决实际问题的能力，使学生培养达到"脑"与"手"相结合的目标。

经过严格遴选，教学团队以西安交通大学人工智能学院教师为主体，并特聘校内其他院系的优秀教师以及国内外知名高校、科研机构和人工智能企业的知名专家加入。例如，"计算机科学与人工智能的数学基础"将由具有数学专业背景，并从事人工智能领域研究的教授讲授；"理论计算机科学的重要思想"是由在国内理论计算机领域享有著名声誉的南京大学教授主讲；"认知心理学基础"由陕西师范大学心理学院知名教授讲授；"人工智能的科学理解"将由在科研与教学有着丰富实践和深刻洞见的资深教授讲授；微软亚洲研究院的专家也将参与"机器学习工具与平台"的授课等。

在当前第四次技术革命的背景下，中国不仅需要培养出更多的人工智能领域人才，更要培养出高层次乃至世界顶尖的人工智能人才。本书作为人工智能人才培养极为关键的本科阶段课程体系的指导，期望通过该课程体系引导的教学工作，使学生能掌握扎实的人工智能基础理论与方法，拥有特色的学科交叉背景，为他们今后能成长为人工智能领域的科学家、工程师，以及相关领域的创新、创业的高层次人才奠定良好的知识与技能的基础。

本书难免还存在一些缺点和不足，殷切希望广大国内同仁和读者批评指正。

西安交通大学人工智能学院

本科专业知识体系建设与课程设置工作组

2019 年 7 月

《人工智能本科专业知识体系与课程设置》
第二版修订工作组

组　长
　　郑南宁（西安交通大学人工智能学院）

副组长
　　辛景民（西安交通大学人工智能学院）

成　员（按姓氏笔画排序）
　　刘妹琴（西安交通大学人工智能学院）
　　刘剑毅（西安交通大学人工智能学院）
　　孙宏滨（西安交通大学人工智能学院）
　　任鹏举（西安交通大学人工智能学院）
　　杜少毅（西安交通大学人工智能学院）
　　汪建基（西安交通大学人工智能学院）
　　杨　勐（西安交通大学人工智能学院）
　　张雪涛（西安交通大学人工智能学院）
　　张　璇（西安交通大学人工智能学院）
　　陈霸东（西安交通大学人工智能学院）
　　徐林海（西安交通大学人工智能学院）
　　魏　平（西安交通大学人工智能学院）

参加第二版修订的其他人员（按姓氏笔画排序）
　　丁　宁（西安交通大学人工智能学院）
　　马永强（西安交通大学人工智能学院）
　　刘龙军（西安交通大学人工智能学院）
　　李宏荣（西安交通大学物理学院）
　　杨　森（西安交通大学物理学院）
　　杨　静（西安交通大学自动化科学与工程学院）
　　张　驰（西安交通大学人工智能学院）
　　陈仕韬（西安交通大学人工智能学院）
　　周三平（西安交通大学人工智能学院）

姜沛林(西安交通大学人工智能学院)
袁泽剑(西安交通大学人工智能学院)
蒋才桂(西安交通大学人工智能学院)
惠　维(西安交通大学计算机科学与技术学院)
葛晨阳(西安交通大学人工智能学院)
薛建儒(西安交通大学人工智能学院)

郑南宁、辛景民、张璇、杨勍、魏平完成了第二版的统稿工作。

西安交通大学人工智能学院
本科专业知识体系建设与课程设置
工 作 组

组 长：
  郑南宁（西安交通大学人工智能学院）

副组长：
  华　　刚（西安交通大学人工智能学院）
  辛景民（西安交通大学人工智能学院）
  Jose C. Principe（美国佛罗里达大学）

成 员（按姓氏笔画排序）：
  王　　乐（西安交通大学人工智能学院）
  兰旭光（西安交通大学人工智能学院）
  任鹏举（西安交通大学人工智能学院）
  刘龙军（西安交通大学人工智能学院）
  刘跃虎（西安交通大学人工智能学院）
  孙宏滨（西安交通大学人工智能学院）
  杜少毅（西安交通大学人工智能学院）
  杨　　勐（西安交通大学人工智能学院）
  汪建基（西安交通大学人工智能学院）
  陈霸东（西安交通大学人工智能学院）
  徐林海（西安交通大学人工智能学院）
  魏　　平（西安交通大学人工智能学院）

参与本书编写的人员还有（按姓氏笔画排序）：
  Georgios N. Yannakakis（马耳他大学数字游戏研究所）
  丁晓军（西安交通大学人文社会科学学院）
  王　　锋（西安交通大学经济与金融学院）
  左炜亮（西安交通大学人工智能学院）
  白惠仁（浙江大学哲学学院）
  刘妹琴（西安交通大学人工智能学院）
  朱晓燕（西安交通大学计算机科学与技术学院）

仲　盛（南京大学计算机科学与技术系）
李延海（西安交通大学生命科学与技术学院）
李　昊（西安交通大学计算机科学与技术学院）
杨建国（西安交通大学电气工程学院）
杨　强（香港科技大学计算机科学与工程系）
杨　静（西安交通大学自动化科学与工程学院）
张元林（西安交通大学人工智能学院）
张　帆（西安交通大学经济与金融学院）
张　驰（西安交通大学人工智能学院）
张志伟（西安交通大学人文社会科学学院）
张　玥（西安交通大学人工智能学院）
张雪涛（西安交通大学人工智能学院）
张　璇（西安交通大学人工智能学院）
宗成庆（中国科学院自动化研究所）
赵晶晶（陕西师范大学心理学院）
姜沛林（西安交通大学人工智能学院）
姚慧敏（西安交通大学人工智能学院）
袁泽剑（西安交通大学人工智能学院）
高彦杰（微软亚洲研究院）
唐亚哲（西安交通大学计算机科学与技术学院）
梅魁志（西安交通大学人工智能学院）
蒋才桂（西安交通大学人工智能学院）
葛晨阳（西安交通大学人工智能学院）
惠　维（西安交通大学计算机科学与技术学院）
薛建儒（西安交通大学人工智能学院）

　　郑南宁、辛景民、张璇、杨勋完成了全书的统稿。
　　注："工科数学分析""线性代数与解析几何""概率统计与随机过程""复变函数与积分变换"的教学大纲在西安交通大学数学与统计学院数学教学中心制定的基础上进行了修订；"大学物理（含实验）"教学大纲在西安交通大学理学院大学物理部、大学物理教学实验中心制定的基础上进行了修订。

CONTENTS 目　　录

第 1 章　人工智能及人才培养定位　　001
1.1　人工智能　　001
1.2　人才培养国家需求　　002
1.3　本科专业人才培养定位　　002

第 2 章　培养方案　　003
2.1　培养目标　　003
2.2　培养方式　　003
2.3　专业知识体系　　004
2.4　专业课程设置　　005
2.5　学期安排　　007
2.6　毕业要求　　009

第 3 章　"数学与统计"课程群　　010
3.1　"工科数学分析"课程大纲　　010
  3.1.1　课程目的和基本内容（Course Objectives and Basic Content）　　010
  3.1.2　课程基本情况（Course Arrangements）　　011
  3.1.3　教学目的和基本要求（Teaching Objectives and Basic Requirements）　　012
  3.1.4　教学内容及安排（Syllabus and Arrangements）　　012
  3.1.5　实验环节（Experiments）　　024
3.2　"线性代数与解析几何"课程大纲　　025
  3.2.1　课程目的和基本内容（Course Objectives and Basic Content）　　025
  3.2.2　课程基本情况（Course Arrangements）　　027
  3.2.3　教学目的和基本要求（Teaching Objectives and Basic Requirements）　　028
  3.2.4　教学内容及安排（Syllabus and Arrangements）　　028

- 3.3 "计算机科学与人工智能的数学基础"课程大纲 035
  - 3.3.1 课程目的和基本内容(Course Objectives and Basic Content) 035
  - 3.3.2 课程基本情况(Course Arrangements) 038
  - 3.3.3 教学目的和基本要求(Teaching Objectives and Basic Requirements) 039
  - 3.3.4 教学内容及安排(Syllabus and Arrangements) 039
- 3.4 "概率统计与随机过程"课程大纲 051
  - 3.4.1 课程目的和基本内容(Course Objectives and Basic Content) 051
  - 3.4.2 课程基本情况(Course Arrangements) 053
  - 3.4.3 教学目的和基本要求(Teaching Objectives and Basic Requirements) 054
  - 3.4.4 教学内容及安排(Syllabus and Arrangements) 054
- 3.5 "复变函数与积分变换"课程大纲 062
  - 3.5.1 课程目的和基本内容(Course Objectives and Basic Content) 062
  - 3.5.2 课程基本情况(Course Arrangements) 063
  - 3.5.3 教学目的和基本要求(Teaching Objectives and Basic Requirements) 064
  - 3.5.4 教学内容及安排(Syllabus and Arrangements) 065
- 3.6 "博弈论"课程大纲 071
  - 3.6.1 课程目的和基本内容(Course Objectives and Basic Content) 071
  - 3.6.2 课程基本情况(Course Arrangements) 073
  - 3.6.3 教学目的和基本要求(Teaching Objectives and Basic Requirements) 074
  - 3.6.4 教学内容及安排(Syllabus and Arrangements) 074

第 4 章 "科学与工程"课程群 085

- 4.1 "现代物理与人工智能"课程大纲 085
  - 4.1.1 课程目的和基本内容(Course Objectives and Basic Content) 085
  - 4.1.2 课程基本情况(Course Arrangements) 089
  - 4.1.3 教学目的和基本要求(Teaching Objectives and Basic Requirements) 090
  - 4.1.4 教学内容及安排(Syllabus and Arrangements) 091

4.1.5　实验环节(Experiments) 104
　4.2　"电子技术与系统"课程大纲 106
　　　4.2.1　课程目的和基本内容(Course Objectives and Basic Content) 106
　　　4.2.2　课程基本情况(Course Arrangements) 109
　　　4.2.3　教学目的和基本要求(Teaching Objectives and Basic Requirements) 109
　　　4.2.4　教学内容及安排(Syllabus and Arrangements) 110
　　　4.2.5　实验环节(Experiments) 122
　4.3　"现代控制工程"课程大纲 124
　　　4.3.1　课程目的和基本内容(Course Objectives and Basic Content) 125
　　　4.3.2　课程基本情况(Course Arrangements) 126
　　　4.3.3　教学目的和基本要求(Teaching Objectives and Basic Requirements) 127
　　　4.3.4　教学内容及安排(Syllabus and Arrangements) 127
　　　4.3.5　实验环节(Experiments) 137
　4.4　"数字信号处理"课程大纲 138
　　　4.4.1　课程目的和基本内容(Course Objectives and Basic Content) 138
　　　4.4.2　课程基本情况(Course Arrangements) 140
　　　4.4.3　教学目的和基本要求(Teaching Objectives and Basic Requirements) 141
　　　4.4.4　教学内容及安排(Syllabus and Arrangements) 141
　　　4.4.5　实验环节(Experiments) 153

第 5 章　"计算机科学与技术"课程群 155

　5.1　"计算机程序设计"课程大纲 155
　　　5.1.1　课程目的和基本内容(Course Objectives and Basic Content) 155
　　　5.1.2　课程基本情况(Course Arrangements) 157
　　　5.1.3　教学目的和基本要求(Teaching Objectives and Basic Requirements) 158
　　　5.1.4　教学内容及安排(Syllabus and Arrangements) 158
　　　5.1.5　实验环节(Experiments) 169
　5.2　"数据结构与算法"课程大纲 172
　　　5.2.1　课程目的和基本内容(Course Objectives and Basic Content) 172
　　　5.2.2　课程基本情况(Course Arrangements) 174

5.2.3 教学目的和基本要求(Teaching Objectives and Basic Requirements) 174
5.2.4 教学内容及安排(Syllabus and Arrangements) 175
5.2.5 实验环节(Experiments) 184
5.3 "计算机体系结构"课程大纲 184
5.3.1 课程目的和基本内容(Course Objectives and Basic Content) 184
5.3.2 课程基本情况(Course Arrangements) 185
5.3.3 教学目的和基本要求(Teaching Objectives and Basic Requirements) 186
5.3.4 教学内容及安排(Syllabus and Arrangements) 186
5.4 "理论计算机科学的重要思想"课程大纲 199
5.4.1 课程目的和基本内容(Course Objectives and Basic Content) 200
5.4.2 课程基本情况(Course Arrangements) 201
5.4.3 教学目的和基本要求(Teaching Objectives and Basic Requirements) 202
5.4.4 教学内容及安排(Syllabus and Arrangements) 202
5.5 "3D计算机图形学"课程大纲 209
5.5.1 课程目的和基本内容(Course Objectives and Basic Content) 209
5.5.2 课程基本情况(Course Arrangements) 210
5.5.3 教学目的和基本要求(Teaching Objectives and Basic Requirements) 211
5.5.4 教学内容及安排(Syllabus and Arrangements) 211
5.5.5 实验环节(Experiments) 216
5.6 "智能感知与移动计算"课程大纲 217
5.6.1 课程目的和基本内容(Course Objectives and Basic Content) 217
5.6.2 课程基本情况(Course Arrangements) 219
5.6.3 教学目的和基本要求(Teaching Objectives and Basic Requirements) 220
5.6.4 教学内容及安排(Syllabus and Arrangements) 220
5.6.5 实验环节(Experiments) 225

第6章 "人工智能核心"课程群 227

6.1 "人工智能概论"课程大纲 227
6.1.1 课程目的和基本内容(Course Objectives and Basic Content) 227

  6.1.2 课程基本情况(Course Arrangements)   228
  6.1.3 教学目的和基本要求(Teaching Objectives and Basic
     Requirements)   229
  6.1.4 教学内容及安排(Syllabus and Arrangements)   230
6.2 "机器学习"课程大纲   233
  6.2.1 课程目的和基本内容(Course Objectives and Basic Content)   234
  6.2.2 课程基本情况(Course Arrangements)   235
  6.2.3 教学目的和基本要求(Teaching Objectives and Basic
     Requirements)   235
  6.2.4 教学内容及安排(Syllabus and Arrangements)   236
6.3 "自然语言处理"课程大纲   242
  6.3.1 课程目的和基本内容(Course Objectives and Basic Content)   243
  6.3.2 课程基本情况(Course Arrangements)   244
  6.3.3 教学目的和基本要求(Teaching Objectives and Basic
     Requirements)   245
  6.3.4 教学内容及安排(Syllabus and Arrangements)   246
  6.3.5 实验环节(Experiments)   252
6.4 "计算机视觉与模式识别"课程大纲   253
  6.4.1 课程目的和基本内容(Course Objectives and Basic Content)   253
  6.4.2 课程基本情况(Course Arrangements)   255
  6.4.3 教学目的和基本要求(Teaching Objectives and Basic
     Requirements)   256
  6.4.4 教学内容及安排(Syllabus and Arrangements)   256
  6.4.5 实验环节(Experiments)   265
6.5 "强化学习"课程大纲   266
  6.5.1 课程目的和基本内容(Course Objectives and Basic Content)   267
  6.5.2 课程基本情况(Course Arrangements)   269
  6.5.3 教学目的和基本要求(Teaching Objectives and Basic
     Requirements)   270
  6.5.4 教学内容及安排(Syllabus and Arrangements)   270
6.6 "生成式 AI 与大语言模型"课程大纲   278
  6.6.1 课程目的和基本内容(Course Objectives and Basic Content)   278
  6.6.2 课程基本情况(Course Arrangements)   279
  6.6.3 教学目的和基本要求(Teaching Objectives and Basic

　　　　　　Requirements) 280
　　　6.6.4 教学内容及安排(Syllabus and Arrangements) 281

第 7 章 "认知与神经科学"课程群　　285

　　7.1 "认知心理学基础"课程大纲　　285
　　　7.1.1 课程目的和基本内容(Course Objectives and Basic Content) 285
　　　7.1.2 课程基本情况(Course Arrangements) 287
　　　7.1.3 教学目的和基本要求(Teaching Objectives and Basic Requirements) 287
　　　7.1.4 教学内容及安排(Syllabus and Arrangements) 288
　　　7.1.5 实验环节(Experiments) 294
　　7.2 "计算神经工程"课程大纲　　295
　　　7.2.1 课程目的和基本内容(Course Objectives and Basic Content) 296
　　　7.2.2 课程基本情况(Course Arrangements) 298
　　　7.2.3 教学目的和基本要求(Teaching Objectives and Basic Requirements) 298
　　　7.2.4 教学内容及安排(Syllabus and Arrangements) 299

第 8 章 "先进机器人技术"课程群　　304

　　8.1 "机器人学基础"课程大纲　　304
　　　8.1.1 课程目的和基本内容(Course Objectives and Basic Content) 304
　　　8.1.2 课程基本情况(Course Arrangements) 306
　　　8.1.3 教学目的和基本要求(Teaching Objectives and Basic Requirements) 306
　　　8.1.4 教学内容及安排(Syllabus and Arrangements) 307
　　　8.1.5 实验环节(Experiments) 312
　　8.2 "多智能体与人机混合智能"课程大纲　　313
　　　8.2.1 课程目的和基本内容(Course Objectives and Basic Content) 313
　　　8.2.2 课程基本情况(Course Arrangements) 315
　　　8.2.3 教学目的和基本要求(Teaching Objectives and Basic Requirements) 315
　　　8.2.4 教学内容及安排(Syllabus and Arrangements) 316
　　8.3 "认知机器人"课程大纲　　324
　　　8.3.1 课程目的和基本内容(Course Objectives and Basic Content) 324

目录

        8.3.2 课程基本情况(Course Arrangements)   326

        8.3.3 教学目的和基本要求(Teaching Objectives and Basic Requirements)   326

        8.3.4 教学内容及安排(Syllabus and Arrangements)   327

  8.4 "先进自动驾驶技术与系统"课程大纲   332

        8.4.1 课程目的和基本内容(Course Objectives and Basic Content)   332

        8.4.2 课程基本情况(Course Arrangements)   334

        8.4.3 教学目的和基本要求(Teaching Objectives and Basic Requirements)   334

        8.4.4 教学内容及安排(Syllabus and Arrangements)   335

第 9 章　"人工智能与社会"课程群　　347

  9.1 "人工智能的科学理解"课程大纲   347

        9.1.1 课程目的和基本内容(Course Objectives and Basic Content)   347

        9.1.2 课程基本情况(Course Arrangements)   349

        9.1.3 教学目的和基本要求(Teaching Objectives and Basic Requirements)   350

        9.1.4 教学内容及安排(Syllabus and Arrangements)   351

  9.2 "人工智能的哲学基础与伦理"课程大纲   357

        9.2.1 课程目的和基本内容(Course Objectives and Basic Content)   357

        9.2.2 课程基本情况(Course Arrangements)   359

        9.2.3 教学目的和基本要求(Teaching Objectives and Basic Requirements)   359

        9.2.4 教学内容及安排(Syllabus and Arrangements)   360

  9.3 "人工智能的社会风险与法律"课程大纲   364

        9.3.1 课程目的和基本内容(Course Objectives and Basic Content)   364

        9.3.2 课程基本情况(Course Arrangements)   366

        9.3.3 教学目的和基本要求(Teaching Objectives and Basic Requirements)   366

        9.3.4 教学内容及安排(Syllabus and Arrangements)   367

第 10 章　"人工智能工具与平台"课程群　　370

  10.1 "深度学习工具与平台"课程大纲   370

        10.1.1 课程目的和基本内容(Course Objectives and Basic

　　　　　　Content) 370
　　　10.1.2 课程基本情况(Course Arrangements) 372
　　　10.1.3 教学目的和基本要求(Teaching Objectives and Basic
　　　　　　Requirements) 372
　　　10.1.4 教学内容及安排(Syllabus and Arrangements) 373
　　　10.1.5 实验环节(Experiments) 380
　10.2 "3D 深度感知"课程大纲 383
　　　10.2.1 课程目的和基本内容(Course Objectives and Basic
　　　　　　Content) 383
　　　10.2.2 课程基本情况(Course Arrangements) 385
　　　10.2.3 教学目的和基本要求(Teaching Objectives and Basic
　　　　　　Requirements) 385
　　　10.2.4 教学内容及安排(Syllabus and Arrangements) 386
　　　10.2.5 实验环节(Experiments) 388
　10.3 "人工智能芯片设计导论"课程大纲 389
　　　10.3.1 课程目的和基本内容(Course Objectives and Basic
　　　　　　Content) 389
　　　10.3.2 课程基本情况(Course Arrangements) 390
　　　10.3.3 教学目的和基本要求(Teaching Objectives and Basic
　　　　　　Requirements) 391
　　　10.3.4 教学内容及安排(Syllabus and Arrangements) 391

第 11 章 "专业综合性实验"课程群　　　　　　　　　　　394

　11.1 "脑信号处理"实验课大纲 394
　　　11.1.1 课程目的和基本内容(Course Objectives and Basic
　　　　　　Content) 394
　　　11.1.2 课程基本情况(Course Arrangements) 395
　　　11.1.3 实验目的和基本要求(Experiment Objectives and Basic
　　　　　　Requirements) 396
　　　11.1.4 实验内容及安排(Experiment Syllabus and Arrangements) 396
　11.2 "游戏 AI 设计与开发"实验课大纲 400
　　　11.2.1 课程目的和基本内容(Course Objectives and Basic Content)
　　　　　　 401
　　　11.2.2 课程基本情况(Course Arrangements) 402

11.2.3 实验目的和基本要求(Experiment Objectives and Basic Requirements) 402
11.2.4 实验内容及安排(Experiment Syllabus and Arrangements) 403

11.3 "机器人导航技术"实验课大纲 405
11.3.1 课程目的和基本内容(Course Objectives and Basic Content) 405
11.3.2 课程基本情况(Course Arrangements) 406
11.3.3 实验目的和基本要求(Experiment Objectives and Basic Requirements) 407
11.3.4 实验内容及安排(Experiment Syllabus and Arrangements) 407

11.4 "创新设计思维"实验课大纲 410
11.4.1 课程目的和基本内容(Course Objectives and Basic Content) 411
11.4.2 课程基本情况(Course Arrangements) 412
11.4.3 实验目的和基本要求(Experiment Objectives and Basic Requirements) 412
11.4.4 实验内容及安排(Experiment Syllabus and Arrangements) 413

后记 417

第 1 章

人工智能及人才培养定位

1.1 人工智能

通过使用机器模仿人类的行为,使机器具有人类的智慧是人类由来已久的梦想和追求。人工智能(Artificial Intelligence,AI)的早期种子可以追溯到古希腊亚里士多德的逻辑推理,而人工智能作为专业术语和一门学问则兴于 1956 年在美国达特茅斯学院(Dartmouth College)召开的"人工智能夏季研讨会"(Dartmouth Summer Research Project on Artificial Intelligence),其目的就是探讨使用机器模仿人类学习以及其他方面的智能。

一般而言,人工智能是指以机器为载体,模拟、延伸和扩展人类或其他生物的智能,使机器能胜任一些通常需要人类智能才能完成的复杂工作。不同种类和程度的智能出现在人和许多生物身上,而人工智能所展示的智能并不局限于这些生物具有的自然智能,它能像人那样思考,也可能超过人的智能。此外,不同的时代、不同的人对这种"复杂工作"的理解是不同的。

在学术上,人工智能通常被划分为符号主义(源于数理逻辑,自顶而下)、连接主义(源于对人脑模拟,人工神经网,自底而上)、行为主义(源于控制论)、贝叶斯方法(基于概率推理)和类推主义(Analogizer)五大学派。人工智能主要通过搜索和优化、逻辑、概率推理、神经网络等工具,解决知识表示、演绎推理、问题求解、学习、感知、规划、运动与操纵、自然语言处理等核心问题。人工智能主要涉及控制理论、计算机科学与工程、数学、统计学、物理学、认知科学、脑科学、神经科学、心理学、语言学、哲学等。

当前,人工智能迎来了第三次发展浪潮,人工智能已成为引领新一轮科技革命和产业变革的战略性技术,人类正走向人工智能时代。以此为契机的人工智能及相关技术的发展和应用对于整个人类的生活、社会、经济和政治都正在产生重大而深远的革命性影响,人工智能已成为国家综合实力与发展的核心竞争力的重要体现。人工智能毫无疑问会改变我们的未来,掌握人工智能技术就意味着价值创造和竞争优势。人工智能是人类历史上最重要的一个演变,人类社会将迎来以有机化学规律演化的生命和无机智慧性的生命形式并存的时代。

1.2 人才培养国家需求

以习近平同志为核心的党中央高度重视人工智能发展。2017年7月,国务院正式发布的《新一代人工智能发展规划》将人工智能上升为我国重大发展战略,明确了"完善人工智能领域学科布局,设立人工智能专业,推动人工智能领域一级学科建设"的重点任务。2018年4月,教育部印发的《高等学校人工智能创新行动计划》指出要加强人工智能领域专业建设和人才培养力度;提出到2020年建立50家人工智能学院、研究院或交叉研究中心,建设100个"人工智能＋X"复合特色专业,编写50本具有国际一流水平的本科生和研究生教材,建设50门人工智能领域国家级精品在线开放课程。2019年以来,教育部公布了2018年度至2022年度普通高等学校本科专业备案和审批结果,现有499所院校获得建设人工智能本科专业的资格。

1.3 本科专业人才培养定位

人工智能多学科交叉、高度复杂、强渗透性的学科特点决定了人工智能的人才培养具有高度挑战性。在探索"人工智能＋X"复合专业培养新模式的同时,应首先立足于根本,把人工智能专业本身建好。人工智能专业培养定位应强调"厚基础""重交叉""宽口径":学生应掌握扎实的数理基础,熟悉人工智能的基本方法及脑认知等交叉学科知识,具备科学家素养、实践动手能力与创新能力,并且拥有较为开阔的产业应用视角与国际前瞻视野。按照此方式培养的学生将具有较强的可塑性和更广阔的发展空间,为学生未来进一步成长为国际一流的人工智能工程师、科学家和企业家奠定知识和能力基础。

从高校发展的层面来讲,人工智能不仅是科学研究的重要内涵,更是大学人才培养和学科建设的新机遇。在当前的人工智能浪潮中,人工智能技术在高等教育、人才培养和各个学科的应用与发展也必将重塑国内外一流大学的格局和地位。各高校可根据发展定位与学科优势特色,探索适合于自身的人工智能专业及"人工智能＋X"复合专业的建设之路。

第 2 章 培养方案

2.1 培养目标

面向国家新一代人工智能发展的重大需求，培养扎实掌握人工智能基础理论、基本方法、应用工程与技术，熟悉人工智能相关交叉学科知识，具备科学素养、实践能力、创新能力、系统思维能力、产业视角与国际视野，未来有潜力成长为国际一流工程师、科学家和企业家，能在我国人工智能学科与产业技术发展中发挥领军作用的优秀人才。

2.2 培养方式

西安交通大学人工智能本科专业的学生选拔遵循"优中选优"的原则，将兴趣、能力与潜力作为选拔与评价的依据，在高考学生、少年班和入校新生中，遴选出数理基础扎实、能力突出、对人工智能兴趣浓厚且具有良好发展潜质的优秀学生进入人工智能专业，并施行动态管理机制。

课程体系建设在参考借鉴国际一流大学课程设置与培养理念的同时，充分把握人工智能学科仍处于高速发展期、渗透性与学科交叉性强的特点，课程设置精练。选用国际一流教材和自编高水平教材，通过讲授基本知识锻炼学习能力与思维方法，让学生拥有自主学习和创造知识的空间。

任课教师的聘用坚持"校内与校外并举、水平与责任并重"的原则，在充分利用校内最优质师资力量的基础上，通过聘任国内外一流教师与海外杰出科学家短期讲学相结合的方式，建设一支学术水平高、责任心强、有热情、肯投入的具有国际化水准的高水平师资队伍，积极探索基于教学评价的教师竞争上岗机制，保证教学水平。

强化并改革实践培养环节,学习国外一流大学培养经验与理念,构建从课内外实验、专业综合性实验、项目实训到导师制科研训练、国际访学、一流企业实习的贯通式实践培养体系,使学生通过参与具体的科研项目,在实践中培养兴趣、锻炼学习能力以及灵活运用所学知识的能力与创新能力。

创新教学方式,采用启发式教学、小组学习、开放式实验与问题研讨等方式,强化学生讨论与课堂互动。通过导师个性化指导和科研项目训练,激发学生好奇心、想象力和批判性思维,培养学生表达能力、发现问题能力和学术判断能力,引导学生发现学术兴趣、选择科研方向、发展学术特长。

加强与人工智能领域领军企业的合作,深化产教融合和校企协同育人。通过设立专项奖学金、开设特色实践课程、提供高质量实训实习基地等方式,将产业的理念、技术、资源整合到培养体系、课程、实训及师资中,最大程度共享和优化配置产教资源,培养高素质和具有产业应用视野的创新人才。

开展国际合作,建立多层次、立体化的高端国际联合培养体系。通过与国际一流大学和研究机构进行学术交流合作、组织高水平国际会议、资助学生国(境)外短/长期交流学习或参加高水平国际会议、邀请国际一流学者访问讲学等,形成国际化培养氛围,拓宽学生的国际化视野。

积极推动人工智能技术在教学与人才培养中的应用。利用新技术升级教学环境、丰富教学手段,改进教学方法,准确掌握教师教学效果与学生掌握情况,以此作为教学水平提升、师资队伍建设与学生个性化指导的抓手,精准发力,促进教学与人才培养模式变革。

2.3 专业知识体系

人工智能本科专业知识体系主要由八大课程群构成,如图2.1所示。

在八大课程群设置中,强调了科学、技术与工程学科交叉、相辅相成,内容设置立足当前、面向未来。其中,"数学与统计"课程群中除了有关数学与统计的通识课程外,还设置了"计算机科学与人工智能的数学基础"和"博弈论"等课程。"科学与工程"课程群包含了"现代物理与人工智能"以及信息学科的工程技术基础——"电子技术与系统""现代控制工程"等课程。"计算机科学与技术"课程群设置了计算机学科核心的"计算机程序设计""数据结构与算法""计算机体系结构"等相关课程。以上三个课程群作为人工智能专业的基础课程群。

图 2.1　人工智能本科专业知识体系：八大课程群

"人工智能核心"课程群由"人工智能概论""机器学习""自然语言处理""计算机视觉与模式识别""生成式 AI 与大语言模型""强化学习"等课程组成，通过不同层次的课程内容启发学生探索人工智能的未来与奥秘。机器人是人工智能在物理现实中应用的载体，具有重要的地位。"先进机器人技术"课程群着重介绍基础及拓展性的机器人相关技术。"人工智能工具与平台"课程群强化了工具与平台在人工智能发展中的重要性，涉及相当广泛的架构、系统、应用等，旨在培养学生应用人工智能技术开发工具和动手的能力。上述三个课程群构成人工智能专业的主干课程群，为本科生奠定稳固的专业核心技术根基。

"认知与神经科学"课程群设置了认知、神经科学与工程相关的基础课程。我们认为要发展新一代人工智能，需要借鉴认知心理学、神经生物学等领域的研究成果，从脑认知和神经科学获得灵感和启示。"人工智能与社会"课程群包含了人工智能的哲学基础与伦理、社会风险与法律等内容，旨在培养学生成为负责任的科学家和工程师。这两个课程群属于人工智能专业的交叉课程群，其目的是为学生奠定基本的认知科学、神经科学、人文社会科学等学科交叉知识。

2.4　专业课程设置

专业课程设置的八大课程群共包含 34 门课程，其中必修 25 门、选修 9 门（完成所需学分须选修其中 5 门）。此外，还特设了"专业综合性实验"课程群，包含必修实验课 4 门，培养学生综合运用所学知识动手解决实际问题的能力。

人工智能本科专业知识体系与课程设置（第 2 版）

课　程　群	课程名称(学分)	学期	必/选修
数学与统计 （必修 6 门,31 学分）	工科数学分析(12)	Ⅰ-1, Ⅰ-2	必修
	线性代数与解析几何(4)	Ⅰ-1	
	计算机科学与人工智能的数学基础(6)	Ⅰ-2, Ⅱ-1	
	概率统计与随机过程(4)	Ⅰ-2	
	复变函数与积分变换(3)	Ⅱ-1	
	博弈论(2)	Ⅱ-2	
科学与工程 （必修 4 门,18 学分）	现代物理与人工智能(7)	Ⅱ-1, Ⅱ-2	必修
	电子技术与系统(5)	Ⅱ-1	
	现代控制工程(3)	Ⅱ-2	
	数字信号处理(3)	Ⅲ-1	
计算机科学与技术 （必修 4 门,9 学分； 选修 1 门,2 学分）	计算机程序设计(2)	Ⅰ-1	必修
	数据结构与算法(3)	Ⅰ-2	
	计算机体系结构(3)	Ⅱ-2	
	理论计算机科学的重要思想(1)	Ⅱ-3	
	3D 计算机图形学(2)	Ⅲ-2	选修 (2 选 1)
	智能感知与移动计算(2)	Ⅲ-2	
人工智能核心 （必修 6 门,14 学分）	人工智能概论(2)	Ⅱ-1	必修
	机器学习(3)	Ⅱ-2	
	自然语言处理(2)	Ⅲ-1	
	计算机视觉与模式识别(3)	Ⅲ-1	
	强化学习(2)	Ⅲ-2	
	生成式 AI 与大语言模型(2)	Ⅲ-2	
先进机器人技术 （必修 2 门,4 学分； 选修 1 门,2 学分）	机器人学基础(2)	Ⅲ-2	必修
	多智能体与人机混合智能(2)	Ⅳ-1	
	认知机器人(2)	Ⅳ-1	选修 (2 选 1)
	先进自动驾驶技术与系统(2)	Ⅳ-1	
认知与神经科学 （必修 2 门,4 学分）	认知心理学基础(2)	Ⅱ-1	必修
	计算神经工程(2)	Ⅲ-1	
人工智能工具与平台 （必修 1 门,2 学分； 选修 1 门,1 学分）	深度学习工具与平台(2)	Ⅱ-2	必修
	3D 深度感知(1)	Ⅳ-1	选修 (2 选 1)
	人工智能芯片设计导论(1)	Ⅳ-1	
人工智能与社会 （选修 2 门,2 学分）	人工智能的科学理解(1)	Ⅲ-1	选修 (3 选 2)
	人工智能的哲学基础与伦理(1)	Ⅲ-1	
	人工智能的社会风险与法律(1)	Ⅲ-1	
专业综合性实验 （必修 4 门,4 学分）	脑信号处理(1)	Ⅲ-1	必修
	游戏 AI 设计与开发(1)	Ⅲ-1	
	机器人导航技术(1)	Ⅲ-2	
	创新设计思维(1)	Ⅳ-1	
合计	93 学分(其中必修 29 门,86 学分；选修 5 门,7 学分)		

2.5　学期安排

学期安排建议如下所示。需要特别说明的是,此表中不含普通高校统一要求的通识教育、集中实践等公共课程,此类课程可根据学校具体要求安排至相应学期。

	第一学期		第二学期		小学期	
	课程名称(学分)	必/选修	课程名称(学分)	必/选修	课程名称(学分)	必/选修
第一学年	工科数学分析Ⅰ(6)	必修	工科数学分析Ⅱ(6)	必修	人工智能前沿系列讲座	
	线性代数与解析几何(4)		计算机科学与人工智能的数学基础Ⅰ(4)			
	计算机程序设计(2)		概率统计与随机过程(4)			
			数据结构与算法(3)			
	必修3门,12学分		必修4门,17学分			

	第一学期		第二学期		小学期	
	课程名称(学分)	必/选修	课程名称(学分)	必/选修	课程名称(学分)	必/选修
第二学年	计算机科学与人工智能的数学基础Ⅱ(2)	必修	现代物理与人工智能Ⅱ(3.5)	必修	理论计算机科学的重要思想(1)	必修
	复变函数与积分变换(3)		现代控制工程(3)			
	现代物理与人工智能Ⅰ(3.5)		计算机体系结构(3)			
	电子技术与系统(5)		机器学习(3)			
	人工智能概论(2)		深度学习工具与平台(2)			
	认知心理学基础(2)		博弈论(2)			
	必修6门,17.5学分		必修6门,16.5学分		必修1门,1学分	

续表

	第一学期		第二学期		小学期	
	课程名称(学分)	必/选修	课程名称(学分)	必/选修	课程名称(学分)	必/选修
第三学年	数字信号处理(3)	必修	强化学习(2)	必修		
	自然语言处理(2)		机器人学基础(2)			
	计算机视觉与模式识别(3)		机器人导航技术(1)			
	计算神经工程(2)		生成式AI与大语言模型(2)			
	脑信号处理(1)		3D计算机图形学(2)	2选1		
	游戏AI设计与开发(1)		智能感知与移动计算(2)			
	人工智能的科学理解(1)	3选2				
	人工智能的哲学基础与伦理(1)					
	人工智能的社会风险与法律(1)					
	必修6门,12学分;选修2门,2学分		必修4门,7学分;选修1门,2学分			

	第一学期		第二学期		小学期	
	课程名称(学分)	必/选修	课程名称(学分)	必/选修	课程名称(学分)	必/选修
第四学年	多智能体与人机混合智能(2)	必修				
	创新设计思维(1)					
	认知机器人(2)	2选1				
	先进自动驾驶技术与系统(2)					
	3D深度感知(1)	2选1				
	人工智能芯片设计导论(1)					
	必修2门,3学分;选修2门,3学分					

注:专业课程合计93学分(其中必修29门,86学分;选修5门,7学分)

2.6 毕业要求

西安交通大学人工智能本科专业毕业需要修满148学分,其中专业课程89学分(必修25门,82学分;选修5门,7学分),专业综合性实验课4门(必修4门,4学分),通识教育公共课程39学分,集中实践类课程(含毕业设计(论文))16学分。

人工智能专业学制为四年,毕业将授予工学学士学位。

第 3 章

"数学与统计"课程群

3.1 "工科数学分析"课程大纲

课程名称：工科数学分析
Course：Mathematical Analysis for Engineering
先修课程：无
Prerequisites：None
学分：12
Credits：12

3.1.1 课程目的和基本内容（Course Objectives and Basic Content）

本课程是人工智能学院本科专业基础必修课。

This course is a basic compulsory course for undergraduates in College of Artificial Intelligence.

本课程介绍了极限、微分、积分、级数等重要的数学工具，并将分析、代数和几何内容进行了有机结合。相关的知识对包括人工智能专业在内的众多工科专业提供了不可缺的高等数学基础，也使学生在数学的抽象性、逻辑性和严谨性等方面受到必要的熏陶和训练。为学生今后增进数学知识、学习人工智能的方法奠定良好的基础，培养学生应用数学知识进行数据分析和建模，以及解决实际问题的意识、兴趣和能力。

本课程的教学，要求学生系统地掌握一元函数微积分学、无穷级数、多元函数微积分学、常微分方程组的基本概念、基本理论和基本方法，同时通过数学实验培养学生的综合素质，即实验动手能力、分析设计能力及团队合作精神，拓展学生思维，激发学生的创新意识。对数学分析的基本思维方法进行必要的训练，逐步提高数学素养以及运算能力、抽象思维能力、逻辑推理能力、空间想象能力、学习能力、分析问题和解决问题的能力，并对现代数学的某些思想方法有所了解，以利于与今后学习现代数学接轨。

This course introduces some important mathematical tools, such as limit, differential, integral and series, and organically combines analysis, algebra and geometry. Relevant knowledge in this course provides an indispensable foundation of advanced mathematics for AI majors and lots of other engineering majors. It lets students take necessary trainings in abstraction, logic and rigor of mathematics, and also lays a good foundation to absorb more mathematics knowledge and learn artificial intelligence in the future, and cultivates readers' awareness, interest and ability to apply mathematics knowledge to solve practical problems.

This course requires students to systematically master the basic concepts, basic theories and basic methods of calculus of unary functions, infinite series, multivariate function calculus, and ordinary differential equations. Meanwhile, the comprehensive quality of students can be cultivated through mathematical experiments. It can be called experimental hands-on ability, analysis design ability and teamwork spirit, and it can also expand student thinking and stimulate students' sense of innovation. Simultaneously, let students have some understanding of thinking methods of modern mathematics, so as to facilitate the future study of modern mathematics.

3.1.2　课程基本情况（Course Arrangements）

课程名称	工科数学分析 Mathematical Analysis for Engineering								
开课时间	一年级		二年级		三年级		四年级		数学与统计
^	秋	春	秋	春	秋	春	秋	春	^
课程定位	本科生"数学与统计"课程群必修课								必修（学分）
学　分	12 学分								^
总学时	216 学时 （授课 192 学时、实验 24 学时）								^
授课学时分配	课堂讲授（190 学时），小组讨论（2 学时）								选修（学分）
先修课程	无								
后续课程									
教学方式	课堂教学、上机教学、课外学习								
考核方式	闭卷考试成绩占 80%，平时作业占 10%，数学实验成绩占 10%								

数学与统计 必修（学分）:
- 工科数学分析(12)
- 线性代数与解析几何(4)
- 计算机科学与人工智能的数学基础(6)
- 概率统计与随机过程(4)
- 复变函数与积分变换(3)
- 博弈论(2)

选修（学分）: /

续表

参考教材	王绵森,马知恩.工科数学分析基础[M].3版.北京:高等教育出版社,2017.
参考资料	李继成.数学实验[M].北京:高等教育出版社,2014.
其他信息	

3.1.3 教学目的和基本要求(Teaching Objectives and Basic Requirements)

(1) 系统掌握一元函数微积分、无穷级数、多元函数微积分、常微分方程组的基本概念、基本理论和基本方法;

(2) 训练数学分析的基本思维方法,提高运算能力、抽象思维能力、逻辑推理能力、空间想象能力,逐步提高数学素养、学习能力、分析问题和解决问题的能力;

(3) 了解现代数学的思想方法,以利于与今后学习现代数学接轨。

3.1.4 教学内容及安排(Syllabus and Arrangements)

第一章 映射、极限、连续(Mapping, Limit and Continuity)

章节序号 Chapter Number	章节名称 Chapters	课时 Class Hour	知 识 点 Key Points
1.1	集合、映射与函数 Set, mapping and function	4	(1) 了解实数集的完备性及确界概念 (2) 理解映射与函数的概念 (1) Understand the completeness and concepts of supremum and infimum (2) Comprehend the concepts of mapping and function
1.2	数列的极限 Limit of sequence	6	(1) 理解数列极限的概念与性质 (2) 了解数列极限收敛性的判别准则 (3) 掌握数列极限的求解方法 (1) Comprehend the concepts and properties of the limit of sequence (2) Understand some criteria for existence of the limit of sequence (3) Master the solution to the limit of sequence

续表

章节序号 Chapter Number	章节名称 Chapters	课时 Class Hour	知识点 Key Points
1.3	函数的极限 Limit of a function	3	(1) 理解函数极限的概念与性质 (2) 掌握两个重要极限 (3) 了解函数极限的存在准则 (4) 掌握函数极限的求解方法 (1) Comprehend the concepts and properties of functional limit (2) Master two important limits of function (3) Understand the existence criteria of function limit (4) Master the solution to the limit of function
1.4	无穷小量和无穷大量 Infinitesimal and infinite quantities	2	(1) 理解无穷小量与无穷大量的概念 (1) Comprehend the concepts of infinitesimal and infinite quantities
1.5	连续函数 Continuous function	8	(1) 理解连续函数的概念与性质 (2) 了解闭区间上连续函数的性质 (3) 了解一致连续的概念 (4) 了解压缩映射原理 (1) Comprehend properties and concepts of continuous function (2) Understand properties of continuous function on a closed interval (3) Understand continuity of elementary functions (4) Understand the principle of compression mapping

第二章 一元函数微分学及其应用(Unary Function Differential Calculus and Its Applications)

章节序号 Chapter Number	章节名称 Chapters	课时 Class Hour	知识点 Key Points
2.1	导数概念 Concept of derivative	2	(1) 理解导数的概念 (1) Comprehend concept of derivative

续表

章节序号 Chapter Number	章节名称 Chapters	课时 Class Hour	知 识 点 Key Points
2.2	求导的基本法则 Fundamental derivatives rules	4	(1) 掌握求导的基本法则 (1) Master fundamental derivatives rules
2.3	函数的微分 The differential of function	6	(1) 理解微分的概念 (2) 了解高阶微分的概念及微分在近似计算中的应用 (1) Comprehend concept of differential (2) Understand applications of the high-order differential in approximate computation
2.4	微分中值定理及其应用 The mean value theorem and its applications	2	(1) 理解微分中值定理 (2) 掌握洛必达法则求不定式的极限 (1) Comprehend the mean value theorem (2) Master L'Hospital's rule to solve the limit of infinitive
2.5	泰勒公式 Taylor formula	2	(1) 了解泰勒定理 (1) Understand Taylor's theorem
2.6	函数性质研究 Function properties study	6	(1) 掌握用导数研究函数单调性及极值的方法 (2) 理解函数极值的概念 (3) 掌握求函数的最大值与最小值的方法 (4) 了解函数凸性的概念 (1) Master the method of using the derivative to study the monotonicity and extremum of the function (2) Comprehend the concept of function extremum (3) Master the method of solving the maximum and minimum values of a function (4) Understand the concept of function convexity

第三章　一元函数积分学及其应用（Unary Function Integrals Calculus and Its Applications）

章节序号 Chapter Number	章节名称 Chapters	课时 Class Hour	知　识　点 Key Points
3.1	定积分的概念与性质 Concept and properties of definite integral	3	（1）理解定积分的概念与性质 （2）了解定积分存在的条件 （1）Comprehend concept and properties of definite integral （2）Understand the conditions for integral exist
3.2	微积分基本公式与基本定理 Basic formulas for indefinite integral and basic theorems	5	（1）理解不定积分的概念与性质 （2）掌握微积分基本公式与基本定理 （1）Comprehend concept and properties of indefinite integral （2）Master basic formulas for indefinite integral and basic theorems
3.3	换元积分法与分部积分法 Integration by substitution and by parts in definite integral	7	（1）掌握换元积分法与分部积分法 （1）Master integration by substitution and by parts in definite integral methods
3.4	定积分的应用 Applications of definite integrals systems	6	（1）掌握建立积分表达式的微元法及用定积分去计算一些几何量（如面积、体积等）和一些物理量（如功、压力、引力和函数的平均值等）的方法 （1）Master the method of establishing the integral expression of the micro-element method and definite integral to calculate some geometric quantities (such as area, volume, etc.) and some physical quantities (such as work, pressure, gravity and the average value of the function, etc.)

章节序号 Chapter Number	章节名称 Chapters	课时 Class Hour	知识点 Key Points
3.5	反常积分 Improper integral	5	（1）理解反常积分的概念 （2）了解反常积分的审敛准则 （3）了解 Γ 函数的概念 (1) Comprehend concept of improper integral (2) Understand the criteria for improper integral (3) Understand the concept of function Γ

第四章 微分方程（Differential Equations）

章节序号 Chapter Number	章节名称 Chapters	课时 Class Hour	知识点 Key Points
4.1	微分方程的基本概念与可分离变量的微分方程 Basic concept of differential equation and separable equations	7	（1）理解常微分方程与常微分方程组的基本概念及其相互关系 （2）掌握变量可分离微分方程和一阶线性微分方程的解法 （3）了解可降阶微分方程的解法 (1) Comprehend the basic concepts of ordinary differential equation and system ordinary differential equation and their relationship (2) Master the solution of variable separable differential equation and first-order linear differential equation (3) Understand the solution of reduced order differential equation

续表

章节序号 Chapter Number	章节名称 Chapters	课时 Class Hour	知 识 点 Key Points
4.2	微分方程的解 Solution of differential equation	8	(1) 理解线性微分方程组的解的性质及解的结构 (2) 掌握常系数线性微分方程组的求解方法 (3) 理解高阶线性微分方程解的结构 (4) 掌握常系数齐次线性微分方程的求解方法 (5) 掌握非齐次项 $f(x)$ 为一些常见类型的(如 $\varphi(t)e^{\mu t}$、$\varphi(x)e^{\mu t}\cos\nu t$、$\varphi(x)e^{\mu t}\sin\nu t$，其中 $\varphi(t)$ 为多项式)的二阶常系数非齐次线性微分方程的特解求解方法 (6) 了解欧拉微分方程的解法及微分方程的幂级数解法 (1) Comprehend properties and structure of solution of linear differential equation (2) Master the solution of linear differential equation with constant coefficients (3) Comprehend structure of solution of higher-order linear differential equation (4) Master solution methods of homogeneous linear differential equation with constant coefficients (5) Master the particular solution method of second-order nonhomogeneous linear differential equation with constant coefficients and the function $f(x)$ includes $\varphi(t)e^{\mu t}$, $\varphi(x)e^{\mu t}\cos\nu t$, $\varphi(x)e^{\mu t}\sin\nu t$, where $\varphi(t)$ is a polynomial (6) Understand solution methods of Euler's differential equation and the power series solution of differential equation
4.3	微分方程的定性分析方法初步 Preliminary analysis of qualitative analysis methods for differential equation	6	(1) 了解自治系统和稳定性的基本概念 (2) 了解判定稳定的李雅普诺夫方法和线性近似系统方法 (1) Understand the basic concepts of autonomous systems and stability (2) Understand the Liapunov method and the linear approximation system method for determining stability

第五章　多元函数微分学及其应用（Multi-variable Function Differential Calculus and Its Applications）

章节序号 Chapter Number	章节名称 Chapters	课时 Class Hour	知　识　点 Key Points
5.1	多元函数的基本概念 The basic concept of multi-variable functions	2	（1）了解 R^n 中点列的极限的概念 （2）了解 R^n 中的开集、闭集、紧集与区域等概念 （3）了解多元连续函数的性质 (1) Understand concept of the limit of a point set sequence in R^n (2) Understand concepts of open set, closed set, tight set and region in R^n (3) Understand properties of multi-variable continuous function
5.2	偏导数 Partial derivative	5	（1）理解多元函数的偏导数的概念 （2）了解方向导数与梯度的概念 （3）掌握多元复合函数的偏导数的求解方法 （4）掌握高阶偏导数的求解方法 （5）掌握由一个方程确定的隐函数的偏导数的计算方法 （6）掌握由方程组所确定的隐函数的偏导数的计算方法 (1) Comprehend concept of partial derivatives of multi-variable function (2) Understand concepts of directional derivative and the gradient (3) Master the solution method of partial derivatives of multi-variable composite (4) Master the solution method of high-order partial derivative (5) Master the calculation method of partial derivative of implicit function determined by an equation (6) Master the calculation method of partial derivative of the implicit function determined by equation system

续表

章节序号 Chapter Number	章节名称 Chapters	课时 Class Hour	知 识 点 Key Points
5.3	全微分 Total differential	9	(1) 理解多元函数的全微分的概念 (2) 掌握多元复合函数的全微分的求解方法 (3) 掌握求解高阶全微分的方法 (4) 掌握由一个方程确定的隐函数的全微分的计算方法 (5) 掌握由方程组确定的隐函数的全微分的计算方法 (1) Comprehend concept of total differential of multi-variable functions (2) Master the calculation method of total differential of multi-variable composite functions (3) Master the solution method of high-order total differential (4) Master the calculation method of total differential of implicit function determined by an equation (5) Master the calculation method of total differential of the implicit function determined by equation systems
5.4	多元函数的泰勒公式与极值问题 Taylor formula of multi-variable function and extreme value	3	(1) 了解多元函数的泰勒公式 (2) 理解无约束极值和有约束极值的概念 (3) 掌握多元函数的极值及一些最大最小值应用问题的求解方法 (1) Understand Taylor formula of multi-variable function (2) Comprehend concepts of unrestricted and constrained extreme value (3) Master the extreme values of multi-variable functions and applications about maximum and minimum values

续表

章节序号 Chapter Number	章节名称 Chapters	课时 Class Hour	知 识 点 Key Points
5.5	多元向量值函数的导数与微分 Derivative and derivation of multivariate vector value function	5	（1）理解向量值函数的导数与微分的概念 （2）掌握向量值函数的导数与微分的求解方法 （1）Comprehend concepts of derivative and differential of vector value function （2）Master the solution method of derivative and differential of vector value function
5.6	多元函数微分学的几何应用 Applications in geometry of the differential for multi-variable function	6	（1）掌握空间曲线的切线与法平面方程的计算方法 （2）掌握曲线弧长的求解方法 （3）掌握曲面的切平面与法线方程的法求解方法 （1）Master the calculation of tangent line and normal plane of a space curve （2）Master the solution method of curve arc length （3）Master the method of solving the tangent plane and the normal equation of the curved surface
5.7	空间曲线的曲率与挠率 Curvature and torsion of space curves	4	（1）掌握空间曲线的切线与法平面方程的求解方法 （2）了解空间曲线的弗莱纳坐标系 （3）弗莱纳标架与弗莱纳公式 （4）掌握求解曲线的曲率和挠率的方法 （1）Master the solution method of tangent and normal plane equation of the space curve （2）Understand the Frenet of the space curve （3）Frenet frame and Frenet formula （4）Master the methods of solving the curvature and torsion of the curve

第六章　多元函数积分学及其应用（Multi-variable Function Integrals Calculus and Its Applications）

章节序号 Chapter Number	章节名称 Chapters	课时 Class Hour	知　识　点 Key Points
6.1	多元函数积分的概念与性质 Concept and properties of multi-variable functions' integral	3	（1）理解多元函数积分的概念与性质 (1) Comprehend concept and properties of multi-variable functions' integral
6.2	二重积分 Double integral	7	（1）理解二重积分的几何意义 （2）掌握二重积分在直角坐标系及极坐标系下的计算方法 （3）了解二重积分在曲线坐标系下的计算方法 (1) Comprehend geometric meaning of double integral (2) Master the calculation of double integral in rectangular and polar coordinates (3) Understand the calculation of double integral in curve coordinates
6.3	三重积分 Triple integral	6	（1）掌握三重积分在直角坐标系、柱面坐标系及球面坐标系下的计算方法 (1) Master the calculation of triple integral in rectangular, cylindrical and spherical coordinates
6.4	重积分的应用 Applications of multiple integral	2	（1）了解重积分的微元法及重积分在几何、物理中的一些应用（如求曲面面积、立体的体积、质量、引力、质心及转动惯量等） (1) Understand the micro-element methods of multiple integral and some applications of multiple integral in geometry and physics (such as surface area, three-dimensional volume, mass, gravity, the center of mass and moment of inertia)

续表

章节序号 Chapter Number	章节名称 Chapters	课时 Class Hour	知 识 点 Key Points
6.5	含参变量的积分与反常重积分 Parametric integral and improper multiple integral	6	（1）了解含参变量的积分与反常重积分的概念 (1) Understand concepts of parametric integral and improper multiple integral
6.6	第一型线积分与面积分 Line integral of a scalar field and surface integrals	6	（1）理解第一型线积分与面积分的概念 （2）掌握第一型线积分与面积分的计算方法 (1) Comprehend concepts of line integral of a scalar field and surface integrals (2) Master the calculation of line integral of a scalar field and surface integrals
6.7	第二型线积分与面积分 Line integral of a vector field and surface integrals	6	（1）理解第二型线积分与面积分的概念 （2）掌握第二型线积分与面积分的计算方法 (1) Comprehend concept of line integral of a vector field and surface integrals (2) Master the calculation of line integral of a vector field and surface integrals
6.8	各种积分的联系及其在场论中的应用 The connection of various integrals and its application in field theory	8	（1）掌握格林公式 （2）理解平面积分与路径无关的条件 （3）了解斯托克斯公式与旋度的概念 （4）了解高斯公式与散度的概念 （5）了解几种重要的特殊向量场 (1) Master Green's formula (2) Comprehend the conditions for surface integrals and path independence (3) Understand concepts of Stokes' formula and curl (4) Understand concepts of Gauss' formula and divergence (5) Understand several important special vector fields

第七章　无穷级数(Infinite Series)

章节序号 Chapter Number	章节名称 Chapters	课时 Class Hour	知　识　点 Key Points
7.1	常数项级数 Series with constant terms	6	(1) 理解无穷级数的基本概念 (2) 了解无穷级数的性质及柯西收敛原理 (1) Comprehend the basic concepts of infinite series with constant terms (2) Understand properties of infinite series and Cauchy principle of convergence
7.2	函数项级数 Series with function terms	4	(1) 理解函数项级数的处处收敛与和函数的概念 (2) 了解函数项级数一致收敛的概念、性质及判别方法 (3) 掌握正项级数的审敛准则 (4) 了解变号级数的审敛准则 (1) Comprehend the concepts of the convergence and sum function of the series with function terms (2) Understand the concepts, properties and discriminant method of uniform convergence of function series (3) Master the criteria for positive series (4) Understand the criteria for series of variable signs
7.3	幂级数 Power series	6	(1) 理解阿贝尔定理 (2) 掌握幂级数收敛区间的求解方法 (3) 了解幂级数的性质 (4) 掌握将函数展开成幂级数的方法 (5) 了解幂级数在近似计算等问题中的简单应用 (1) Comprehend Abel's theorem (2) Master the solution method of convergence interval of the power series (3) Understand properties of the power series (4) Master the solution method of expanding a function into the power series (5) Understand the simple application of the power series in approximate calculation problems

续表

章节序号 Chapter Number	章节名称 Chapters	课时 Class Hour	知识点 Key Points
7.4	傅里叶级数 Fourier series	6	（1）掌握欧拉-傅里叶公式及狄利克雷定理 （2）掌握将函数展开为傅里叶级数的方法 （3）了解傅里叶级数的复数形式 (1) Master the Euler-Fourier formula and Dirichlet theorem (2) Master the solution method of expanding a function into the Fourier series (3) Understand the plural form of the Fourier series

3.1.5 实验环节（Experiments）

序号 Num.	实验内容 Experiment Content	课时 Class Hour	知识点 Key Points
1	基于 MATLAB 软件的计算方法 MATLAB based numerical solution methods	24	（1）迭代法 （2）最优化方法 （3）数据拟合 （4）数据插值 （5）数值积分 （6）微分方程的数值解方法 (1) Iterative method (2) Optimization method (3) Data fitting (4) Data interpolation (5) Numerical integration (6) Numerical solution of differential equation

大纲制定者：西安交通大学数学与统计学院数学教学中心

大纲修订者：杜少毅教授（西安交通大学人工智能学院）、汪建基副教授（西安交通大学人工智能学院）

大纲审定：西安交通大学人工智能学院本科专业知识体系建设与课程设置第二版修订工作组

3.2 "线性代数与解析几何"课程大纲

课程名称：线性代数与解析几何
Course：Linear Algebra and Analytic Geometry
先修课程：无
Prerequisites：None
学分：4
Credits：4

3.2.1 课程目的和基本内容（Course Objectives and Basic Content）

本课程是人工智能学院本科专业基础必修课。

This course is a basic compulsory course for undergraduates in College of Artificial Intelligence.

本课程的内容对近些年计算机技术的快速发展和人工智能领域的技术进步都有着重要的理论支撑，如计算机视觉与图像处理本质上就可看作是一种向量、矩阵或几何的运算。同时，本课程在教学中精简了内容，淡化了繁杂的运算技巧。这样可以使学生在掌握必要理论知识的同时能有更充足的时间进行应用实践，为将来在计算机科学和人工智能等领域的学习奠定重要的理论基础。

本课程力求将线性代数与解析几何融为一体，与数学分析的内容相互渗透，并为数学分析的多元部分提供必要的代数与几何基础。通过本课程的教学，使学生系统地获取线性代数与空间解析几何的基本知识、基本理论与基本方法，提高运用所学知识分析和解决问题的能力，并为学习相关课程及进一步学习现代数学奠定必要的数学基础。课堂教学中，注重将数学建模思想融入理论课教学，培养学生应用线性代数知识解决实际问题的能力和创新意识。

本课程的内容主要包括：行列式、矩阵、几何向量及其应用、n 维向量与线性方程组、线性空间与欧氏空间、特征值与特征向量、二次曲面与二次型、线性变换等。课程的第一章引入行列式并讨论了其基本性质和计算方法。第二章主要介绍了矩阵的基本概念及其运算。第三章首先介绍了向量的概念及它的线性运算和乘法运算，并引入向量坐标的概念将向量运算转化为代数运算,然后利用向量研究平面和空间直线问题。第四章不仅讨论了向量相关的基本理论,还利用矩阵和向量等工具完整地解决线性方程组的求解问题。第五章介绍了线性空间与欧氏空间的基本概念,并讨论了它们的基本性质和基本结构。第六章介绍了特征值与特征向量的概念、性质与计算,然后讨论了矩阵对角化的问题和特征值的典型应用实例。第七章主要讨论了二次型相关理论。第八章介绍了线性变换的基本知识,包括线性变化的基本概念、线性变换的矩阵表示等。

This course has supported various important theoretical progresses to the rapid development of computer and artificial intelligence technologies in recent years. For example, computer vision and image processing can essentially be viewed as vector, matrix or geometric operation. At the same time, this course simplifies the content and weakens the complicated operation skills in teaching, so that students master the necessary theoretical knowledge and take more time to practice, which would lay an important theoretical foundation for the future study of computer science and artificial intelligence.

This course seeks to integrate linear algebra and analytic geometry together, infiltrate the contents of mathematical analysis, and provide the necessary algebraic and geometric basis for the multivariate part of mathematical analysis. By studying this course, students should systematically acquire the basic knowledge, theory and methods of linear algebra and spatial analytic geometry, improve their ability to analyze and solve problems with the knowledge they have learned, and lay the necessary mathematical foundation for learning related courses and further studying modern mathematics. In classroom teaching, this course takes the thinking of mathematical modelling into theoretical teaching to cultivate students' ability of innovative consciousness and the ability to solve practical problems with linear algebra knowledge.

The content of this course mainly includes determinant, matrix, geometric vector with applications, n-dimensional vector and system of linear equations, linear space and Euclidean space, eigenvalue and eigenvector, quadratic surface and quadratic form, and linear transformation, etc. Chapter 1 introduces the determinant and discusses its basic properties and calculation methods. The basic concepts and operations of matrix are introduced in Chapter 2. In Chapter 3, the concept of vector and its linear and multiplication operations are firstly introduced. Moreover, the concept of vector coordinates helps to transform vector operations into algebraic operations, and then the planar and spatial straight line problems can be studied by vectors. In Chapter 4, the basic theories of vector correlation are discussed, and the methods to solve the system of linear equations by using tools such as matrix and vector are also introduced. Chapter 5 introduces the concepts of linear space and Euclidean space, and discusses their basic properties and basic structure. Chapter 6 introduces the concept, properties and calculations of eigenvalues and eigenvectors, and then discusses the problems of matrix diagonalization and typical application examples of eigenvalues. Chapter 7 focuses on the theory of quadratic correlation. Chapter 8 introduces the basics of linear transformation, including the basic concepts of linear variation, matrix representations of linear transformations.

3.2.2 课程基本情况(Course Arrangements)

课程名称	线性代数与解析几何 Linear Algebra and Analytic Geometry								
开课时间	一年级		二年级		三年级		四年级		数学与统计
	秋	春	秋	春	秋	春	秋	春	
课程定位	本科生"数学与统计"课程群必修课								必修 (学分)
学 分	4学分								
总学时	64学时 (授课64学时、实验0学时)								
授课学时 分配	课堂讲授(62学时), 小组讨论(2学时)								选修 (学分)
先修课程	无								
后续课程									
教学方式	课堂教学、作业、自学								
考核方式	期中闭卷考试成绩占30%,平时作业占10%,期终闭卷考试成绩占60%								
参考教材	魏战线,李继成.线性代数与解析几何[M].北京:高等教育出版社,2015.								
参考资料	魏战线.线性代数辅导与典型题解析[M].北京:高等教育出版社,2018.								
其他信息									

必修(学分) 栏目:
- 工科数学分析(12)
- 线性代数与解析几何(4)
- 计算机科学与人工智能的数学基础(6)
- 概率统计与随机过程(4)
- 复变函数与积分变换(3)

选修(学分) 栏目:
- 博弈论(2)
- /

3.2.3 教学目的和基本要求(Teaching Objectives and Basic Requirements)

(1) 系统地掌握行列式、矩阵、几何向量及其应用、n维向量与线性方程组、线性空间与欧氏空间(初步)、特征值与特征向量、二次曲面与二次型、线性变换(初步)的基本知识、基本理论与基本方法;

(2) 提高学生的运算能力;

(3) 训练学生的逻辑推理能力、抽象思维能力和空间想象能力;

(4) 能够运用所获取的知识去分析和解决问题。

3.2.4 教学内容及安排(Syllabus and Arrangements)

第一章 行列式(Determinant)

章节序号 Chapter Number	章节名称 Chapters	课时 Class Hour	知 识 点 Key Points
1.1	行列式的定义与性质 Definition and properties of determinant	2	(1) 2阶行列式与一类二元线性方程组的解 (2) n阶行列式的定义 (3) 行列式的基本性质 (1) Solution of the 2nd order determinant and a kind of bivariate linear equations (2) Definition of nth-order determinant (3) Main properties of determinant
1.2	行列式的计算 The calculation of determinant	2	(1) 上三角行列式的转换与计算 (2) 降阶法的应用 (1) Conversion and calculation of the upper triangular determinant (2) Application of the reduced order method
1.3	克莱姆法则 Cramer's law	1	(1) 克莱姆法则的定理、推论以及应用 (1) Theorem, inference and applications of Cramer's law

第二章 矩阵（Matrix）

章节序号 Chapter Number	章节名称 Chapters	课时 Class Hour	知　识　点 Key Points
2.1	矩阵及其运算 Matrix with operations	2	（1）矩阵的概念 （2）矩阵的代数运算 （3）矩阵的转置 （4）方阵的行列式 (1) Concept of matrix (2) Algebraic operation of matrix (3) Transpose of matrix (4) Determinant of square matrix
2.2	逆矩阵 Inverse matrix		（1）逆矩阵 （2）伴随矩阵的定义、定理和推论 (1) Inverse matrix (2) Definition of adjoint matrix with its theorem and inference
2.3	分块矩阵及其运算 Partitioned matrix and operations	1	（1）子矩阵 （2）分块矩阵 (1) Submatrix (2) Partitioned matrix
2.4	初等变换与初等矩阵 Elementary transformation and elementary matrix	1	（1）初等变换与初等矩阵 （2）阶梯形矩阵 （3）再论可逆矩阵 (1) Elementary transformation and elementary matrix (2) Echelon form (3) Re-discussion on reversible matrix
2.5	矩阵的秩 Rank of matrix	1	（1）矩阵的秩的定义和相关推论 (1) The definition and related inferences of rank of matrix

第三章 几何向量及其应用(Geometric Vector with Applications)

章节序号 Chapter Number	章节名称 Chapters	课时 Class Hour	知 识 点 Key Points
3.1	向量及其线性运算 Vector and linear operations	3	(1) 向量的基本概念 (2) 向量的线性运算 (3) 向量共线、共面的充要条件 (4) 空间坐标系与向量的坐标 (1) Basic concept of vector (2) Linear operation of vector (3) Necessary and sufficient conditions for vector collinearity and coplanarity (4) Spatial coordinate system and coordinate of vector
3.2	数量积、向量积、混合积 Quantitative product, vector product, and triple product	2	(1) 两个向量的数量积(内积、外积) (2) 两个向量的向量积(内积、外积) (3) 混合积 (1) Quantitative product of two vectors (inner product, outer product) (2) Vector product of two vectors (inner product, outer product) (3) Triple product
3.3	平面和空间直线 Plane and space line	3	(1) 平面的方程 (2) 两个平面的位置关系 (3) 空间直线的方程 (4) 两条直线的位置关系 (5) 直线与平面的位置关系 (6) 距离 (1) Plane equation (2) Positional relationship between two planes (3) Equation of space line (4) Positional relationship between two straight lines (5) Positional relationship between line and plane (6) Distance

第四章　n 维向量与线性方程组（n-Dimensional Vectors and Systems of Linear Equations）

章节序号 Chapter Number	章节名称 Chapters	课时 Class Hour	知　识　点 Key Points
4.1	消元法 Elimination method	2	（1）n 元线性方程组 （2）消元法 （3）线性方程组的解 （4）数域 (1) System of linear equations with n variables (2) Elimination method (3) Solution of system of linear equations (4) Number field
4.2	向量组的线性相关性 Linear correlation of vector groups	3	（1）n 维向量及其线性运算 （2）线性表示与等价向量 （3）线性相关与线性无关 (1) n-dimensional vector and linear operation (2) Linear representation and equivalent vector (3) Linear correlation and independence
4.3	向量组的秩 Rank of vector group	2	（1）向量组的极大无关组与向量组的秩 （2）向量组的秩与矩阵的秩的关系 (1) Maximum independent group of vector group and rank of vector group (2) The relationship between rank of vector group and rank of matrix
4.4	线性方程组的 解的结构 The structure of solutions of linear equations	3	（1）齐次线性方程组 （2）非齐次线性方程组 (1) Homogeneous linear equations (2) Nonhomogeneous linear equations

第五章 线性空间与欧氏空间（Linear Space and Euclidean Space）

章节序号 Chapter Number	章节名称 Chapters	课时 Class Hour	知 识 点 Key Points
5.1	线性空间的基本概念 Basic concept of linear space	5	(1) 线性空间的定义 (2) 线性空间的基本性质 (3) 线性子空间的定义 (4) 基、维数和向量的坐标 (5) 基变换与坐标变换 (6) 线性空间的同构 (7) 子空间的交与和 (1) Definition of linear space (2) Basic properties of linear space (3) Definition of linear subspace (4) Coordinates of bases, dimensions and vectors (5) Base transformation and coordinate transformation (6) Isomorphism of linear spaces (7) Intersection and sum of subspaces
5.2	欧氏空间的基本概念 Basic concept of Euclidean space	5	(1) 内积及其基本性质 (2) 范数和夹角 (3) 标准正交基及其基本性质 (4) 格拉姆-施密特正交化方法 (5) 正交矩阵 (6) 矩阵的 QR 分解 (7) 正交分解和最小二乘法 (1) Inner product and its basic properties (2) Norm and angle (3) Standard orthogonal basis and its basic properties (4) Gram-Schmidt orthogonalization method (5) Orthogonal matrix (6) QR decomposition of the matrix (7) Orthogonal decomposition and least squares

第六章 特征值与特征向量（Eigenvalues and Eigenvectors）

章节序号 Chapter Number	章节名称 Chapters	课时 Class Hour	知 识 点 Key Points
6.1	矩阵的特征值与特征向量 Eigenvalues and eigenvectors of matrix	2	(1) 特征值与特征向量的定义 (2) 特征方程、特征多项式与特征子空间的定义 (1) Definition of eigenvalues and eigenvectors (2) Definition of characteristic equations, characteristic polynomials and feature subspaces
6.2	相似矩阵与矩阵的相似对角化 Similar matrix and similar diagonalization of matrix	4	(1) 相似矩阵 (2) 矩阵可对角化的条件 (3) 实对称矩阵的对角化 (1) Similar matrix (2) Condition of matrix diagonalization (3) Diagonalization of real symmetric matrix
6.3	应用举例 Application examples	2	(1) 一类常系数线性微分方程组的求解 (2) 斐波那契数列与递推关系式的矩阵解法 (1) Solving a class of linear differential equations with constant coefficients (2) Matrix solutions of Fibonacci sequence and recursion relation

第七章 二次曲面与二次型（Quadric Surface and Quadric Form）

章节序号 Chapter Number	章节名称 Chapters	课时 Class Hour	知 识 点 Key Points
7.1	曲面与空间曲线 Surface and space curve	3	(1) 曲面与空间曲线的方程 (2) 柱面、锥面、旋转面 (3) 5 种典型的二次曲面 (4) 4 种曲面在坐标面上的投影 (5) 空间区域的简图 (1) Equations of surface and space curve (2) Cylinder, tapered surface, rotating surface (3) Five quadric surfaces typically (4) Projections of four kinds of surfaces on the coordinate plane (5) Sketch of the space area

续表

章节序号 Chapter Number	章节名称 Chapters	课时 Class Hour	知识点 Key Points
7.2	实二次型 Real quadratic form	5	（1）二次型及其矩阵表示 （2）二次型的标准型 （3）合同变换与惯性定理 （4）正定二次型 （5）二次曲面的标准方程 (1) Quadratic form and its matrix representation (2) Standard Quadratic form (3) Congruent transformation and inertia theorem (4) Positive definite quadratic form (5) Standard equation of quadric

第八章 线性变换（Linear Transformation）

章节序号 Chapter Number	章节名称 Chapters	课时 Class Hour	知识点 Key Points
8.1	线性变换及其运算 Linear transformation and its operations	4	（1）线性变换的定义及其基本性质 （2）核与值域 （3）线性变换的运算 (1) Definition and basic properties of linear transformation (2) Core and range (3) Operations of linear transformation
8.2	线性变换的矩阵表示 Matrix representation of linear transformation	4	（1）线性变换的矩阵 （2）线性算子在不同基下的矩阵之间的关系 (1) Matrix of linear transformation (2) The relationship between matrices of linear operators with different bases

　　大纲制定者：西安交通大学数学与统计学院数学教学中心

　　大纲修订者：杜少毅教授（西安交通大学人工智能学院）、汪建基副教授（西安交通大学人工智能学院）

　　大纲审定：西安交通大学人工智能学院本科专业知识体系建设与课程设置第二版修订工作组

3.3 "计算机科学与人工智能的数学基础"课程大纲

课程名称：计算机科学与人工智能的数学基础
Course：Math Foundation of Computer Science and Artificial Intelligence
先修课程：工科数学分析、线性代数与解析几何
Prerequisites：Mathematical Analysis for Engineering, Linear Algebra and Analytic Geometry
学分：6
Credits：6

3.3.1 课程目的和基本内容（Course Objectives and Basic Content）

本课程是人工智能学院本科专业基础必修课。
This course is a basic compulsory course for undergraduates in College of Artificial Intelligence.

为了加强学生关于计算机和人工智能学科的数学基础，特开设本课程，其主要目的有：

（1）人工智能相关的研究与实践需要诸多数学知识作为基础，而已开设的其他数学类课程：工科数学分析、线性代数与解析几何、概率统计与随机过程、复变函数与积分变换及博弈论，虽已为相关领域的学习打下良好基础，但仍有部分内容尚未涉及（如数值计算与优化理论等），这些内容将在本课程中进行介绍；

（2）部分内容（例如矩阵分析等），虽在其他开设课程中已有涉及，但人工智能方向的研究与应用需要更加深入地了解这些内容，本课程将对这部分内容进行更深入和更有针对性的介绍。

"计算机科学和人工智能的数学基础"课程所包含的内容主要分为以下6部分：数理逻辑、集合论与组合分析、图论初步、矩阵与刚体运动、数值计算和优化理论初步。我们将其中部分内容命名为"初步"是因为对它们单独进行系统介绍都可能超过4学时，而本课程也并非仅仅对它们进行概念的介绍、浅尝辄止，而是对这些部分与人工智能学习非常密切的内容进行深入的介绍。其中，数理逻辑包含命题逻辑和谓词逻辑；

集合论与组合分析包含了集合的基本概念与运算、组合分析初步以及可数集 & 不可数集 & 康托集等内容；图论初步包括图的基本概念、特殊的图及树等内容；矩阵与刚体运动包含了矩阵基础、应用回归分析及三维刚体运动等内容；数值计算包括数值计算的数学基础、非线性方程的数值解法、线性方程组的数值解法、函数插值与逼近方法等内容；优化理论初步则包含了优化基础、无约束优化及约束优化等内容。

通过对上述内容的学习，为人工智能学院本科生进一步学习和实践打下扎实的数学基础。其中，数理逻辑不仅是人工智能三大学派之一的符号主义学派的理论基石，也是逻辑电路设计等课程的基础；集合论与组合分析初步可以帮助学生更好地用集合进行表达与分析，并为学习概率论奠定基础；图论初步部分为学生的编程学习以及学习数据结构等课程都有重要的帮助；矩阵与刚体理论的学习可以帮助学生更好地利用矩阵这一重要工具分析和解决在学习和实践中碰到的具体问题；数值计算为学生在实际中利用计算机解决各种数学问题打下基础；而优化理论初步所介绍的方法可以使学生在遇到实际问题时学会如何对问题更好地进行建模与优化求解。

课程采用集中授课与小组学习相结合的模式，并辅之以小组讨论、日常作业等教学手段，加强学生对数学基础的认识，为日后更好地利用数学知识解决在计算机及人工智能学科中遇到的问题奠定基础。课程还将通过大作业和算法编程实现等实践环节进一步加强学生独立分析问题、解决问题的能力，培养综合设计及创新能力，培养实事求是、严肃认真的科学作风和良好的实验习惯，为今后的学习和工作打下良好的基础。

To further strengthen the students' mathematical foundation on computer science and artificial intelligence, the course is offered specially. Its main purposes include the following two points.

(1) Although other mathematics courses, including Mathematical Analysis for Engineering, Linear Algebra and Analytic Geometry, Probability Statistics and Stochastic Processes, Complex Variable Function and Integral Transform, Game Theory, have laid a good foundation for students on the study of related fields, there are still some contents that have not been covered, such as numerical computation and optimization theory. These contents will be included in the course "Math Foundation of CS and AI".

(2) Some contents, such as matrix analysis, have already been introduced in other courses, but research and application in the field of artificial intelligence require a deeper understanding of these contents. This cource will provide move in-depth and

targeted introduction to this part.

The main contents of this course include the following six parts: Mathematical Logic, Set Theory and Combination Analysis, Graph Theory, Matrix and Rigid Body Motion, Numerical Computation, Optimization Theory. The part of Mathematical Logic includes two chapters: propositional logic and predicate logic; The part of Set Theory and Combination Analysis includes basic concepts and operations of sets, combination analysis, and countable set & uncountable set & Cantor Set; The part of Graph Theory includes basic concepts of graphs, special graphs, and trees; The part of Matrix and Rigid Body Motion includes Matrix foundation, applied regression analysis, and rigid body motion in three-dimensional space; The part of Numerical Computation includes four chapters: mathematics basis of numerical computation, numerical solutions of nonlinear equations, numerical solutions of linear equations, and function interpolation and approximation methods; The part of Optimization Theory includes basis of optimization, unconstrained optimization, and constrained optimization.

By studying the above contents, it lays a solid mathematical foundation for the further study and practice of undergraduates in AI College. Mathematical Logic is not only the basis of the follow-up contents in this course, but also is the basis for the study of proposition representation and reasoning, logic circuit design, etc. The study of Set Theory and Combination Analysis can help students better use sets for expression and analysis, and lay the foundation of probability theory. The knowledge in Graph Theory provides some good ideas in programming, and it also is a basis to learn the course "Data Structures and Algorithms". Matrix is an important tool which can help students to analyze and solve practical problems well. Numerical Computation lays a foundation for students to solve various mathematical problems by computers. The methods introduced in Optimization Theory can help students think about how to solve problems arising in practice.

The course adopts the group learning supplemented by group discussion, daily homework and other teaching methods, to strengthen the students' understanding of the mathematical knowledge and methods, which is the mathematical foundation for better use of mathematics knowledge to solve problems of computer science and artificial intelligence in the future. The course also further strengthen students' ability to analyze problems and solve problems independently via large course assignments and algorithm programming, which can train comprehensive design and innovation ability. Moreover, this course will cultivate realistic, serious scientific style and good experimental habits, which can lay a good foundation for future work.

3.3.2 课程基本情况(Course Arrangements)

课程名称	计算机科学与人工智能的数学基础 Math Foundation of Computer Science and Artificial Intelligence							
开课时间	一年级		二年级		三年级		四年级		数学与统计
^^	秋	春	秋	春	秋	春	秋	春
课程定位	本科生"数学与统计"课程群必修课								必修 (学分)
学 分	6学分							
总学时	96学时 (授课96学时,实验0学时)							
授课学时分配	课堂讲授(96学时)								选修 (学分)
先修课程	工科数学分析、线性代数与解析几何							
后续课程	人工智能概论、机器学习							
教学方式	课堂教学、课后作业							
考核方式	笔试成绩占70%,平时成绩(作业、大作业、上机实验等)占20%,考勤占10%							
参考教材	1. 耿素云.离散数学[M].北京:清华大学出版社,2013. 2. 李桂成.计算方法[M].北京:电子工业出版社,2018. 3. 孙文瑜,徐成贤,朱德通.最优化方法[M].北京:高等教育出版社,2010. 4. 钱颂迪,等.运筹学[M].北京:清华大学出版社,2018.							
参考资料	1. Lehman E,Leighton F T,Meyer A R. Mathematics for Computer Science[M]. Cambridge:MIT Press,2016. 2. 张贤达.矩阵分析与应用[M].北京:清华大学出版社,2016. 3. 何晓群,刘文卿.应用回归分析[M].北京:中国人民大学出版社,2015. 4. 高立.数值最优化方法[M].北京:北京大学出版社,2018. 5. Boyd S,Vandenberghe L.凸优化[M].王书宁,许鋆,黄晓霖,译.北京:清华大学出版社,2018. 6. 高翔,张涛,等.视觉SLAM十四讲:从理论到实践[M].北京:电子工业出版社,2019.							
其他信息								

数学与统计 必修(学分):
- 工科数学分析(12)
- 线性代数与解析几何(4)
- 计算机科学与人工智能的数学基础(6)
- 概率统计与随机过程(4)
- 复变函数与积分变换(3)
- 博弈论(2)

选修(学分):/

3.3.3 教学目的和基本要求(Teaching Objectives and Basic Requirements)

(1) 掌握命题逻辑中的命题符号化、命题公式及分类、等值验算、范式与基本的逻辑推理方法,了解全功能集;

(2) 深入理解谓词逻辑中的合式公式及解释,并学会利用谓词逻辑等值式求前束范式;

(3) 熟悉集合的基本概念、基本运算与集合元素的计数方法,学会利用组合分析方法对集合或多重集中的元素进行计数,了解基于递推方程的算法复杂度分析方法;

(4) 理解无限集的势和可数集,了解不可数集和常见集合的势,了解康托集;

(5) 了解图的基本概念并学会图的矩阵表示方法,掌握一些常见的特殊图并了解其重要的应用实例,掌握树的概念与基本分析方法;

(6) 熟练掌握主成分分析方法,掌握矩阵的奇异分解与 K-SVD 算法,掌握稀疏矩阵方程求解的常用方法;

(7) 熟悉矩阵与向量的求导法则,并学会利用求导法则解决实际问题,掌握基于帽子矩阵的多元线性回归方法;

(8) 熟练掌握群、旋转矩阵、变换矩阵、特殊正交群和特殊欧氏群等概念,学会应用旋转向量与四元数;了解李群和李代数的基本概念及作用在其上的映射变换关系;

(9) 掌握二分法、牛顿迭代法、弦截法和迭代法等非线性方程的数值解法,理解高斯消元法、矩阵分解法和迭代法等线性方程组的数值解法;

(10) 熟悉拉格朗日插值、牛顿插值、埃尔米特插值和样条插值等多项式插值和分段插值的方法,掌握最佳一致逼近和最佳平方逼近理论;

(11) 理解最优化问题,掌握凸集、凸函数和凸优化的概念,学会使用黄金分割法和二分法等最优化方法;

(12) 掌握最速下降法、牛顿法和共轭梯度法等无约束优化方法;了解等式约束优化、不等式约束优化和二次规划的基本理论;

(13) 熟悉使用 C 语言和 MATLAB 进行数值计算和优化方法的实现。

3.3.4 教学内容及安排(Syllabus and Arrangements)

第一部分 数理逻辑(Mathematical Logic)

第一章 命题逻辑(Propositional Logic)

章节序号 Chapter Number	章节名称 Chapters	课时 Class Hour	知　识　点 Key Points
1.1	命题符号化及联结词 Symbolization of propositions & connectives	2	（1）命题及其真值 （2）联结词（否定联结词、合取联结词、析取联结词、蕴涵联结词、等价联结词） (1) Propositions and their real values (2) Connectives (negation connectives, conjunction connectives, disjunction connectives, conditional connectives, biconditional connectives)
1.2	命题公式及分类 Propositional formula & classification	2	（1）命题公式及其赋值 （2）真值表 （3）重言式、矛盾式、可满足式 (1) Propositional formula & assignment (2) Truth table (3) Tautology, contradiction, satisfactable formula
1.3	等值验算 Equivalent deduction		（1）置换规则 (1) Replacement rule
1.4	范式 Normal form	2	（1）析取范式、合取范式 （2）主析取范式、主合取范式 (1) Disjunctive normal form and conjunctive normal form (2) Principal disjunctive normal form and principal conjunctive normal form
1.5	联结词全功能集 Set of fully capable connectives	2	（1）联结词全功能集 (1) Set of fully capable connectives
1.6	推理理论 Reasoning theory		（1）前提、推理、结论 (1) Premise, logical deduction, conclusion

第二章　谓词逻辑(Predicate Logic)

章节序号 Chapter Number	章节名称 Chapters	课时 Class Hour	知　识　点 Key Points
2.1	谓词逻辑基本概念 Basic concept of predicate logic	2	(1) 个体词、谓词 (2) 存在量词、全称量词 (3) 特性谓词 (1) Individual term, predicate (2) Existential quantifier, universal quantifier (3) Characteristic predicate
2.2	谓词逻辑合式公式及解释 Well-formed formula in predicate logic and its interpretation	2	(1) 合式公式 (2) 逻辑有效式、矛盾式、可满足式 (1) Well-formed formula (2) Tautology, contradiction, satisfactable formula
2.3	谓词逻辑等值式与前束范式 Logical equivalence and prenex normal in predicate logic	2	(1) 等值式 (2) 前束范式 (1) Logical equivalence (2) Prenex normal

第二部分　集合论与组合分析(Set Theory and Combination Analysis)

第三章　集合的基本概念和运算(Basic Concepts and Operations of Set)

章节序号 Chapter Number	章节名称 Chapters	课时 Class Hour	知　识　点 Key Points
3.1	集合的基本概念 Basic concepts of set	2	(1) 子集、空集、幂集 (1) Subset, empty set, power set
3.2	集合的基本运算 Basic operations of set		(1) 并集、交集、补集 (2) 对称差 (3) 文氏图 (1) Union set, intersection set, complementary set (2) Symmetric difference (3) Venn diagram

续表

章节序号 Chapter Number	章节名称 Chapters	课时 Class Hour	知识点 Key Points
3.3	集合中元素的计数 Cardinality of set	2	(1) 包含排斥原理 (1) Principle of inclusion and exclusion

第四章 组合分析初步(Combinatorial Analysis)

章节序号 Chapter Number	章节名称 Chapters	课时 Class Hour	知识点 Key Points
4.1	加法法则和乘法法则 Sum rule and product rule	2	(1) 加法法则 (2) 乘法法则 (1) Sum rule (2) Product rule
4.2	基本排列组合的计数方法 Counting method of permutation and combination		(1) 排列、组合 (2) 多重集 (1) Permutation, combination (2) Multiple set
4.3	递推方程的求解与应用 Solution and application of recursive equation	2	(1) 迭代 (2) 递推方程 (1) Iteration (2) Recursive equation

第五章 可数集、不可数集、康托集(Countable Set, Uncountable Set, Cantor Set)

章节序号 Chapter Number	章节名称 Chapters	课时 Class Hour	知识点 Key Points
5.1	映射、对等与可数集 Mapping, counter and countable set	2	(1) 映射、满射、单射、双射 (2) 可数集 (1) Mapping, surjection, injection, bijection (2) Countable set

续表

章节序号 Chapter Number	章节名称 Chapters	课时 Class Hour	知识点 Key Points
5.2	不可数集、集合的势 Uncountable set, cardinality of set,	2	(1) 康托闭集定理、不可数集、集合的势、伯恩斯坦定理 (1) Cantor's intersection theorem, uncountable set, cardinality of set, Bernstein's theorem
5.3	康托集 Cantor set		(1) 康托集 (1) Cantor set

第三部分　图论初步(Graph Theory)
第六章　图的基本概念(Basis Concepts of Graph)

章节序号 Chapter Number	章节名称 Chapters	课时 Class Hour	知识点 Key Points
6.1	无向图和有向图 Undirected graph and digraph	2	(1) 无向图、有向图 (2) 顶点、边、握手定理、图的同构 (1) Undirected graph, digraph (2) Vertex, edge, handshake theorem, graph isomorphism
6.2	通路、回路和图的连通性 Pathway, cycle and connectivity of graph	2	(1) 通路、回路、简单通路、简单回路、初级通路、初级回路 (2) 连通、可达、点割集、边割集 (1) Path, cycle, simple path, simple cycle, primary path, primary cycle (2) Connectivity, reachability, vertex cut set, edge cut set
6.3	图的矩阵表示 Matrix representation of graph	2	(1) 关联矩阵、邻接矩阵、可达矩阵 (1) Incidence matrix, adjacency matrix, reachability matrix
6.4	最短路径、关键路径和着色 Shortest path, critical path and coloring	2	(1) 最短路径、关键路径 (2) 着色 (1) Shortest path, critical path (2) Coloring

第七章 特殊的图(Special Graphs)

章节序号 Chapter Number	章节名称 Chapters	课时 Class Hour	知识点 Key Points
7.1	二部图 Bipartite graph	2	(1) 二部图 (1) Bipartite graph
7.2	欧拉图 Euler graph		(1) 欧拉图 (1) Euler graph
7.3	哈密顿图 Hamilton graph	2	(1) 哈密顿图 (1) Hamilton graph
7.4	平面图 Plane graph		(1) 平面图,欧拉公式 (1) Plane graph, Euler's formula

第八章 树(Tree)

章节序号 Chapter Number	章节名称 Chapters	课时 Class Hour	知识点 Key Points
8.1	无向树及生成树 Undirected tree and spanning tree	2	(1) 无向树、生成树 (1) Undirected tree, spanning tree
8.2	根树及其应用 Root tree and its applications	2	(1) 根数、二叉树 (2) 最佳前缀码 (1) Root tree, binary tree (2) Best prefix code

第四部分 矩阵与刚体运动(Matrix and Rigid Body Motion)

第九章 矩阵基础(Matrix Foundation)

章节序号 Chapter Number	章节名称 Chapters	课时 Class Hour	知识点 Key Points
9.1	特征分析 Eigen analysis	3	(1) 特征值与特征向量、特征多项式 (2) 主成分分析 (1) Eigenvalues and eigenvectors, eigen polynomial (2) Principal component analysis

续表

章节序号 Chapter Number	章节名称 Chapters	课时 Class Hour	知 识 点 Key Points
9.2	矩阵的求导 Derivation of matrix	2	(1) 函数矩阵的求导 (2) 雅可比矩阵与梯度矩阵 (1) Derivation of function matrix (2) Jacobian matrix and gradient matrix
9.3	矩阵的奇异分解 Singular decomposition of matrix	1	(1) 奇异分解 (1) Singular decomposition
9.4	广义逆矩阵与 最小二乘法 Least square method	2	(1) 广义逆矩阵 (2) 最小二乘法 (1) Generalized inverse matrix (2) Least square method
9.5	稀疏矩阵 Sparse matrix	3	(1) 稀疏表征 (2) 稀疏矩阵方程的求解 (3) K-SVD算法 (1) Sparse representation (2) Solution of equations with sparse matrix (3) K-SVD algorithm

第十章 应用回归分析初步(Applied Regression Analysis)

章节序号 Chapter Number	章节名称 Chapters	课时 Class Hour	知 识 点 Key Points
10.1	回归问题概述与 线性回归 Overview of regression problems	2	(1) 变量间的统计关系、回归分析 (2) Statistical relations among variables, regression analysis
10.2	线性回归 Linear regression		(1) 一元线性回归 (2) 多元线性回归 (3) 帽子矩阵求解法 (1) Simple regression (2) Multivariate linear regression (3) Hat-matrix method for linear regression

章节序号 Chapter Number	章节名称 Chapters	课时 Class Hour	知 识 点 Key Points
10.3	多元相关性与条件不相关性 Multivariate correlation and conditional uncorrelation	2	(1) 无符号相关系数与无符号不相关系数 (2) 条件不相关系数 (1) Unsigned correlation coefficient and unsigned uncorrelation coefficient (2) Conditional uncorrelation coefficient
10.4	非线性回归 Nonlinear regression	1	(1) 多项式回归 (2) 其他非线性回归模型 (1) Polynomial regression (2) Other nonlinear regression models

第十一章 三维空间刚体运动(Rigid Body Motion in Three-Dimensional Space)

章节序号 Chapter Number	章节名称 Chapters	课时 Class Hour	知 识 点 Key Points
11.1	代数系统 Algebraic system		(1) 群、环、域 (1) Group, ring, field
11.2	刚体运动 Rigid body motion	4	(1) 旋转矩阵与特殊正交群 SO(3) (2) 变换矩阵与特殊欧式群 SE(3) (3) 旋转向量与四元数 (1) Rotation matrix and special orthogonal group SO(3) (2) Transformation matrix and special euclidean group SE(3) (3) Rotation vector and quaternion
11.3	李群与李代数 Lie group and Lie algebra	3	(1) 李群、李代数 (2) SO(3)上的指数映射 (3) SE(3)上的指数映射 (1) Lie group, Lie algebra (2) Exponential map of SO(3) (3) Exponential map of SE(3)

第五部分　数值计算（Numerical Computation）
第十二章　数值计算的数学基础（Mathematics Basis of Numerical Computation）

章节序号 Chapter Number	章节名称 Chapters	课时 Class Hour	知　识　点 Key Points
12.1	数值算法概论 Introduction to numerical algorithm	4	(1) 数值解与逼近解的概念 (1) Concepts of numerical solution and approximate solution
12.2	向量和矩阵范数 Norms of vector and matrix		(1) 范数的定义 (1) Norm definition
12.3	误差 Error		(1) 误差的定义 (1) Definition of error

第十三章　非线性方程的数值解法（Numerical Solutions of Nonlinear Equations）

章节序号 Chapter Number	章节名称 Chapters	课时 Class Hour	知　识　点 Key Points
13.1	非线性方程问题 Nonlinear equation problem	2	(1) 非线性方程问题 (1) Nonlinear equation problem
13.2	二分法 Dichotomy method		(1) 二分法 (1) Dichotomy method
13.3	牛顿迭代法 Newton's iterative method		(1) 牛顿迭代法 (1) Newton's iterative method
13.4	弦截法 Chord section methods	2	(1) 单点弦截法、双点弦截法 (1) Single-point chord section method, double-point chord section method
13.5	迭代法 Iterative methods		(1) 不动点迭代方法 (2) 收敛性质、收敛阶 (1) Fixed point iteration method (2) Convergence property, order of convergence

第十四章　线性方程组的数值解法（Numerical Solutions of Linear Equations）

章节序号 Chapter Number	章节名称 Chapters	课时 Class Hour	知　识　点 Key Points
14.1	线性方程组问题 Linear equations problem		（1）线性方程问题 （1）Linear equations problem
14.2	高斯消元法 Gauss elimination methods	3	（1）高斯消元法 （2）主元素高斯消元法 （3）高斯-约当消元法 （1）Gauss elimination method （2）Principal element Gauss elimination method （3）Gauss-Jordan elimination method
14.3	矩阵分解法 Matrix decomposition methods		（1）矩阵三角分解法 （2）乔列斯基分解法 （1）Matrix triangular decomposition method （2）Cholesky decomposition method
14.4	误差分析 Error analysis	3	（1）不适定问题、病态问题 （2）病态方程组、条件数 （1）Ill-posed problem, ill-conditioned problem （2）Ill-conditioned equation system, condition number
14.5	迭代法 Iterative methods		（1）雅可比迭代法、高斯-赛德尔迭代法 （2）迭代法的收敛性 （1）Jacobi iterative method, Gauss-Seidel iterative method （2）Convergence of iterative method

第十五章　函数插值与逼近方法（Function Interpolation and Approximation Methods）

章节序号 Chapter Number	章节名称 Chapters	课时 Class Hour	知　识　点 Key Points
15.1	函数插值与逼近 Function interpolation and approximation	1	（1）插值的基本概念 （2）逼近的基本概念 （1）Concepts of interpolation （2）Concepts of approximation

续表

章节序号 Chapter Number	章节名称 Chapters	课时 Class Hour	知识点 Key Points
15.2	插值方法 Interpolation methods	4	(1) 拉格朗日插值、分段线性插值、牛顿插值、埃尔米特插值、样条插值 (1) Lagrange interpolation, piecewise linear interpolation, Newton interpolation, Hermite interpolation, spline interpolation
15.3	函数的内积与正交多项式 Innerproduct of functions and orthogonal polynomials	3	(1) 函数的内积 (2) 正交多项式 (1) Innerproduct of functions (2) Orthogonal polynomials
15.4	函数最佳逼近 Optimal approximation of functions		(1) 最佳一致逼近 (2) 最佳平方逼近 (1) Optimal consistent approximation (2) Optimal square approximation

第六部分　优化理论初步（Optimization Theory）

第十六章　优化基础（Basis of Optimization）

章节序号 Chapter Number	章节名称 Chapters	课时 Class Hour	知识点 Key Points
16.1	最优化问题 Optimization problem	1	(1) 最优化问题 (2) 约束优化、无约束优化 (3) 线性规划、二次规划 (1) Optimization problem (2) Constrained optimization, unconstrained optimization (3) Linear programming, quadratic programming
16.2	凸集、凸函数和凸优化 Convex set, convex function and convex optimization	2	(1) 凸集、凸函数 (2) 凸优化 (1) Convex set, convex function (2) Convex optimization
16.3	最优化方法 Optimal method		(1) 最优化算法和收敛性 (2) 黄金分割法、二分法 (1) Optimization algorithm and convergence (2) Golden section method, dichotomy method

第十七章 无约束优化(Unconstrained Optimization)

章节序号 Chapter Number	章节名称 Chapters	课时 Class Hour	知 识 点 Key Points
17.1	无约束优化问题 Unconstrained optimization problem	1	(1) 无约束优化问题 (2) 最小二乘法 (1) Unconstrained optimization problem (2) Least square method
17.2	最速下降法 Steepest descent method	2	(1) 无约束优化方法 (2) 最速下降法 (1) Unconstrained optimization method (2) Steepest descent method
17.3	牛顿法 Newton method		(1) 牛顿法 (2) Newton method
17.4	共轭梯度法 Conjugate gradient method		(1) 共轭梯度法 (2) Conjugate gradient method

第十八章 有约束优化(Constrained Optimization)

章节序号 Chapter Number	章节名称 Chapters	课时 Class Hour	知 识 点 Key Points
18.1	有约束优化 Constrained optimization	2	(1) 等式约束优化 (2) 不等式约束优化 (1) Equality constrained optimization (2) Inequality constrained optimization
18.2	点集配准实例 Examples of point set registration	2	(1) 点集配准问题 (2) 几何变换及其代数表达式 (3) 迭代最近点算法 (4) 场景重建与定位 (1) Point set registration problem (2) Geometric transformation and its algebraic expression (3) Iterative closest point algorithm (4) Scene reconstruction and localization

大纲指导者：郑南宁教授（西安交通大学人工智能学院）

大纲制定者：杜少毅教授（西安交通大学人工智能学院）、汪建基副教授（西安交通大学人工智能学院）

大纲审定：西安交通大学人工智能学院本科专业知识体系建设与课程设置第二版修订工作组

3.4 "概率统计与随机过程"课程大纲

课程名称：概率统计与随机过程
Course：Probability Theory and Stochastic Process
先修课程：工科数学分析、线性代数与解析几何
Prerequisites：Mathematical Analysis for Engineering, Linear Algebra and Analytic Geometry
学分：4
Credits：4

3.4.1 课程目的和基本内容（Course Objectives and Basic Content）

本课程是人工智能学院本科专业基础必修课。

This course is a basic compulsory course for undergraduates in College of Artificial Intelligence.

本课程为计算机科学与人工智能提供了重要的数理统计基础。人工智能的相关方法大多涉及数据分析问题，其中不确定性几乎是不可避免的。因此，引入随机变量并建立相关的理论、模型和方法是人工智能的一个重要理论基础。本课程包含概率论、数理统计和随机过程三部分内容。其中第一章到第四章介绍了概率论中的基本概念及基本原理：随机事件与概率、随机变量及其概率分布、随机变量的数字特征、极限定理等；第五章到第七章介绍了数理统计的基本概念及经典方法：参数估计、假设检验等；第八章和第九章介绍了随机过程的基本知识以及平稳过程等。

课程通过对概率论和数理统计基本知识的学习，要求学生理解并掌握随机事件与概率的基本概念和基本计算方法，理解并掌握随机变量及概率分布的概念及基本性质，掌握随机变量的数学特征的基本概念和计算方法，了解大数定律的基本原理，会用

中心极限定理求近似概率,了解数理统计的基本概念,掌握参数估计及假设检验的基本理论和方法,熟悉随机过程(包括复的)的概论,理解平稳过程的概念、相关函数的性质,了解各态历经性的判定,掌握谱密度的概念、性质和计算方法,了解平稳时间序列的概念、线性模型及模型识别,会进行有关的参数估计并会用这些方法解决一些工程和经济管理中遇到的实际问题。

概率统计与随机过程是从数量方面研究随机现象统计规律性的一门学科,它在人工智能、模式识别、计算机视觉、经济管理、金融投资、保险精算、企业管理等众多领域都有广泛的应用。学习和正确运用概率统计方法已成为对工科类大学生的基本要求。使学生掌握处理随机现象的基本思想和方法,培养他们运用概率统计知识分析和解决实际问题的能力,并为学习后继课程和继续深造打好基础。

This course provides an important mathematical statistics foundation for computer science and artificial intelligence. Most artificial intelligence methods involve data analysis, where uncertainty is almost inevitable. Therefore, the introduction of random variables and the establishment of related theories, models and methods are important theoretical basis of artificial intelligence. This course consists of three parts: Probability Theory, Mathematical Statistics, and Stochastic Process. Chapter 1 to Chapter 4 introduce the basic concepts and principles of probability theory, such as random events and probability, random variables and their probability distribution, digital characteristics of random variables, limit theorem, etc. Chapter 5 to Chapter 7 introduce the basic concepts and classical methods of mathematical statistics, such as parameter estimation, hypothesis test, etc. Chapter 8 to Chapter 9 introduce the basic knowledge of stochastic process and stationary process.

Through studying the basic knowledge of probability theory and mathematical statistics, this course requires students to understand and master the basic concepts and calculation methods of random events and probability. Understand and master the concepts and basic properties of random variables and probability distribution. Grasp the basic concepts and calculation methods of the mathematical characteristics of random variables. Understand the basic principles of the law of large numbers. The approximate probability can be obtained by using the central limit theorem. Understand the basic concepts of mathematical statistics. Grasp the basic theory and method of parameter estimation and hypothesis test. Be familiar with general knowledge of stochastic process, including complex ones. Understand the concept of stationary process and the properties of correlation function. Understand the determination of ergodicity of states. Grasp the concept, properties and calculation methods of spectral density. Understand the concept of stationary time series, linear model and model recognition. The relevant parameters can be estimated and these methods can be used to solve some practical problems encountered in engineering and economic management.

Probabilistic statistics and stochastic process are disciplines that study the statistical regularity of stochastic phenomena in quantity. It is widely used in many fields, such as artificial intelligence, pattern recognition, computer vision, economic management, financial investment, insurance actuarial, enterprise management. Learning and correctly using probability and statistics methods have become the basic requirements for students major in engineering. It is a basic theoretical course for students to master the basic ideas and methods of dealing with random phenomena, to train their abilities to analyze and solve practical problems by using probability and statistics knowledge, and to lay a good foundation for subsequent courses and further studies.

3.4.2 课程基本情况（Course Arrangements）

课程名称	概率统计与随机过程 Probability Theory and Stochastic Process								
开课时间	一年级		二年级		三年级		四年级		数学与统计
	秋	春	秋	春	秋	春	秋	春	
课程定位	本科生"数学与统计"课程群必修课								必修 （学分）	工科数学分析(12)
学　分	4学分								线性代数与解析几何(4)
总学时	64学时 （授课64学时、实验0学时）								计算机科学与人工智能的数学基础(6)
授课学时 分配	课堂讲授(62学时)， 大作业讨论(2学时)								概率统计与随机过程(4)
									复变函数与积分变换(3)
									选修 （学分）	博弈论(2)
									/
先修课程	工科数学分析、线性代数与解析几何							
后续课程								
教学方式	课堂教学、综合大作业							
考核方式	期中考试成绩占30%，期末考试成绩占50%，平时作业占10%，实验成绩占10%							
参考教材	1. 施雨,李耀武.概率论与数理统计应用[M].西安：西安交通大学出版社,2015. 2. 魏平,王宁,符世斌.概率论与数理统计教程[M].西安：西安交通大学出版社,2007.							
参考资料	魏平.概率论与数理统计综合辅导[M].西安：西安交通大学出版社,2007.							
其他信息								

3.4.3 教学目的和基本要求(Teaching Objectives and Basic Requirements)

(1) 理解随机事件与概率的基本概念,掌握其基本计算方法;
(2) 掌握随机变量及概率分布的概念及基本性质;
(3) 熟悉随机变量的数学特征的基本概念,掌握其计算方法;
(4) 了解大数定律的基本原理,会用中心极限定理求近似概率;
(5) 理解数理统计的基本概念,掌握参数估计及假设检验的基本理论和方法;
(6) 熟悉随机过程的概论,理解平稳过程的概念、相关函数的性质;
(7) 了解各态历经性的判定,掌握谱密度的概念、性质和计算方法;
(8) 掌握平稳时间序列的概念、线性模型及模型识别;
(9) 会进行有关的参数估计并会用这些方法解决一些工程和经济管理中的实际问题。

3.4.4 教学内容及安排(Syllabus and Arrangements)

第一章 随机事件与概率(Random Events and Probability)

章节序号 Chapter Number	章节名称 Chapters	课时 Class Hour	知识点 Key Points
1.1	随机事件 Random events	1	(1) 随机现象与随机试验 (2) 样本空间与随机事件 (3) 事件的关系与运算 (1) Random phenomena and random experiments (2) Sample space and random events (3) The relation and operation of events
1.2	概率 Probability	1	(1) 概率的古典定义 (2) 概率的统计定义 (3) 概率的公理化定义 (4) 概率的性质 (1) Classical definition of probability (2) Statistical definition of probability (3) Axiomatic definition of probability (4) The properties of probability

续表

章节序号 Chapter Number	章节名称 Chapters	课时 Class Hour	知 识 点 Key Points
1.3	古典概率的计算 The calculation of classical probability	2	(1) 古典概率的计算方法 (1) The calculating method of classical probability
1.4	条件概率， 事件的独立性 Conditional probability, event independence	4	(1) 条件概率与乘法定理 (2) 全概率公式与贝叶斯公式 (3) 事件的独立性 (1) Conditional probability and multiplication theorem (2) Total probability formula and Bayesian formula (3) Independence of events

第二章　随机变量及概率分布（Random Variables and Probability Distribution）

章节序号 Chapter Number	章节名称 Chapters	课时 Class Hour	知 识 点 Key Points
2.1	一维随机变量 One-dimensional random variable	4	(1) 随机变量与分布函数 (2) 离散型随机变量 (3) 连续型随机变量 (1) Random variables and distribution functions (2) Discrete random variables (3) Continuous random variables
2.2	二维随机变量 Two-dimensional random variable	4	(1) 二维随机变量与联合分布函数 (2) 二维离散型随机变量 (3) 二维连续型随机变量 (1) Two-dimensional random variables and joint distribution function (2) Two-dimensional discrete random variables (3) Two-dimensional continuous random variables

续表

章节序号 Chapter Number	章节名称 Chapters	课时 Class Hour	知识点 Key Points
2.3	条件分布 Conditional distribution	1	(1) 条件分布律 (2) 条件概率密度 (1) Conditional distribution law (2) Conditional probability density
2.4	随机变量的相互独立性 Interdependence of random variables	1	(1) 随机变量的相互独立性 (1) Interdependence of random variables
2.5	随机变量函数的概率分布 Probability distribution of functions of random variables	2	(1) 一维随机变量的函数的概率分布 (2) 二维随机变量的函数的概率分布 (1) Probability distribution of functions of one-dimensional random variables (2) Probability distribution of functions of two-dimensional random variables

第三章 随机变量的数字特征(Digital Characteristics of Random Variables)

章节序号 Chapter Number	章节名称 Chapters	课时 Class Hour	知识点 Key Points
3.1	数学期望 Mathematical expectation	2	(1) 数学期望的定义 (2) 随机变量函数的数学期望 (3) 数学期望的性质 (1) Definition of mathematical expectation (2) Mathematical expectations of functions of random variables (3) The properties of mathematical expectation

续表

章节序号 Chapter Number	章节名称 Chapters	课时 Class Hour	知识点 Key Points
3.2	方差 Variance	2	(1) 方差和标准差 (2) 方差的性质 (1) Variance and standard deviation (2) The properties of variance
3.3	协方差与相关系数,矩 Covariance and correlation coefficient, moment	2	(1) 协方差与相关系数 (2) 矩 (3) 协方差矩阵 (1) Covariance and correlation coefficient (2) Moment (3) Covariance matrix

第四章 大数定律及中心极限定理(Law of Large Numbers and Central Limit Theorem)

章节序号 Chapter Number	章节名称 Chapters	课时 Class Hour	知识点 Key Points
4.1	大数定律 Law of large number	1.5	(1) 切比雪夫不等式 (2) 切比雪夫大数定律 (3) 贝努利大数定律 (1) Chebyshev inequality (2) Chebyshev's law of large number (3) Bernoulli law of large number
4.2	中心极限定理 Central limit theorem	1.5	(1) 独立同分布的中心极限定理 (2) 不同分布的中心极限定理 (1) Central limit theorem of independent and identical distribution (2) Central limit theorem of different distributions

第五章　数理统计的基本概念(Basic Concept of Mathematical Statistics)

章节序号 Chapter Number	章节名称 Chapters	课时 Class Hour	知　识　点 Key Points
5.1	总体与样本 Population and sample	1	(1) 总体及分布 (2) 样本 (1) Population and distribution (2) Sample
5.2	样本分布 Sample distribution	1	(1) 样本频数分布与频率分布 (2) 频率直方图 (3) 经验分布函数 (1) Sample frequency distribution and frequency distribution (2) Frequency histograms (3) Empirical distribution function
5.3	统计量 Statistic	1	(1) 统计量概念 (2) 几个常用的统计量 (1) Concept of statistics (2) Several commonly used statistics
5.4	抽样分布 Sampling distribution	2	(1) 几个常用的重要分布 (2) 分位数 (3) 正态总体的抽样分布 (1) Several commonly used important distributions (2) Quantiles (3) Sampling distribution of normal population

第六章　参数估计(Parameter Estimation)

章节序号 Chapter Number	章节名称 Chapters	课时 Class Hour	知　识　点 Key Points
6.1	点估计 Point estimation	2	(1) 矩估计法 (2) 极大似然估计法 (1) Moment estimation method (2) Maximum likelihood estimation

续表

章节序号 Chapter Number	章节名称 Chapters	课时 Class Hour	知 识 点 Key Points
6.2	估计量的评选标准 Criteria for selection of estimators	1	(1) 无偏性 (2) 有效性 (3) 相合性 (1) Unbiased (2) Effectiveness (3) Consistency
6.3	区间估计 Interval estimation	2	(1) 双侧区间估计 (2) 单侧区间估计 (1) Bilateral interval estimation (2) Unilateral interval estimation
6.4	正态总体参数的区间估计 Interval estimation of normal population parameters	1	(1) 单个总体 $N(\mu,\sigma^2)$ 的情形 (2) 两个总体 $N(\mu_1,\sigma_1^2)$ 和 $N(\mu_2,\sigma_2^2)$ 的情形 (1) The case of a single population (2) The case of two populations

第七章 假设检验(Hypothesis Test)

章节序号 Chapter Number	章节名称 Chapters	课时 Class Hour	知 识 点 Key Points
7.1	假设检验的基本概念 Basic concepts of hypothesis test	1	(1) 假设检验的基本原理 (2) 假设检验的一般步骤 (1) Basic principles of hypothesis test (2) General steps of hypothesis test
7.2	正态总体参数的假设检验 Hypothesis test of normal population parameters	2	(1) 单个总体 $N(\mu,\sigma^2)$ 的情形 (2) 两个总体 $N(\mu_1,\sigma_1^2)$ 和 $N(\mu_2,\sigma_2^2)$ 的情形 (1) The case of a single population (2) The case of two populations

续表

章节序号 Chapter Number	章节名称 Chapters	课时 Class Hour	知 识 点 Key Points
7.3	单边假设检验 Unilateral hypothesis test	1	(1) 单边假设 (1) Unilateral hypothesis
7.4	参数假设的大样本检验 Large sample Test of parametric hypothesis	1	(1) 参数假设的大样本检验方法 (1) Large sample test method for parametric hypothesis
7.5	总体分布的假设检验 Hypothesis test of population distribution	1	(1) 分布拟合检验 (2) 皮尔逊定理 (3) χ^2 拟合检验法 (1) Distribution fitting test (2) Pearson theorem (3) χ^2 fitting test method

第八章 随机过程的基本知识(Basic Knowledge of Stochastic Process)

章节序号 Chapter Number	章节名称 Chapters	课时 Class Hour	知 识 点 Key Points
8.1	随机过程的概念 The concept of stochastic process	2	(1) 随机过程的概念和记号 (1) Concept and notation of stochastic process
8.2	随机过程的概率特征 Probabilistic characteristics of stochastic process	2	(1) 有限维分布函数族 (2) 随机过程的数字特征 (3) 两个随机过程的不相关与相互独立 (1) Finite dimensional distribution function family (2) Digital characteristics of stochastic process (3) Uncorrelated and independent of two random processes

续表

章节序号 Chapter Number	章节名称 Chapters	课时 Class Hour	知 识 点 Key Points
8.3	随机过程的基本类型 Basic types of stochastic process	2	(1) 按参数集与状态空间分类 (2) 按过程的性质特点分类 (1) Classification by parameter set and state space (2) Classification according to the nature and characteristics of the process
8.4	泊松过程与布朗运动 Poisson process and Brownian motion	2	(1) 泊松过程的定义与性质 (2) 布朗运动 (1) Definition and properties of Poisson process (2) Brownian motion

第九章 平稳过程(Stationary Process)

章节序号 Chapter Number	章节名称 Chapters	课时 Class Hour	知 识 点 Key Points
9.1	平稳过程概念 The concept of stationary process	2	(1) 平稳过程的概念 (1) The concept of stationary process
9.2	相关函数的性质 Properties of correlation function	2	(1) 自相关函数的性质 (2) 互相关函数的性质 (1) Properties of auto-correlation function (2) Properties of cross-correlation function
9.3	平稳过程的谱密度 Spectral density of stationary process	2	(1) 相关过程的谱分解 (2) 谱密度的物理意义 (3) 谱密度与互谱密度的性质 (4) 相关函数与谱密度之间的变换 (1) Spectral decomposition of related process (2) Physical significance of spectral density (3) Properties of spectral density and cross-spectral density (4) Transform between correlation function and spectral density

章节序号 Chapter Number	章节名称 Chapters	课时 Class Hour	知 识 点 Key Points
9.4	各态的历经性 Ergodicity of states	2	(1) 各态历经性概念 (2) 各态历经定理 (3) 各态历经的应用 (1) The concept of ergodicity of states (2) Ergodic theorems of states (3) Applications of ergodic states

大纲制定者：西安交通大学数学与统计学院数学教学中心

大纲修订者：杜少毅教授（西安交通大学人工智能学院）、汪建基副教授（西安交通大学人工智能学院）

大纲审定：西安交通大学人工智能学院本科专业知识体系建设与课程设置第二版修订工作组

3.5 "复变函数与积分变换"课程大纲

课程名称：复变函数与积分变换
Course：Complex Analysis and Integral Transformation
先修课程：工科数学分析、线性代数与解析几何
Prerequisites：Mathematical Analysis for Engineering，Linear Algebra and Analytic Geometry
学分：3
Credits：3

3.5.1 课程目的和基本内容（Course Objectives and Basic Content）

本课程是人工智能学院本科专业基础必修课。

This course is a basic compulsory course for undergraduates in College of Artificial Intelligence.

本课程为数字信号处理等专业课打好基础，培养学生的数学素质，提高其应用数学知识解决实际问题的能力，也为计算机科学与人工智能的学习提供了重要的理论基

础。本课程旨在使学生初步掌握复变函数与积分变换的基本理论和方法,为学习有关后继课程和进一步扩大数学知识面奠定必要的基础。本课程的内容包括:复数与复变函数、复变函数的导数及其性质,复变函数的积分及其性质,解析函数的性质(包括高阶导数公式)、幂级数和罗伦级数的展开,孤立奇点的分类(包括无穷远点),留数及其应用,共形映射的概念及性质(特别要掌握双线性映射以及几个初等函数定义的映射所具有的性质),傅里叶变换及其性质,拉普拉斯变换及其应用。

This course lays a foundation for major courses such as digital signal processing and etc. , which cultivates students' mathematical quality and improves the students' ability to apply mathematics knowledge to solve practical problems. It also provides an important theoretical foundation for the study of computer science and artificial intelligence. The course is offered to make students grasp the basic theories and methods of complex analysis and integral transformation, and lay a necessary foundation for learning the subsequent courses and further expanding mathematical knowledge. The content of this course includes: complex and complex function, derivative of complex function and its properties, integral of complex function and its properties, properties of analytic functions (including higher derivative formulas), expansion of power series and Loren series, classification of isolated singularities (including infinite points), residual number and its applications, the concept and properties of conformal mapping (in particular, students should master the properties of bilinear mappings and mappings defined by several elementary functions), Fourier transform and its properties, Laplace transform and its applications.

3.5.2 课程基本情况(Course Arrangements)

课程名称	复变函数与积分变换 Complex Analysis and Integral Transformation								
开课时间	一年级		二年级		三年级		四年级		数学与统计
	秋	春	秋	春	秋	春	秋	春	
课程定位	本科生"数学与统计"课程群必修课								必修 (学分)	工科数学分析(12)
学分	3学分								线性代数与解析几何(4)
									计算机科学与人工智能的数学基础(6)
总学时	48学时 (授课48学时、实验0学时)								概率统计与随机过程(4)
									复变函数与积分变换(3)
									博弈论(2)
授课学时 分配	课堂讲授(40学时), 小组讨论(8学时)								选修 (学分)	/

续表

先修课程	工科数学分析、线性代数与解析几何
后续课程	数字信号处理
教学方式	课堂教学、大作业、小组讨论
考核方式	闭卷考试成绩占80%，平时成绩占20%
参考教材	1. 王绵森.复变函数[M].北京：高等教育出版社，2008. 2. 张元林.积分变换[M].北京：高等教育出版社，2004.
参考资料	王绵森.复变函数学习辅导与习题选解[M].北京：高等教育出版社，2004.
其他信息	

3.5.3 教学目的和基本要求（Teaching Objectives and Basic Requirements）

（1）掌握复数的各种表示方法及其运算，了解区域的概念，了解复球面与无穷远点的概念，理解复变函数的基本概念，了解复变函数的极限和连续性的概念；

（2）理解复变函数的导数及复变函数解析的概念，掌握复变函数解析的充要条件，了解调和函数与解析函数的关系，会从解析函数的实（虚）部求其虚（实）部，了解指数函数、三角函数、双曲函数、对数函数及幂函数的定义及它们的主要性质（包括在单值域中的解析性）；

（3）了解复变函数积分的定义及性质，会求复变函数的积分，理解柯西积分定理，掌握柯西积分公式和解析函数的高阶导数公式，了解解析函数无限次可导的性质；

（4）理解复数项级数收敛、发散及绝对收敛等概念，了解幂级数收敛的概念，会求幂级数的收敛半径，了解幂级数在收敛圆内的一些基本性质，理解泰勒定理，了解 e^z，$\sin z$，$\cos z$，$\ln(1+z)$，$(1+z)^\mu$ 的马克劳林展开式，并会利用它们将一些简单的解析函数展开为幂级数，理解洛朗定理及孤立奇点的分类（包括无穷远点），会用间接方法将简单的函数在其孤立奇点附近展开为洛朗级数；

（5）熟悉留数概念，掌握极点处留数的求法（包括无穷远点），掌握留数定理，掌握用留数求围道积分的方法，会用留数求一些实变函数的积分；

（6）掌握解析函数导数的几何意义及共形映射的概念，掌握线性映射的性质和分式性映射的保圆性及保对称性，了解函数 $w=z^\alpha$（α 为正有理数）、$w=e^z$ 和有关映射的性质，会求一些简单区域（例如平面、半平面、角形域、圆、带形域等）之间的共形映射；

(7) 理解傅里叶变换的概念,掌握傅里叶变换的性质,了解傅里叶变换的基本应用;

(8) 熟悉拉普拉斯变换的概念,掌握拉普拉斯变换的性质,了解拉普拉斯变换的基本应用。

3.5.4 教学内容及安排(Syllabus and Arrangements)

第一章 复数与复变函数(Complex Number and Complex Function)

章节序号 Chapter Number	章节名称 Chapters	课时 Class Hour	知 识 点 Key Points
1.1	复数的表示与运算 Representation and operation of complex number	3	(1) 区域的概念 (2) 复球面与无穷远点的概念 (1) The concept of region (2) The concept of complex sphere and infinite point
1.2	复变函数的基本概念 Basic concept of complex function		(1) 复变函数的极限 (2) 复变函数的连续性 (1) Limit of complex function (2) Continuity of complex function

第二章 解析函数及其在平面场中的应用(Analytic Function and Its Application in Plane Field)

章节序号 Chapter Number	章节名称 Chapters	课时 Class Hour	知 识 点 Key Points
2.1	解析函数的概念 The concept of analytic function	4	(1) 复变函数的导数 (2) 复变函数解析的概念 (3) 复变函数解析的充要条件 (1) Derivative of complex function (2) The concept of analysis of complex function (3) Necessary and sufficient condition of analysis of complex function

续表

章节序号 Chapter Number	章节名称 Chapters	课时 Class Hour	知　识　点 Key Points
2.2	解析函数的性质 Properties of analytic function		(1) 调和函数与解析函数的关系 (2) 从解析函数的实(虚)部求其虚(实)部 (3) 指数函数、三角函数、双曲函数、对数函数及幂函数的定义及它们的主要性质(包括在单值域中的解析性) (1) The relation between harmonic function and analytic function (2) Finding the real (virtual) part of analytic function from the real(virtual) part (3) Definitions of exponential function, trigonometric function, hyperbolic function, logarithmic function and power function and their main properties (including analyticity in single value domain)

第三章　复变函数的积分(Integral of Complex Function)

章节序号 Chapter Number	章节名称 Chapters	课时 Class Hour	知　识　点 Key Points
3.1	复变函数积分的定义 Definition of complex function integral	1	(1) 复变函数积分的概念 (2) 复变函数积分的性质 (1) The concept of complex function integral (2) Properties of complex function integral
3.2	复变函数积分的公式 Formula for integral of complex function	1	(1) 复变函数积分的求解 (1) Solving the integral of complex function
3.3	柯西积分定理 Cauchy integral theorem	1	(1) 柯西积分定理的定义 (2) 柯西积分定理的性质 (1) Definition of Cauchy integral theorem (2) Properties of Cauchy integral theorem

续表

章节序号 Chapter Number	章节名称 Chapters	课时 Class Hour	知　识　点 Key Points
3.4	柯西积分公式 Cauchy integral formula	1	(1) 柯西积分的求解 (1) Solution of Cauchy integral
3.5	解析函数的 高阶导数公式 The formula of higher order derivative of analytic function	1	(1) 解析函数的高阶导数求解 (1) Solution of higher order derivative of analytic function

第四章　复变函数项级数(Term Series of Complex Function)

章节序号 Chapter Number	章节名称 Chapters	课时 Class Hour	知　识　点 Key Points
4.1	复变函数项级数 Term series of complex function	1	(1) 复数项级数收敛 (2) 复数项级数发散 (3) 复数项级数绝对收敛 (1) Convergence of complex series (2) Divergence of complex series (3) Absolute convergence of complex series
4.2	幂级数收敛 Convergence of power series	1	(1) 幂级数的收敛半径 (2) 幂级数在收敛圆内的基本性质 (1) Convergence radius of power series (2) Basic properties of power series in convergent circle
4.3	泰勒定理 Taylor theorem	1	(1) 泰勒展开式 (2) 马克劳林展开式 (3) 解析函数展开为幂级数 (1) Taylor expansion (2) Marklaurin expansion (3) Analytic function expands to power series

续表

章节序号 Chapter Number	章节名称 Chapters	课时 Class Hour	知 识 点 Key Points
4.4	洛朗定理 Laurent theorem	2	(1) 洛朗定理 (2) 孤立奇点的分类 (3) 用间接方法将简单的函数在其孤立奇点附近展开为洛朗级数 (1) Laurent theorem (2) Classification of isolated singularities (3) Simple functions are expanded to Laurent series near their isolated singularities via indirect method

第五章 留数及其应用(Residue and Its Application)

章节序号 Chapter Number	章节名称 Chapters	课时 Class Hour	知 识 点 Key Points
5.1	留数的概念 The concept of residue	1	(1) 留数的定义 (2) 极点处留数的求法 (1) Definition of residue (2) Solution of residual number at pole
5.2	留数定理 Residue theorem	2	(1) 留数定理 (2) 留数求围道积分的方法 (1) Residue theorem (2) The method of finding contour integral by residual number
5.3	留数的应用 Applications of residue	2	(1) 用留数求一些实变函数的积分 (1) Solving integrals of some real analysis by residual number

第六章 共形映射(Conformal Mapping)

章节序号 Chapter Number	章节名称 Chapters	课时 Class Hour	知识点 Key Points
6.1	共形映射的概念 The concept of conformal mapping	1	(1) 解析函数导数的几何意义 (2) 共形映射的概念 (1) The geometric meaning of derivative of analytic function (2) The concept of conformal mapping
6.2	线性映射和分式性映射的性质 Properties of linear mapping and fractional mapping	2	(1) 线性映射的性质 (2) 分式性映射的保圆性 (3) 分式性映射的保对称性 (4) 函数 $w=z^\alpha$(α 为正有理数)、$w=e^z$ 和有关映射的性质 (1) Properties of linear mapping (2) Roundness preservation of fractional mapping (3) Symmetry preservation of fractional mapping (4) Properties of mappings of $w=z^\alpha$ (α is a positive rational number), $w=e^z$ and other functions
6.3	共形映射的求解 Solution of conformal mapping	1	(1) 求解简单区域(例如平面、半平面、角形域、圆、带形域等)之间的共形映射 (1) Solving conformal mappings between simple domains (e.g. plane, half plane, angular domain, circle, band domain, etc.)

第七章 傅里叶变换(Fourier Transform)

章节序号 Chapter Number	章节名称 Chapters	课时 Class Hour	知识点 Key Points
7.1	傅里叶变换的概念 The concept of Fourier transform	1	(1) 傅里叶变换的定义 (1) Definition of Fourier transform

续表

章节序号 Chapter Number	章节名称 Chapters	课时 Class Hour	知识点 Key Points
7.2	傅里叶变换的性质（一） Properties of Fourier transform(1)	2	（1）傅里叶变换的性质（一） (1) Properties of Fourier transform(1)
7.3	傅里叶变换的性质（二） Properties of Fourier transform(2)	2	（1）傅里叶变换的性质（二） (1) Properties of Fourier transform(2)
7.4	傅里叶变换的应用（一） Applications of Fourier transform(1)	1	（1）傅里叶变换的应用（一） (1) Applications of Fourier transform(1)
7.5	傅里叶变换的应用（二） Applications of Fourier transform(2)	1	（1）傅里叶变换的应用（二） (1) Applications of Fourier transform(2)

第八章　拉普拉斯变换（Laplace Transform）

章节序号 Chapter Number	章节名称 Chapters	课时 Class Hour	知识点 Key Points
8.1	拉普拉斯变换的概念 Concept of Laplace transform	1	（1）拉普拉斯变换的概念 (1) The concept of Laplace transform
8.2	拉普拉斯变换的性质（一） Properties of Laplace transform(1)	2	（1）拉普拉斯变换的性质（一） (1) Properties of Laplace transform(1)
8.3	拉普拉斯变换的性质（二） Properties of Laplace transform(2)	1	（1）拉普拉斯变换的性质（二） (1) Properties of Laplace transform(2)
8.4	拉普拉斯变换的应用（一） Applications of Laplace transform(1)	1	（1）拉普拉斯变换的应用（一） (1) Applications of Laplace transform(1)

续表

章节序号 Chapter Number	章节名称 Chapters	课时 Class Hour	知识点 Key Points
8.5	拉普拉斯变换的应用(二) Applications of Laplace transform(2)	2	(1) 拉普拉斯变换的应用(二) (1) Applications of Laplace transform(2)

大纲制定者：西安交通大学数学与统计学院数学教学中心

大纲修订者：杜少毅教授(西安交通大学人工智能学院)、汪建基副教授(西安交通大学人工智能学院)

大纲审定：西安交通大学人工智能学院本科专业知识体系建设与课程设置第二版修订工作组

3.6 "博弈论"课程大纲

课程名称：博弈论

Course：Game Theory

先修课程：工科数学分析、概率统计与随机过程

Prerequisites：Mathematical Analysis for Engineering，Probability Theory and Stochastic Process

学分：2

Credits：2

3.6.1 课程目的和基本内容(Course Objectives and Basic Content)

本课程是人工智能学院本科专业基础必修课。

This course is a basic compulsory course for undergraduates in College of Artificial Intelligence.

该课程介绍有关决策主体的行为产生相互作用时，各决策主体之间的最优策略

选择以及策略均衡的知识体系。博弈论不仅是现代经济学的一个标准分析工具,而且在管理学、政治学、国际关系学、军事战略等学科有着广泛的应用。随着博弈论的不断发展和完善,该理论逐渐被应用到电力系统、人工智能等工程设计领域。近几年,算法博弈论迅速发展,并与多智能系统研究融合,其普及程度已逐渐追赶上人工智能的发展。博弈论的思维模式和分析方法将会对人工智能领域的研究起到重要的推动作用。

该课程以博弈的信息结构、博弈过程和博弈方式为主线,按照博弈的类型,用 8 章的内容系统介绍以下 7 种博弈的基本原理和分析方法:完全信息静态博弈、完全且完美信息动态博弈、重复博弈、完全但不完美信息动态博弈、不完全信息静态博弈、不完全信息动态博弈、博弈学习和进化博弈。第一章是该课程的导论,主要通过与日常生活密切相关的博弈游戏介绍博弈的基本特征和博弈的分类。第二章到第八章分别介绍上述 7 种博弈。

该课程主要采用课堂授课模式,并辅之以小组讨论、行为经济学实验等教学手段。在课堂授课中,将始终以博弈实例或博弈模型分析为基本手段,帮助学生通过实例分析掌握博弈论的基本概念、基本原理和分析方法,最终使得学生在自己的知识体系中构建起博弈论的理论框架,在思维习惯上培养学生用博弈思想分析决策问题的思维模式,在掌握的分析工具中学会应用博弈分析方法。在小组讨论中,将训练学生用博弈论的基本理论和方法分析解决实际决策问题的能力,并引导学生逐渐将博弈论的思想融入人工智能领域的学习和研究中。

This course introduces the knowledge system about optimal strategy choice and strategic equilibrium among players when the behavior of players interacts. Game Theory is not only a standard analytical tool for modern economics, but also has a wide range of applications in Management Science, Politics, International Relations, Military Strategy, and other disciplines. With the continuous development and improvement of Game Theory, this theory has been used in engineering design fields such as power systems and artificial intelligence. In recent years, algorithm game theory has developed rapidly and merged with multi-intelligent systems, and its popularity has gradually caught up with the development of AI. The thinking mode and analysis method of game theory will play an important role in promoting the research in the field of AI.

Based on the main logical line of information structures, game processes and game modes, according to the types of game, the basic principles and analysis methods of seven kinds of games are introduced in eight chapters. The theories of seven games are as follows: games with complete information, dynamic games with complete and perfect information, repeated games, dynamic games with complete but imperfect information, static games with incomplete information, dynamic games with incomplete information, game learning and evolutionary games. Chapter 1 is an introduction to this course, which introduces the basic characteristics of the game and the classification of games mainly through games closely related to daily life. The seven types of games are introduced in Chapter 2 to Chapter 8 respectively.

The course mainly adopts classroom teaching mode, supplemented by group discussion, behavioral economics experiment, and other teaching methods. In classroom teaching, we will use game examples or game models analysis as the basic means to help students obtain the basic concepts, basic principles and analysis methods of Game Theory. By teaching this course, we expect to enable students to construct a theoretical framework of Game Theory in their own knowledge system, to train students' thinking mode of using Game Theory to analyze decision-making problems in their thinking habits, and to teach students can use game analysis tool in their tool boxes. In group discussions and behavioral economics experiments, students will be trained to use the basic theory and methods of Game Theory to analyze and solve practical decision-making problems, and guide students to gradually incorporate the idea of game theory into the study and research of AI.

3.6.2 课程基本情况（Course Arrangements）

课程名称	博弈论 Game Theory									
开课时间	一年级		二年级		三年级		四年级		数学与统计	
	秋	春	秋	春	秋	春	秋	春	必修 (学分)	工科数学分析(12)
课程定位	本科生"数学与统计"课程群必修课									线性代数与解析几何(4)
学　分	2 学分									计算机科学与人工智能 的数学基础(6)
总学时	32 学时 (授课 32 学时、实验 0 学时)									概率统计与随机过程(4)
										复变函数与积分变换(3)
										博弈论(2)
授课学时 分配	课堂讲授(32 学时)									选修 (学分)	/

续表

先修课程	工科数学分析、概率统计与随机过程
后续课程	
教学方式	课堂教学、小组讨论
考核方式	闭卷考试成绩占70%，小组讨论占5%，平时成绩占25%
参考教材	谢识予.经济博弈论[M].上海：复旦大学出版社，2018.
参考资料	1. 葛泽慧，于艾琳，赵瑞，等.博弈论入门[M].北京：清华大学出版社，2018. 2. Gibbons M.博弈论基础[M].高峰，译.北京：中国社会科学出版社，1999. 3. Maschler M，Solan E，Zamir S.博弈论[M].赵世永，译.北京：格致出版社，2018.
其他信息	

3.6.3 教学目的和基本要求（Teaching Objectives and Basic Requirements）

（1）理解博弈论的基本概念、博弈的特征和博弈的分类；

（2）熟练掌握完全信息静态博弈的基本分析思路和方法、纳什均衡、混合策略和混合策略纳什均衡；

（3）熟悉完全且完美信息动态博弈的表示法和特点、子博弈和子博弈完美纳什均衡、逆推归纳法；

（4）了解有限次重复博弈和无限次重复博弈；

（5）理解完全但不完美信息动态博弈以及完美贝叶斯均衡；

（6）熟练掌握不完全信息静态博弈和贝叶斯纳什均衡；

（7）了解不完全信息动态博弈、声明博弈和信号博弈；

（8）掌握有限理性博弈、博弈学习模型和演化博弈论。

3.6.4 教学内容及安排（Syllabus and Arrangements）

第一章 导论（Introduction）

章节序号 Chapter Number	章节名称 Chapters	课时 Class Hour	知识点 Key Points
1.1	从游戏到决策理论 From game to decision theory	1	（1）博弈论的学习目的及学习内容 （1）Learning purpose and contents of game theory

续表

章节序号 Chapter Number	章节名称 Chapters	课时 Class Hour	知 识 点 Key Points
1.2	典型的博弈例子 Typical game examples	1	(1) 囚徒困境 (2) 双寡头竞价博弈 (3) 田忌赛马 (4) 古诺模型 (5) 空城计中的博弈 (1) Prisoners' dilemma (2) Duopoly bidding game (3) Tianji horse racing (4) Cournot model (5) Game in empty city planning
1.3	博弈的特征和 博弈的分类 Characteristics and classification of game	1	(1) 博弈方 (2) 博弈策略 (3) 博弈过程 (4) 博弈得益 (5) 博弈信息结构 (6) 博弈决策方式 (7) 博弈方式 (8) 博弈理论结构 (1) Game players (2) Game strategy (3) Game process (4) Game payoff (5) Information structure of game (6) Decision-making mode of game (7) Game mode (8) Structure of game theory

第二章　完全信息静态博弈(Game with Complete Information)

章节序号 Chapter Number	章节名称 Chapters	课时 Class Hour	知识点 Key Points
2.1	博弈的基本分析思路和方法 Basic analytical thinking and methods of game	1	(1) 上策均衡 (2) 严格下策反复消去法 (3) 划线法 (4) 箭头法 (1) Dominant strategy equilibrium (2) Iterated elimination of weakly dominated strategies (3) Line-drawing method (4) Arrow method
2.2	纳什均衡 Nash equilibrium		(1) 纳什均衡的定义 (2) 纳什均衡与严格下策反复消去法 (3) 纳什均衡的一致预测性质 (1) Definition of Nash equilibrium (2) Nash equilibrium and iterated elimination of weakly dominated strategies (3) Uniform prediction of Nash equilibrium
2.3	无限策略博弈分析和反应函数 Infinite strategy game analysis and response function	1	(1) 古诺模型 (2) 反应函数 (3) 伯特兰德寡头模型 (4) 公共资源问题 (5) 反应函数的问题 (1) Cournot model (2) Response function (3) Bertrand oligopoly model (4) Public resources problem (5) Problem of response function

续表

章节序号 Chapter Number	章节名称 Chapters	课时 Class Hour	知识点 Key Points
2.4	混合策略和混合策略纳什均衡 Mixed strategy and mixed strategy Nash equilibrium	2	(1) 严格竞争博弈和混合策略的引进 (2) 多重均衡博弈和混合策略 (3) 混合策略和严格下策反复消去法 (4) 混合策略反应函数 (1) Strict competition game and introduction of mixed strategy (2) Multiple equilibrium game and mixed strategy (3) Mixed strategy and iterated elimination of weakly dominated strategies (4) Mixed strategy response function
2.5	纳什均衡的存在性 The existence of Nash equilibrium		(1) 纳什定理 (2) 纳什定理的意义和扩展 (1) Nash theorem (2) Significance and extension of Nash theorem
2.6	纳什均衡的选择和分析方法扩展 Selection and analysis method extension of Nash equilibrium	1	(1) 帕累托和风险上策均衡 (2) 聚点和相关均衡 (3) 共谋和防共谋均衡 (1) Pareto dominant strategy equilibrium and risk dominant strategy equilibrium (2) Focus equilibrium and relevance equilibrium (3) Collusion and anti-collusion equilibrium

第三章 完全且完美信息动态博弈(Dynamic Game with Complete and Perfect Information)

章节序号 Chapter Number	章节名称 Chapters	课时 Class Hour	知识点 Key Points
3.1	动态博弈的表示法和特点 Representation and characteristics of dynamic game	2	(1) 动态博弈的阶段和扩展形表示 (2) 动态博弈的基本特点 (1) Stage and extensive-form of dynamic game (2) Basic characteristics of dynamic game

续表

章节序号 Chapter Number	章节名称 Chapters	课时 Class Hour	知 识 点 Key Points
3.2	策略的可信性和纳什均衡的问题 The credibility of strategy and problem of a Nash equilibrium		(1) 选择和策略的可信性问题 (2) 纳什均衡的问题 (3) 逆推归纳法 (1) Selection and credibility of a strategy (2) Problem of Nash equilibrium (3) Reverse induction
3.3	子博弈和子博弈完美纳什均衡 Subgame and subgame perfect Nash equilibrium		(1) 子博弈与子博弈完美纳什均衡 (1) Subgame and subgame perfect Nash equilibrium
3.4	经典的动态博弈模型 Classical dynamic game model	1	(1) 斯塔克博格模型 (2) 劳资博弈 (3) 议价博弈 (4) 委托人-代理人理论 (1) Stackelberg model (2) The game between labor unions and firms (3) Bargaining game (4) Principal-agent theory
3.5	有同时选择的动态博弈模型 Dynamic game model with simultaneous selection	1	(1) 标准模型 (2) 间接融资和挤兑风险 (3) 国际竞争和最优关税 (4) 有同时选择的委托人-代理人关系 (1) Standard model (2) Indirect financing and risk of bank run (3) International competition and optimal tariff (4) Principal-agent relationship with simultaneous choice

续表

章节序号 Chapter Number	章节名称 Chapters	课时 Class Hour	知　识　点 Key Points
3.6	动态博弈分析的问题和扩展讨论 Problems and extended discussion of dynamic game analysis	1	（1）逆推归纳法的问题 （2）颤抖手均衡和顺推归纳法 （3）蜈蚣博弈 (1) Problems of inverse induction (2) Trembling hand equilibrium and progressive induction (3) Centipede game

第四章　重复博弈（Repeated Game）

章节序号 Chapter Number	章节名称 Chapters	课时 Class Hour	知　识　点 Key Points
4.1	重复博弈引论 Introduction to repeated game		（1）重复博弈的定义和意义 （2）重复博弈的基本概念 (1) Definition and significance of repeated game (2) Basic concepts of repeated game
4.2	有限次重复博弈 Finite repeated game	2	（1）两人零和博弈的有限次重复博弈 （2）唯一纯策略纳什均衡博弈的有限次重复博弈 （3）多个纯策略纳什均衡博弈的有限次重复博弈 （4）有限重复博弈的民间定理 (1) Finite repeated game of two-person zero-sum game (2) Finite repeated game with unique pure strategy Nash equilibrium (3) Finite repeated game with multiple pure strategy Nash equilibriums (4) The folk theorems of finite repeated game

续表

章节序号 Chapter Number	章节名称 Chapters	课时 Class Hour	知识点 Key Points
4.3	无限次重复博弈 Infinite repeated game	2	(1) 两人零和博弈的无限次重复博弈 (2) 唯一纯策略纳什均衡博弈的无限次重复博弈 (3) 无限次重复古诺模型 (4) 有效工资率 (1) Infinite repeated game of two-person zero-sum game (2) Infinite repeated game with unique pure strategy Nash equilibrium (3) Infinite repeated Cournot model (4) Effective wage rate

第五章 完全但不完美信息动态博弈（Dynamic Game with Complete but Imperfect Information）

章节序号 Chapter Number	章节名称 Chapters	课时 Class Hour	知识点 Key Points
5.1	不完美信息动态博弈 Dynamic game with imperfect information	2	(1) 概念和例子 (2) 不完美信息动态博弈的表示 (3) 不完美信息动态博弈的子博弈 (1) Concepts and examples (2) Representation of dynamic game with imperfect information (3) Subgame of dynamic game with imperfect information
5.2	完美贝叶斯均衡 Perfect Bayesian equilibrium		(1) 完美贝叶斯均衡的定义 (2) 关于判断形成的进一步理解 (1) Definition of perfect Bayesian equilibrium (2) Further understanding of judgment formation

续表

章节序号 Chapter Number	章节名称 Chapters	课时 Class Hour	知识点 Key Points
5.3	单一价格二手车交易的博弈 The game of used vehicle trading with single price	2	(1) 单一价格二手车交易的博弈模型 (2) 均衡的类型 (3) 博弈模型的纯策略完美贝叶斯均衡 (4) 博弈模型的混合策略完美贝叶斯均衡 (1) The game model of used vehicle trading with single price (2) Types of equilibriums (3) Perfect Bayesian equilibrium with pure strategy of the game model (4) Perfect Bayesian equilibrium with mixed strategy of the game model
5.4	双价二手车交易的博弈 The game of two-price used car trading		(1) 双价二手车交易博弈模型 (2) 博弈模型的均衡 (1) The game model of two-price used car trading (2) Equilibrium of game model
5.5	有退款保证的双价二手车交易 Two-price used car trading with refund guarantee		(1) 有退款保证的双价二手车交易 (1) Two-price used car trading with refund guarantee

第六章 不完全信息静态博弈(Static Game with Incomplete Information)

章节序号 Chapter Number	章节名称 Chapters	课时 Class Hour	知识点 Key Points
6.1	不完全信息静态博弈和贝叶斯纳什均衡 Incomplete information static game and Bayesian Nash equilibrium	2	(1) 问题和例子 (2) 不完全信息静态博弈的一般表示 (3) 海萨尼转换 (4) 贝叶斯纳什均衡 (1) Problems and examples (2) General representation of incomplete information static game (3) Harsanyi conversion (4) Bayesian Nash equilibrium

续表

章节序号 Chapter Number	章节名称 Chapters	课时 Class Hour	知 识 点 Key Points
6.2	暗标拍卖 Sealed-bid auction		（1）暗标拍卖 (1) Sealed-bid auction
6.3	双方报价拍卖 Bid auction	2	（1）双方报价拍卖的贝叶斯纳什均衡条件 （2）线性策略均衡 (1) Conditions of Bayesian Nash equilibrium of bid auction (2) Linear strategic equilibrium

第七章 不完全信息动态博弈（Dynamic Game with Incomplete Information）

章节序号 Chapter Number	章节名称 Chapters	课时 Class Hour	知 识 点 Key Points
7.1	不完全信息动态博弈及其转换 Dynamic game with incomplete information and its conversion	2	（1）不完全信息动态博弈问题 （2）类型和海萨尼转换 (1) Problems of dynamic game with incomplete information (2) Types and Harsanyi conversion
7.2	声明博弈 Declarations game		（1）声明和信息传递 （2）离散型声明博弈 （3）连续型声明博弈 (1) Declarations and information transfer (2) Discrete declarations game (3) Continuous declarations games

续表

章节序号 Chapter Number	章节名称 Chapters	课时 Class Hour	知 识 点 Key Points
7.3	信号博弈 Signaling game	2	（1）行为传递的信息和信号机制 （2）信号博弈模型和完美贝叶斯均衡 （3）股权换投资博弈 （4）劳动市场信号博弈 (1) Information transmitted by behaviors and signal mechanism (2) Signaling game model and perfect Bayesian equilibrium (3) Game of exchanging stock right for investment (4) Signaling game in labor market
7.4	不完全信息工会厂商谈判 Negotiation between labor union and firms under the condition of incomplete information		（1）不完全信息工会厂商谈判 (1) Negotiation between labor and firms under the condition of incomplete information

第八章 博弈学习和进化博弈论（Game Learning and Evolutionary Game）

章节序号 Chapter Number	章节名称 Chapters	课时 Class Hour	知 识 点 Key Points
8.1	有限理性和博弈分析 Limited rationality and game analysis	1	（1）有限理性问题 （2）有限理性博弈分析方法 (1) The problem of limited rationality (2) Analysis methods of game with limited rationality
8.2	博弈学习模型 Game learning model	1	（1）最优反应动态 （2）虚拟行动 （3）博弈学习模型小结 (1) Optimal response dynamics (2) Virtual action (3) Summary of game learning model

续表

章节序号 Chapter Number	章节名称 Chapters	课时 Class Hour	知 识 点 Key Points
8.3	进化博弈论 Evolutionary game theory	1	（1）生物进化博弈论 （2）经济中的进化博弈论 (1) Game theory of biology evolutionary (2) Evolutionary game theory in economy

大纲制定者：王锋教授(西安交通大学经济与金融学院)、张帆教授(西安交通大学经济与金融学院)

大纲审定：西安交通大学人工智能学院本科专业知识体系建设与课程设置第二版修订工作组

第 4 章

"科学与工程"课程群

4.1 "现代物理与人工智能"课程大纲

课程名称：现代物理与人工智能
Course：Physics for Artificial Intelligence
先修课程：工科数学分析
Prerequisites：Mathematical Analysis for Engineering
学分：7
Credits：7

4.1.1 课程目的和基本内容（Course Objectives and Basic Content）

本课程是人工智能学院本科专业基础必修课。

This course is a basic compulsory course for undergraduates in College of Artificial Intelligence.

物理学是自然科学中最基本的学科，从理论和实验两个角度研究各层次的物质基本结构及其运动规律以及各层次间相互作用的学科，其定律通常用数学语言来表达。物理学通常细分为两大类：经典物理学和现代物理学。其中，现代物理学则是 20 世纪上半叶发展起来的物理学的一个分支，用后牛顿的新物理学概念研究微观粒子和高速宏观物理现象的理论和方法体系，其基石是 20 世纪初的两个里程碑突破：量子理论和相对论。现代物理学比经典物理学更广泛，能够解释尺度非常小的物质微小基本成分和非常接近光速的极高速度的物理现象。经典物理定律通常不适用或仅作为现代物理定律的近似值适用。

物理学与人工智能之间存在着深刻的交叉和互动。物理学为人工智能提供了理论基础和实践工具，对人工智能的发展产生着重大而深远的影响，而人工智能则推动了物理学的发展，帮助我们更好地理解和控制物理世界。物理学是以解决具体问题为中心的学科，注重实证、实验和建模。将这种思维方式引入人工智能领域，可以提高学生们解决问题的能力，让他们能够更好地理解和使用人工智能技术。物理学往往需要运用抽象的思维和创新的观念来解决问题，这种创新思维对人工智能领域来说非常重要，有助于学生发现新的问题和解决方案。

当前，许多人工智能系统都采用了基于物理的模型。量子计算、统计力学等物理概念已经可用于优化机器学习算法；在计算机视觉中，基于光学的物理模型可用于解决光照、阴影和反射等问题；在自然语言处理中，声学模型帮助我们理解和模拟声音信号的产生和传播；在机器人学中，机械学模型可用于设计和控制机器人。这些模型都源自于物理学的深入研究。特别地，物理学为人工智能提供了一些重要的理论基础。例如，统计力学为理解和设计机器学习算法提供了理论基础，量子力学为构建未来的超级计算机（量子计算机）提供了基础。信息论和复杂度理论为理解和设计高效的人工智能算法和系统提供了工具。此外，物理学在计算硬件领域的发展对人工智能也产生了深远影响。例如，光子计算利用光子的波动性质和速度，实现了比传统电子计算机更高效、更快速的信息处理。神经形态计算则通过新的计算基材模拟生物神经网络的结构和功能，提供了一种新的、更接近人脑的计算方式。

另外，人工智能也在不断推动着对物理学理论和实践的创新。当前，人工智能正在帮助物理学家设计实验、处理和分析大量的实验数据，甚至发现新的物理法则。例如，人工智能已经成功应用于粒子物理实验中，帮助科学家在海量数据中寻找新粒子的迹象。人工智能在理解复杂物理系统机理的研究中发挥着重要作用，许多物理系统的行为复杂到人类无法直接理解，人工智能可以在这些情况下帮助研究人员建立复杂物理系统的行为模型。例如，神经网络可以模拟复杂的量子系统，深度学习可以帮助我们理解复杂的非线性动力学。通过对量子系统的深入学习，人工智能可以把我们对量子力学的理解推进到一个新的阶段。因此，人工智能不仅可以帮助我们理解和利用已有的物理理论，也可能推动我们对物理世界的新的理解。

人工智能与物理学的交融正在开创一种全新的研究范式。"现代物理与人工智能"课程的教学大纲特别设计用于人工智能本科专业学生，与普通大学物理课程不同，该课程在注重与经典物理知识衔接的基础上，学习和探究先进物理理论在人工智能领域的实际应用和研究。本课程包括经典物理学概述及其发展、量子力学与量子计算、相对论与电磁学、物态与凝聚态物理、统计物理和机器学习、非线性动力学与复杂系统、物理模拟与虚拟现实、现代物理实验和项目等主题，更深入地探讨物理学在现代科学技术和人工智能领域中的应用，培养学生运用物理学知识解决人工智能实际问题的能力，为探究人工智能未来发展的新方向奠定现代物理知识基础，为学生未来在人工智能领域的前沿研究和工作做好准备。

对于人工智能本科专业的学生来说，本课程具有很大的实用性和前瞻性。掌握现代物理学中的相关知识可以为人工智能的发展提供重要的理论基础和应用工具，使学生建立跨学科视角，进而理解物理学和人工智能这两个看似截然不同的领域之间存在的深刻的联系，帮助学生在日益竞争激烈的人工智能领域中取得优势，将来在现代物理学和人工智能交叉领域研究的蓬勃发展中取得更多新的发现和创新。

Physics is the most basic subject in natural sciences, studying the fundamental structure of matter at various levels and their laws of motion, as well as the interaction between different levels, both from theoretical and experimental perspectives. Its laws are often expressed in mathematical language. Physics is usually divided into two main categories: classical physics and modern physics. Modern physics emerged in the first half of the 20th century as a branch of physics, mainly researching microscopic particles and high-speed macro physical phenomena using post-Newtonian concepts. Its foundation lies in two milestone breakthroughs at the beginning of the 20th century: quantum theory and relativity. Modern physics is more extensive than classical physics and can explain physical phenomena at very small scales and at speeds close to the speed of light. The laws of classical physics are often inapplicable or only approximate within the laws of modern physics.

There is a profound intersection and interaction between physics and artificial intelligence. Physics provides theoretical foundations and practical tools for AI and significantly impacts its development, while AI advances physics, helping us better understand and control the physical world. Additionally, physics is centered around problem-solving, focusing on empiricism, experimentation, and modeling. Introducing this thinking into AI can enhance students' problem-solving abilities, enabling them to better understand and use AI technologies. Also, physics often requires abstract thinking and innovative concepts to solve problems. This innovation is vital in AI, helping students identify new problems and solutions.

Currently, many AI systems employ physics-based models. For example, quantum computing and statistical mechanics have been applied to optimize machine learning algorithms; optical physics models are used in computer vision to address issues of lighting, shadows, and reflections; acoustics models aid in understanding and simulating sound signals in natural language processing; and mechanical models are used in designing and controlling robots in robotics. These models stem from deep research in physics. Particularly, physics offers essential theoretical foundations for AI, such as statistical mechanics for understanding and designing machine learning algorithms, and quantum mechanics for building future supercomputers (quantum computers). Information theory and complexity theory provide tools for efficient AI algorithms and systems. Moreover, advancements in physical computing hardware deeply impact AI, with technologies like photonic computing and neuromorphic computing offering more efficient and brain-like computation methods.

On the other hand, AI continuously drives innovation in physics theory and practice. Currently, AI helps physicists design experiments, process and analyze vast amounts of experimental data, and even discover new physical laws. For instance, AI has been successfully applied in particle physics experiments, helping scientists find traces of new particles in massive data. AI plays a critical role in understanding complex physical systems, where human comprehension is limited, and can assist researchers in modeling complex systems using neural networks and deep learning. Through intensive study of quantum systems, AI might usher in a new stage in our understanding of quantum mechanics. Therefore, AI can help us understand and utilize existing physical theories and possibly foster new insights into the physical world.

The integration of AI and physics is forging a new research paradigm. The course "Physics for Artificial Intelligence" is specially designed for undergraduate AI students. Unlike standard university physics courses, this course emphasizes the study and exploration of advanced physics theories' practical applications and research in AI, building upon classical physics knowledge. Topics include overviews of classical physics and the development, quantum mechanics and quantum computing, relativity and electromagnetism, physics of state and condensed matter, statistical physics and machine learning, nonlinear dynamics and complex systems, physics simulations and virtual reality, modern physics experiments and projects, and more. This approach explores physics applications in modern science, technology, and AI, nurturing students' ability to use physics in solving AI problems, laying the foundation for exploring new AI directions, and preparing students for cutting-edge research and work in AI.

Therefore, the course "Physics for Artificial Intelligence" holds great practicality and foresight for AI undergraduate students. Mastering relevant knowledge in modern physics can provide essential theoretical foundations and tools for AI development, allowing students to establish interdisciplinary perspectives and understand the profound connections between physics and AI. This understanding can help students gain an edge in the increasingly competitive AI field, leading to further discoveries and innovations in the flourishing intersection of modern physics and AI.

4.1.2 课程基本情况(Course Arrangements)

课程名称	现代物理与人工智能 Physics for Artificial Intelligence						
开课时间	一年级	二年级	三年级	四年级	\multicolumn{3}{c}{科学与工程}
	秋 春	秋 春	秋 春	秋 春			
课程定位	本科生"科学与工程"课程群必修课				必修 (学分)	现代物理与人工智能(7)	
学 分	7学分					电子技术与系统(5)	
总 学 时	128学时 (授课96学时、实验32学时)					现代控制工程(3)	
						数字信号处理(3)	
授课学时 分配	课堂讲授(96学时)				选修 (学分)	/	
先修课程	工科数学分析						
后续课程							
教学方式	课堂教学、作业、讨论、实验						
考核方式	闭卷考试成绩占70%,作业成绩和考勤占30%						
参考教材	Serway R A,Moses C J,Moyer C A. 近代物理学[M]. 3版. 北京:清华大学出版社,2008.						
参考资料	1. Feynman R P,Leighton R B,Matthew S. 费曼物理学讲义[M]. 上海:上海科学技术出版社,2005. 2. 吴百诗. 大学物理[M]. 北京:科学出版社,2007.						
其他信息							

4.1.3　教学目的和基本要求（Teaching Objectives and Basic Requirements）

（1）理解先进物理理论并能将其应用于人工智能技术；

（2）培养物理实验和数值模拟能力；

（3）能将物理学原理应用于人工智能算法和模型；

（4）有能力阅读物理和人工智能领域的专业文献；

（5）"经典物理学概述及其发展"章节从力、热、光、电、磁等方面对经典物理学关于宏观物质运动描述的主要思想体系、方法和主要规律进行概述，初步介绍从经典物理学延伸到微观、宇观世界探索途径的基本思想；

（6）"量子力学与量子计算"章节致力于为学生提供量子力学和量子计算的全面引导，从基本的物理原理到现代量子计算技术的实际应用，再到量子技术与人工智能领域的交叉和整合，可以提供深入了解现代计算技术和物理原理的途径；

（7）"相对论与电磁学"章节有助于学生全面理解狭义相对论、电磁波方程及其与电磁学的相对论性描述，并强调与人工智能领域的交叉和整合；

（8）"物态与凝聚态物理"章节主要培养学生对物态与凝聚态物理的基本理解，并分析其与人工智能未来发展的交叉和互动，促进跨学科思维；

（9）"统计物理和机器学习"章节旨在培养学生对统计物理和机器学习之间相互联系和交叉影响的理解，强调基本理论的掌握，同时也关注当前的研究前沿和实际应用；

（10）"非线性动力学与复杂系统"章节主要介绍非线性动力学和复杂系统的基本概念和原理，及其更广泛的人工智能和工程问题中的应用，增进学生对复杂系统和模式的理解，并为进一步研究提供基础；

（11）"物理模拟与虚拟现实"章节既强调理论基础，也关注实际应用和技能训练，旨在使学生掌握物理建模和模拟的基础知识，以及虚拟现实（包括虚拟环境的构建和用户交互技术）的基本概念，同时，配合实验和项目实践，让学生在虚拟现实开发环境中进行物理模拟的实验，增强学生实际动手能力，加深对物理模拟在虚拟现实中的应用理解，培养学生在人工智能、虚拟现实和物理模拟等交叉领域的综合素质；

（12）"现代物理实验和项目"利用虚拟环境下的物理模拟、动手实践操作等方式，通过宏观手段观测微观量，帮助学生理解量子力学、相对论的基本数学模型、光学基本原理、凝聚态物理等知识点，培养学生掌握现代物理实验的基本操作和分析方法，并通过交叉项目的实践，培养跨学科的思维能力和创新精神。

4.1.4 教学内容及安排(Syllabus and Arrangements)

第一章 经典物理学概述及其发展(Overview of Classical Physics and the Development)

阐述经典物理的基本体系和发展,内容包含经典物理发展至现代物理及其在新技术领域应用,主要线索为物理学研究的对象,分为宏观、微观和宇观世界的基本运动形式、主要思想、方法和部分规律。该章内容以引导学生对物理学整体脉络的了解为主,其中所出现的公式主要用于内容介绍的需要,学习中不要求对定量计算的掌握,但在课堂教学中将以示例讨论加强学生对一些重要的公式和定理的理解。该章内容特别强调中物理学发展对新技术革命的基础作用,旨在引导和激发学生在相关方向的思考和继续探索。

章节序号 Chapter Number	章节名称 Chapters	课时 Class Hour	知识点 Key Points
1.1	机械运动概述 Overview of mechanical movement	2	(1) 测量与运动的量化,机械运动的分类,平动、转动、振动与波动、流体运动模型及其描述方法 (2) 机械运动的主要规律及其应用方法 (3) 惯性系与力学体系的局限性 (1) Measure and quantify motion, classification of mechanical motion; moving, rotation, vibration and wave, fluid motion models, basic methods (2) Main laws and application methods of mechanical motion (3) Inertial frames and limitations of mechanical systems

续表

章节序号 Chapter Number	章节名称 Chapters	课时 Class Hour	知识点 Key Points
1.2	热物理概述 Overview of thermal physics	2	(1) 热运动及热力学系统模型,温度与热力学第零定律,热机与热力学第一、第二定律,冷却与第三定律 (2) 热力学系统的微观模型,分布与统计模型,热力学第二定律的统计诠释,热力学熵 (3) 现代统计物理思想,熵与信息基础 (1) Thermal motion and thermodynamic system models, temperature and zero law of thermodynamics, heat engine and the first and the second laws of thermodynamics, cooling and the third law (2) Micro-model of thermal system, distribution and statistical models, statistical interpretation of the second law of thermodynamics, thermodynamic entropy (3) Introduction to modern statistical physics, Entropy and information
1.3	电磁运动规律 Electricity and magnetism	2	(1) 电荷与电场,电流(运动电荷)与磁场,静电场与稳恒磁场的基本规律,场的散度与旋度 (2) 法拉第电磁感应定律,麦克斯韦方程组,电磁波与光,光的干涉与测量,偏振光(选讲) (1) Charge and electric field, current (moving charge) and magnetic field, basic laws of electrostatic field and constant magnetic field, divergence and curl of field (2) Faraday's law of electromagnetic induction, Maxwell's equations, electromagnetic waves and light, interference of light and precise measurement, polarized light (Selected)

续表

章节序号 Chapter Number	章节名称 Chapters	课时 Class Hour	知识点 Key Points
1.4	原子与电子 Atoms and electrons	2	(1) 黑体辐射与普朗克的能量子,原子的探索与玻尔模型,能级与光谱,光电效应与爱因斯坦的光辐射理论,光的波粒二象性 (2) 电子自旋、泡利不相容原理、元素周期表,能带与半导体,超导及其微观模型简介 (3) 粒子物理简介 (1) Blackbody radiation and Planck's energy quanta, exploration of atoms and Bohr model, energy levels and spectra, photoelectric effect and Einstein's theory of light radiation, wave particle duality of light (2) Introduction to electron spin, Pauli incompatibility principle, periodic table of elements, energy bands and semiconductors, superconductivity and its microscopic models (3) Introduction to particle physics
1.5	相对论与宇观世界引论 Introduction to relativity and cosmic world	2	(1) 运动的相对性和相对时空观,相对论基本思想 (2) 黎曼几何与弯曲时空,爱因斯坦引力场方程,广义相对论与实验观测,宇宙学模型简介 (1) Transforms of motion and relative spacetime, basic idea of Relativity (2) Riemannian geometry and curved spacetime, Einstein's gravitational field equation, general relativity and experimental observations, introduction to cosmological models

第二章 量子力学与量子计算(Quantum Mechanics and Quantum Computing)

对于人工智能本科专业的学生,量子力学与量子计算的基础知识可以提供深入了解现代计算技术和物理原理的途径。本章的内容致力于为学生提供对量子力学和量子计算的全面引导,从基本的物理原理到现代量子计算技术的实际应用,并介绍了量子技术与人工智能领域的交叉融合。

章节序号 Chapter Number	章节名称 Chapters	课时 Class Hour	知识点 Key Points
2.1	量子力学的波动理论 Wave theory of quantum mechanics	6	（1）经典物理的局限性 （2）从光的波粒二象性到德布罗意物质波 （3）波函数及其波恩诠释 （4）薛定谔方程及量子力学的波动理论 （5）简单应用 (1) Limitations of classical physics (2) Wave particle duality of light and de Broglie matter waves (3) Wave functions with the Bonn interpretation (4) Schrödinger equation and wave theory of quantum mechanics (5) Some examples
2.2	量子态及量子力学基础理论 Fundamental theory of quantum states and quantum mechanics	6	（1）狄拉克算符和表象理论 （2）海森堡的量子跃迁（矩阵）理论 （3）量子力学的数学工具：希尔伯特空间，算符和本征值问题 （4）量子态、测量和叠加：量子比特和量子态，叠加原理，纯态与混合态 （5）测不准原理：测不准原理的表述和理解，测不准关系在量子力学中的作用 (1) Dirac operator and representation theory (2) Heisenberg's quantum transition (matrix) theory (3) Mathematical tools of quantum mechanics: Hilbert spaces, operators, and eigenvalue problems (4) Quantum state, measurement and superposition: qubit and quantum state, superposition principle, pure states and mixed states (5) Uncertainty principle: expression and understanding of the uncertainty principle, and role of the uncertainty relationship in quantum mechanics

续表

章节序号 Chapter Number	章节名称 Chapters	课时 Class Hour	知识点 Key Points
2.3	量子计算和信息 Quantum computing and information	6	(1) 量子信息的基本概念 (2) 量子比特的定义：基本量子逻辑门，量子电路的设计和模拟 (3) 量子算法：Shor's 算法，Grover's 算法 (4) 量子纠缠和量子通信：纠缠态和量子隐形传输 (5) 量子技术在人工智能领域的应用：量子计算机的现有技术和限制，量子机器学习和量子优化算法等 (1) Notion of quantum information (2) Definition of qubits：basic quantum logic gates，design and simulation of quantum circuits (3) Quantum algorithms：Shor's algorithm，Grover's algorithm (4) Quantum entanglement and quantum communication：entangled states and quantum teleportation (5) Applications of quantum technology in artificial intelligence：existing technologies and limitations of quantum computers，quantum machine learning and quantum optimization algorithms，etc.

第三章 相对论与电磁学(Relativity and Electromagnetism)

学生通过本章内容的学习能对狭义相对论、电磁波方程及其与电磁学的相对论性描述有全面的理解，并强调了与人工智能领域的交叉融合。

章节序号 Section Number	章节名称 Sections	课时 Class Hour	知识点 Key Points
3.1	狭义相对论 Special relativity	6	(1) 相对论的历史背景和动机：力学相对性原理推广的困难，光速不变与迈克尔逊-莫雷实验，两条基本假设 (2) 洛伦兹变换：洛伦兹变换的数学形式和物理解释，事件、间隔和光锥概念 (3) 相对论动力学：能量-动量关系，相对论下的动量守恒和能量守恒 (4) 示例分析和习题讨论：时间膨胀和长度收缩的概念和公式，习题解析 (1) Historical background and motivation of relativity: difficulties in generalizing the principle of relativity in mechanics, constant of speed of light and Michelson-Morley experiment, two basic assumptions (2) Lorentz transform: mathematical form and physical explanation of Lorentz transformation, concepts of event, interval and light cone (3) Relativistic dynamics: energy-momentum relationship, conservation of momentum and conservation of energy under relativity (4) Example analysis and exercise discussion: concept and formula of time dilation and length contraction, and the analysis of exercises
3.2	真空中的电磁波方程 Equations of electromagnetic waves in a vacuum	4	(1) 电磁场的基本方程：麦克斯韦方程组的引入和基本解释 (2) 真空中的电磁波方程：平面电磁波的解，偏振和波动方向 (3) 电磁波的应用：通信、医疗等领域的实际应用示例 (1) Basic equation of electromagnetic field: introduction and basic interpretation of Maxwell's equations (2) Electromagnetic wave equation in vacuum: solution of planar electromagnetic waves, polarization and wave direction (3) Applications of electromagnetic waves: examples of practical applications in communications, medical and other fields

续表

章节序号 Section Number	章节名称 Sections	课时 Class Hour	知识点 Key Points
3.3	狭义相对论与电磁学应用：电磁场的相对论性描述 Special relativity and electromagnetism applications: relativistic description of electromagnetic fields	4	(1) 相对论变换下的电磁场：电场和磁场在相对运动中的变换关系、充电粒子的动力学和辐射 (2) 相对论性效应在工程技术中的应用：GPS系统中的相对论修正等 (3) 习题讨论与解析：电磁场的相对论性描述的例题分析 (4) 与人工智能交叉的可能性：相对论和电磁学在人工智能领域的潜在应用，推荐阅读、开放性问题探讨 (1) Electromagnetism under relativity relativistic transformation: transformation relationship between electric and magnetic fields in relative motion, dynamics and radiation of charged particles (2) Applications of relativistic effects in engineering technology: relativistic corrections in GPS systems, etc. (3) Discussions and analysis of exercises: example analysis of relativistic descriptions of electromagnetic fields (4) Possibilities of intersection with artificial intelligence: potential applications of relativity and electromagnetism in the field of artificial intelligence, recommended reading, open-ended questions

第四章 物态与凝聚态物理(Physics of State and Condensed Matter)

本章内容可以帮助人工智能专业的本科生全面了解物态与凝聚态物理的基本概念和理论，并引导他们深入探索和思考其在现代科学技术，特别是在人工智能领域的潜在应用。在安排课程时，可以根据学生的基础和兴趣对某些主题进行深入讲解。还可适当增加实验演示和计算机模拟等，帮助学生更直观地理解和掌握物态与凝聚态物理的基本概念和理论。

章节序号 Chapter Number	章节名称 Chapters	课时 Class Hour	知识点 Key Points
4.1	物态的分类和相变 Classification and phase transition of matter states	4	(1) 物态的分类：固态、液态、气态及等离子态，不同物态的特性和微观解释 (2) 相变的基本概念：一级和二级相变的概念和区别，相变的热力学描述，相图和临界点 (3) 相变的深入理解：兰道理论，临界现象，实验中的相变现象观察 (1) Classification of states of matter: solid, liquid, gaseous and plasma, characteristics and microscopic interpretation of different states of matter (2) Basic concepts of phase transition: concepts and differences between primary and secondary phase transitions, thermodynamic description of phase transitions, phase diagrams and critical points (3) In-depth understanding of phase transitions: Landau theory, critical phenomena, observation of phase transition phenomena in experiments
4.2	晶体与晶格振动 Crystal and lattice vibration	3	(1) 晶体结构与晶格对称性：晶格的周期性与对称性 (2) 晶体的结合：原子键合与晶体结合种类 (3) 晶格振动：一维晶体的振动，格波与声子 (1) Crystal structure and lattice symmetry: periodicity and symmetry of lattice (2) Crystal bonding: types of atomic bonding and crystal bonding (3) Lattice vibration: vibration of one-dimensional crystals, lattice waves and phonons

续表

章节序号 Chapter Number	章节名称 Chapters	课时 Class Hour	知识点 Key Points
4.3	凝聚态物理和电子能带理论 Condensed matter physics and electron energy band theory	3	(1) 凝聚态物理概述：凝聚态系统的普遍性质，现代凝聚态物理的重要性 (2) 电子能带理论基础：布洛赫定理，能带结构 (3) 导体、半导体和绝缘体：能带理论的应用，导体、半导体和绝缘体的分类 (1) Overview of condensed matter physics: general properties of condensed matter systems and importance of modern condensed matter physics (2) Fundamentals of electron energy band theory: Bloch's Theorem, band structure (3) Conductors, semiconductors, and insulators: application of energy band theory, classification of conductors, semiconductors, and insulators
4.4	凝聚态物理的前沿主题与AI的交叉互动 Cutting-edge topics in condensed matter physics and their intersecting interactions with AI	4	(1) 物质的磁性与超导电性 (2) 低维电子系统与相变：拓扑绝缘体、石墨烯等，二维体系中的相变 (3) 相变材料与神经形态计算：相变存储基本原理，神经形态计算架构与机理 (1) Magnetic and superconductive properties (2) Low-dimensional electronic systems and phase transitions: topological insulators, graphene, etc., phase transitions in two-dimensional systems (3) Phase change materials and neuromorphic computation: basic principles of phase change storage, architecture and mechanism of neuromorphic computation

第五章 统计物理与机器学习(Statistical Physics and Machine Learning)

本章的内容安排旨在培养学生对统计物理和机器学习之间相互联系和交叉影响的理解，强调基本理论的掌握，同时也关注当前的研究前沿和实际应用。

章节序号 Chapter Number	章节名称 Chapters	课时 Class Hour	知识点 Key Points
5.1	统计物理基础 Fundamentals of statistical physics	6	(1) 微观态与宏观态：微观态的统计描述，宏观态的热力学量 (2) 分布定律：麦克斯韦速度分布，波色-爱因斯坦统计，费米-狄拉克统计 (3) 熵和自由能：热力学第二定律，熵的定义和性质，自由能和平衡态 (1) Micro and macroscopic states: Statistical description of microscopic states, thermodynamic quantities in macroscopic states (2) Law of distribution: Maxwell velocity distribution, Bose-Einstein statistics, Fermi-Dirac statistics (3) Entropy and free energy: second law of thermodynamics, definition and properties of entropy, Free energy and equilibrium
5.2	机器学习中的统计物理方法 Statistical physics methods in machine learning	6	(1) 机器学习与统计物理的关联：玻尔兹曼机的结构和训练，随机过程和马尔可夫链，神经网络与伊辛模型等统计物理模型的对应 (2) 信息论与机器学习：信息熵和KL散度，互信息和特征选择 (1) Correlation between machine learning and statistical physics: structure and training of Boltzmann machines, stochastic processes and Markov chains, correspondence of neural networks to statistical physics models such as the Issing model (2) Information theory and machine learning: information entropy and KL divergence, mutual information and feature selection

第六章 非线性动力学与复杂系统（Nonlinear Dynamics and Complex Systems）

非线性动力学与复杂系统是一个非常丰富和有趣的主题，这个领域的研究对许多科学和工程学科都有深远的影响。在人工智能本科专业中，探讨这些概念可以增进对复杂系统和模式的理解，并为进一步研究提供基础。通过本章内容的学习，学生将能够理解和分析非线性动力学和复杂系统的基本原理，并能够将这些概念应用到更广泛的人工智能和工程问题中。为加深学生对本章内容的理解，需要通过讲解、示例、数值模拟、小组讨论和项目作业等教学方法来实现。

章节序号 Chapter Number	章节名称 Chapters	课时 Class Hour	知识点 Key Points
6.1	非线性系统的基本概念 Basic concepts of nonlinear systems	1	(1) 非线性系统的定义和特性 (2) 非线性方程示例 (3) 线性系统与非线性系统的区别 (1) Definition and characteristics of nonlinear systems (2) Examples of nonlinear equations (3) Differences between linear and nonlinear systems
6.2	一维非线性映射 One-dimensional nonlinear mapping	2	(1) 递归关系和迭代映射 (2) 周期与不动点 (3) 逻辑斯谛映射 (4) 通向混沌：倍周期分叉过程 (1) Recursive relationships and iterative mapping (2) Periodicity and fixed point (3) Logistic mapping (4) Route to chaos: period-doubling bifurcation
6.3	分形与分形维数 Fractals and fractal dimension	3	(1) 迭代、自相似性与分形 (2) 常见的分形集 (3) 维度的扩展：分形维数 (4) 迭代函数系统 (1) Iteration, self-similarity, and fractals (2) Fractal sets (4) Extension of dimension: fractal dimension (3) Iterated function system

续表

章节序号 Chapter Number	章节名称 Chapters	课时 Class Hour	知识点 Key Points
6.4	非线性微分方程和振动系统 Nonlinear differential equations and vibration systems	2	（1）非线性振动器的方程 （2）平衡点分析和稳定性 （3）限制周期和奇异吸引子 (1) Equations for nonlinear vibrators (2) Equilibrium point analysis and stability (3) Period-limiting and singular attractors
6.5	混沌现象 Chaos	2	（1）混沌与蝴蝶效应 （2）洛伦兹方程和奇异吸引子 （3）自然界中的混沌及其应用 (1) Chaos and the butterfly effect (2) Lorenz equation and singular attractors (3) Chaos in nature with applications
6.6	复杂网络和小世界网络 Complex and small-world networks	2	（1）复杂网络的基本理论 （2）小世界网络和无标度网络 （3）网络中的同步和控制 （4）复杂网络在人工智能和其他领域的应用 (1) Basic theory of complex networks (2) Small-world and scale-free networks (3) Synchronization and control in the network (4) Applications of complex networks in artificial intelligence and other fields

续表

章节序号 Chapter Number	章节名称 Chapters	课时 Class Hour	知识点 Key Points
6.7	同步现象 Synchronization phenomenon	2	（1）同步现象的数学描述 （2）耦合振动器的动态行为 （3）Kuramoto 模型 （4）同步在科学和工程中的应用 (1) Mathematical description of synchronization phenomenon (2) Dynamic behavior of coupled vibrators (3) Kuramoto model (4) Simultaneous applications in science and engineering
6.8	复杂系统的综合分析和示例研究 Comprehensive analysis and case studies of complex systems	2	（1）复杂系统理论的综合分析 （2）复杂系统的预测和控制 （3）示例讨论：例如生态系统、交通系统的建模和金融市场的分析等 (1) Comprehensive analysis of complex systems theory (2) Prediction and control of complex systems (3) Example discussions：e.g. ecosystems, modelling of transportation systems, analysis of financial markets, etc.

第七章 物理模拟与虚拟现实(Physical Simulation and Virtual Reality)

通过本章的学习,使学生掌握物理建模和模拟的基础知识,并了解如何在虚拟现实中应用这些技能。本章的内容安排既强调理论基础,也关注实际应用和技能训练,有助于培养学生在人工智能、虚拟现实和物理模拟等交叉领域的综合素质。可以配合实验和项目实践,增强学生的实际动手能力。

章节序号 Chapter Number	章节名称 Chapters	课时 Class Hour	知识点 Key Points
7.1	物理建模基础 Fundamentals of physical modeling	6	（1）微分方程和模拟方法：微分方程的数值解法，时间积分和空间离散化 （2）网格方法和粒子系统：网格和粒子表示，碰撞和约束处理 (1) Differential equations and simulation methods: numerical solutions of differential equations, time integration and spatial discretization (2) Mesh methods and particle systems: mesh and particle representation, collision and constraint handling
7.2	虚拟现实中的物理模拟 Physical simulation in virtual reality	6	（1）刚体动力学模拟：刚体运动方程，碰撞检测和响应 （2）流体动力学模拟：Navier-Stokes方程，布料模拟技术 （3）材质与光照模拟：纹理映射，材质渲染，光源模型，光照算法 (1) Rigid body dynamics simulation: rigid body equations of motion, collision detection and response (2) Fluid dynamics simulation: Navier-Stokes equation, fabric simulation technology (3) Materials and lighting simulation: texture mapping, material rendering, light source models, lighting algorithms

4.1.5 实验环节（Experiments）

支撑实验内容的教学，需要创建量子力学虚拟实验室，通过利用 Unity 的 VR 开发能力和 AI 的智能辅助，构建一个能够帮助学生更好地理解相对论和量子力学的虚拟实验环境，创建场景模型并确保物理行为符合量子力学和相对论的基本数学模型，为学生展示并让他们体验狭义相对论的效应，例如时间膨胀、长度收缩、同时性相对性等。通过本章内容的学习和实验，学生不仅能够掌握现代物理实验的基本操作和分析方法，还可以深入了解量子计算的前沿技术，并通过交叉项目的实践，培养跨学科的思维能力和创新精神。虚拟实验室环节更是充分利用了现代科技手段，让学生在不依赖物理实验室的条件下也能进行深入的实验探索。

序号 Num.	实验内容 Experiment Content	课时 Class Hour	知识点 Key Points
1	虚拟现实中的物理模拟 Simulation for modern physics using virtual reality	8	（1）虚拟环境下的相对论效应观察实验：时间膨胀、长度收缩、同时性的相对性 （2）混沌与分形现象的仿真实验 （3）布料材质模拟实验 (1) Observation of relativity effects in virtual environment: time expansion effect, length contraction effect, relativity of simultaneity (2) Simulation experiments of chaos and fractal phenomena (3) Simulation experiment of fabric material
2	现代光学实验 Modern optics experiments	8	（1）光速测定实验 （2）波粒二象性观察：双缝干涉实验，光电效应实验 (1) Speed of light measurement experiment (2) Wave-particle duality observation: double-slit interference experiment, photoelectric effect experiment
3	量子计算实验与交叉项目设计 Quantum computing experiments and inter-disciplinary project design	8	（1）费米-狄拉克分布实验：通过宏观手段测量复杂的微观量 （2）量子搜索算法设计实验 （3）学生团队选择项目主题，例如 AI 与量子计算结合、量子计算的前景讨论 (1) The Fermi-Dirac distribution experiment: measuring complex microscopic facts through macroscopic means (2) Designing of quantum search algorithms (3) Student teams choose project topics, such as AI combined with quantum computing, discussion of the prospects of quantum computing

续表

序号 Num.	实验内容 Experiment Content	课时 Class Hour	知识点 Key Points
4	凝聚态物理实验 Experiments on condensed matter physics	8	（1）弗兰克—赫兹实验：验证原子的定态能级 （2）霍尔效应实验：观察导电材料中的电流与磁场相互作用 (1) The Frank-Hertz experiment: verification of the steady-state energy levels of atoms (2) The hall effect experiment: observing the interaction between current and magnetic field in conductive materials

大纲指导者：郑南宁教授（西安交通大学人工智能学院）

大纲制定者：郑南宁教授（西安交通大学人工智能学院）、辛景民教授（西安交通大学人工智能学院）、李宏荣教授（西安交通大学物理学院）、杨森教授（西安交通大学物理学院）、刘剑毅副教授（西安交通大学人工智能学院）、杨勐副教授（西安交通大学人工智能学院）、汪建基副教授（西安交通大学人工智能学院）、蒋才桂教授（西安交通大学人工智能学院）

大纲审定：西安交通大学人工智能学院本科专业知识体系建设与课程设置第二版修订工作组

4.2 "电子技术与系统"课程大纲

课程名称：电子技术与系统
Course：Electronic Technology and System
先修课程：工科数学分析
Prerequisites：Mathematical Analysis for Engineering
学分：5
Credits：5

4.2.1 课程目的和基本内容（Course Objectives and Basic Content）

本课程是人工智能学院本科专业必修课。课程由三部分组成，第一部分是电子器件与模拟电路,第二部分是数字电路与系统,第三部分是计算机体系结构基础。

This course is a compulsory course for undergraduates in College of Artificial Intelligence. It consists of three parts: Electronic Devices and Analog Circuits, Digital Circuit and System, Computer Architecture Basics.

第一部分旨在通过对常见电子器件和模拟电路设计方法的学习,帮助学生建立关于模拟器件与电路的知识框架。具体内容包括电路基础、晶体管、运算放大器与负反馈、模数与数模转换器等。课程采用课堂讲授、仿真设计、讨论和实物实验等教学手段,训练学生用基本理论和方法分析解决实际问题的能力,初步掌握模拟电路设计的基本知识和技能。此外,课程通过一个完整的心电测量实验,帮助学生了解模拟信号的获取和处理过程,为今后的工作打下良好的基础。

Part 1 aims to introduce common electronic devices and basic analog circuit design methods, to help students construct the knowledge framework of analog devices and circuits. Contents include circuit basics, transistor, operational amplifier and feedback, analog-to-digital and digital-to-analog converters. The teaching methods includes class teaching, simulation experiment, discussion and practical experiments, to develop the ability to analyze and address practical problems with basic theories and methods, and make students to master the basic knowledge and skills on analog circuit design. In addition, this course also includes a complete ECG signal measure experiment to help students get familiar with the analog signal sampling and processing flow, and lay a solid foundation for their future work.

第二部分旨在通过对数字系统基本理论、设计方法和应用技术的学习,使学生掌握现代数字系统设计与验证过程中涉及的基本原理、方法、编程语言与工具。具体内容包括数制系统与计算机编码方式、布尔代数与组合逻辑、组合逻辑电路、时序逻辑电路、数字系统的寄存器传输级设计、基本处理器结构等。同时,本课程还将讲授 SystemVerilog 硬件描述语言,并使用该语言描述组合逻辑、时序逻辑电路,完成数字系统设计。课程强调基础,通过对数字系统的抽象和硬件描述语言的掌握,使学生能够解决复杂数字系统和与实际硬件系统相关的问题。此外,课程还安排 System Verilog 编程和数字系统设计实验,通过自主实验学习培养学生对数字电路特别是 System Verilog 编程的掌握能力。

Part 2 aims to introduce the basic theories, design methods and application techniques of digital system, to make students master the theories, methods, programming language and tools in design and verification of modern digital systems.

Contents include digital number systems and data encoding in computer, Boolean algebra and combinational logic, combinational logic circuit, sequential logic circuit, register-transfer level design of digital systems, basic processor organization. This course also introduces SystemVerilog hardware description language, and make students design combinational and sequential logic circuits with SystemVerilog. Emphasis is on the fundamentals: the levels of abstraction and hardware description language methods that allow students to cope with hugely complex systems, and connections to practical hardware implementation problems. In addition, this course also arranges three off-class experiments, to develop the interests and practical skills on real digital system design through self-directed experiments.

　　第三部分旨在通过对计算机体系结构基本方法与技术的学习,使学生了解计算机系统设计的基础知识,建立复杂电子系统设计的框架性概念。具体内容包括包括指令集、汇编语言、计算机程序编译过程、微体系结构性能分析、单周期处理器、多周期处理器、流水线处理器、高级微体系结构技术、高速缓存、虚拟内存。课程强调系统思维,通过对微处理器设计的讲解与分析,使学生能够掌握用数字电路知识设计处理器以及其他复杂数字系统的基本方法,掌握处理器通过体系结构设计提升数字系统乃至数模混合系统性能与能效的基本思路,是对前面两部分电路教学内容的综合与提升。此外,课程也安排了两项课程实验,帮助同学进一步深入理解计算机系统性能量化分析与优化设计的方法,提高嵌入式计算机系统编程能力。

　　Part 3 aims to enable students to understand the basic knowledge of computer system design and establish framework concepts for complex electronic system design through the study of basic methods and techniques of computer architecture. Content include instruction sets, assembly language, program compilation process, microarchitecture performance analysis, single-cycle processor, multi-cycle processor, pipelined processor, advanced microarchitecture technology, cache, and virtual memory. The course emphasizes system-level thinking, through the explanation and analysis of microprocessor design, so that students can master the basic methods of designing processors and other complex digital systems with digital circuit knowledge, master the basic ideas about how processors improve the performance and energy efficiency of digital systems and even digital-analog hybrid systems through architecture design, which is the synthesis part of the above two parts of circuit teaching content. In addition, two experiments are arranged to help students further understand the methods of quantitative analysis and optimization of computer system performance, and improve their programming ability on embedded computer system.

4.2.2 课程基本情况(Course Arrangements)

课程名称	电子技术与系统 Electronic Technology and System							
开课时间	一年级		二年级		三年级		四年级	
	秋	春	秋	春	秋	春	秋	春
课程定位	本科生"科学与工程"课程群必修课							
学　　分	5学分							
总 学 时	88学时 (授课72学时,实验16学时)							
授课学时分配	课堂讲授(72学时)							
先修课程	工科数学分析							
后续课程	数字信号处理、计算机体系结构							
教学方式	课堂讲授、大作业与实验、小组讨论							
考核方式	闭卷考试成绩占70%,实验成绩占20%,考勤占10%							
参考教材	1. 杨建国.电子器件与模拟电路(轻印教材),2017. 2. Harris D M.数字设计和计算机体系结构[M].陈俊颖,译.北京：机械工业出版社,2007.							
参考资料	1. Wakerly J F. Digital Design Principles and Practices[M]. New York：Pearson Education,2017. 2. Palnitkar S. Verilog HDL 数字设计与综合[M].夏宇闻,胡燕祥,刁岚松,等译.北京：电子工业出版社,2009.							
其他信息								

科学与工程	
必修 (学分)	现代物理与人工智能(7)
	电子技术与系统(5)
	现代控制工程(3)
	数字信号处理(3)
选修 (学分)	/

4.2.3 教学目的和基本要求(Teaching Objectives and Basic Requirements)

第一部分　电子器件与模拟电路(Electronic Devices and Analog Circuits)

(1) 掌握电路基本元件和电路分析方法、BJT和FET晶体管的放大原理及其仿真软件的使用；

(2) 了解晶体管的开关原理；

(3) 掌握运算放大器的简化模型；

(4) 掌握以运放为核心的负反馈原理,掌握用虚短虚断法分析运放电路的基本方法；

(5) 了解差动放大原理以及共模抑制比概念；

(6) 了解滤波原理,掌握一阶和二阶滤波器的分析方法和设计实践,了解运放电路的带宽和其他性能指标;

(7) 理解采样定理,了解模数转换和数模转换基本原理及其应用场景;

(8) 通过一个心电测量实例,了解一个完整的模拟电路系统。

第二部分 数字系统结构与设计(Structure and Design of Digital System)

(1) 了解数字系统设计的基本概念和过程;

(2) 理解数制的表示、运算,掌握计算机的编码方式;

(3) 掌握布尔代数与组合逻辑的基本概念与分析优化方法;

(4) 理解基本组合逻辑、算数组合元件与状态机的结构;

(5) 掌握时序逻辑元件的基本结构与设计方法;

(6) 掌握 SystemVerilog 硬件描述语言并用该语言进行数字系统设计;

(7) 掌握数字系统高级模块的设计方法。

第三部分 计算机体系结构基础(Computer Architecture Basics)

(1) 掌握指令集架构的基本概念,了解典型的指令集及特点;

(2) 以 ARM 指令集为例,掌握常用指令、寻址模式与基本汇编编程;

(3) 了解计算机程序编译、汇编、连接与装载的过程;

(4) 掌握微体系结构性能分析基本方法;

(5) 掌握单周期、多周期与流水线处理器的特点,了解高级微体系结构技术;

(6) 掌握高速缓存、虚拟内存等存储架构设计的基本方法。

4.2.4 教学内容及安排(Syllabus and Arrangements)

绪论 (Introduction)

章节序号 Chapter Number	章节名称 Chapters	课时 Class Hour	知 识 点 Key Points
0.1	绪论 Introduction	2	(1) 电子系统层级 (2) 电子系统发展历史回顾 (3) 模拟电路、数字电路与体系结构的关系 (1) Hierarchy of electronic system (2) History of electronic system (3) Relationship of analog circuit, digital circuit and architecture

第一部分 电子器件与模拟电路(Electronic Devices and Analog Circuits)

第一章 电路基础(Foundation of Circuits)

章节序号 Chapter Number	章节名称 Chapters	课时 Class Hour	知 识 点 Key Points
1.1	电路基础知识-元件 Basic elements	0.5	(1) 电阻、电容、电感、电源 (2) 二极管、三极管、伏安特性、线性与非线性 (1) Resistors, capacitors, inductors, source (2) Diodes, triodes, voltammetric characteristics, linearity and nonlinearity
1.2	电路基础知识-定理 Basic theorems	0.5	(1) 电阻电路的等效变换、基尔霍夫定律 (2) 叠加原理、戴维南和诺顿定理 (1) Equivalent transformation of resistive circuits, Kirchhoff's law (2) Superposition principle, Thevenin and Norton's theorems
1.3	电路基础知识-方法 Basic methods	1	(1) 时域、频域分析方法 (1) Analysis methods of time domain and frequency domain

第二章 晶体管(Transistors)

章节序号 Chapter Number	章节名称 Chapters	课时 Class Hour	知 识 点 Key Points
2.1	晶体管原理 Principles of transistor	2	(1) BJT和MOSFET的伏安特性、受控机理 (1) Voltammetric characteristics and mechanism of BJT and MOSFET
2.2	晶体管的基本应用 Basic applications of transistor	1.5	(1) 共射极放大电路 (2) 共集电极放大电路 (3) 开关 (4) 与非门 (1) Common emitter amplifying circuit (2) Common collector amplifying circuit (3) Switch (4) NAND gate

续表

章节序号 Chapter Number	章节名称 Chapters	课时 Class Hour	知　识　点 Key Points
2.3	仿真软件介绍 Introduction to simulation software	0.5	(1) 对一个单级晶体管放大电路的仿真 (1) Simulation of a single-stage transistor amplifying circuit

第三章　运放和负反馈（Operational Amplifier and Feedback）

章节序号 Chapter Number	章节名称 Chapters	课时 Class Hour	知　识　点 Key Points
3.1	运算放大器和负反馈原理 Operational amplifier (OPA) and feedback theory	2	(1) 运算放大器的理想模型 (2) 负反馈方框图 (3) 反馈系数 (4) 负反馈对电路性能的影响 (1) Ideal model of OPA (2) Negative feedback block diagram (3) Feedback coefficient (4) The influence of negative feedback on circuit performance
3.2	运算放大器的基本应用 Basic applications of OPA	1	(1) 比例器 (2) 加法器 (3) 跟随器 (1) Inverting and noninverting amplifier (2) Adder (3) Follower
3.3	差动放大和共模抑制 Differential amplifier and CMR	1	(1) 减法器 (2) 共模和差模 (3) 共模抑制比 (1) Subtractor (2) Common mode and differential mode (3) Common mode rejection ratio

续表

章节序号 Chapter Number	章节名称 Chapters	课时 Class Hour	知 识 点 Key Points
3.4	运算放大器及其电路的性能指标 Performance specifications of OPA and their circuits	1	(1) 失调电压 (2) 偏置电流 (3) 至轨特性 (4) 压摆率 (1) Offset voltage (2) Bias current (3) Rail-to-rail characteristics (4) Slew rate
3.5	运算放大器电路的频率特性 Frequency characteristics of OPA circuit	1	(1) GBW 模型 (2) 幅频和相频特性 (3) 闭环频率特性 (4) 负反馈电路稳定性 (1) GBW model (2) Amplitude-frequency and phase-frequency characteristics (3) Closed loop frequency characteristics (4) Robustness analysis of feedback circuit
3.6	运算放大器组成的滤波器 Filter composed of OPA	2	(1) 一阶低通和高通 (2) 二阶 SK 和 MFB 低通 (1) First-order low-pass and high-pass (2) Second-order SK and MFB low-pass

第四章 模数和数模转换器(Analog to Digital and Digital to Analog Converter)

章节序号 Chapter Number	章节名称 Chapters	课时 Class Hour	知 识 点 Key Points
4.1	数据采集和数据恢复 Data collection and data recovery	0.5	(1) 数据采集和奈奎斯特定理 (1) Data acquisition and Nyquist theorem

续表

章节序号 Chapter Number	章节名称 Chapters	课时 Class Hour	知 识 点 Key Points
4.2	模数转换器 Analog to digital convertor (ADC)	1.5	(1) 闪速模数转换器 (2) 逐次逼近模数转换器 (3) 流水线模数转换器 (4) Σ-Δ 型模数转换器 (1) FLASH ADC (2) SAR ADC (3) Pipeline ADC (4) Sigma-delta ADC
4.3	数模转换器 Digital to analog convertor (DAC)	2	(1) 数模转换器 (2) 模数、数模转换器的应用 (1) Digital to analog convertor (2) Applications of ADC and DAC

第五章　ECG 前端电路分析(ECG Front-end Circuit Analysis)

章节序号 Chapter Number	章节名称 Chapters	课时 Class Hour	知 识 点 Key Points
5.1	ECG 前端分析 ECG front-end analysis	4	(1) 右腿驱动电路 (2) 仪表放大器 (3) 贝塞尔滤波器 (4) 模数转换器 (5) 基准和供电 (1) Driven-Right-Leg (DRL) circuit (2) Instrumentation amplifier (3) Bessel filter (4) ADC (5) Reference and power supply

第二部分 数字电路与系统(Digital Circuit and System)
第六章 组合逻辑设计(Combinational Logic Design)

章节序号 Chapter Number	章节名称 Chapters	课时 Class Hour	知 识 点 Key Points
6.1	布尔等式与布尔代数 Boolean equation and Boolean algebra	2	(1) 基本术语 (2) 与或表达式 (3) 或与表达式 (4) 一元定理 (5) 多元定理 (6) 真值表设计 (7) 表达式简化方法 (1) Basic terminology (2) Sum-of-products form (3) Product-of-sums form (4) Theorems of one variable (5) Theorems of several variables (6) Truth table design (7) Simplifying equation
6.2	逻辑门设计与多级组合逻辑 Logic gate design and multi-level combinatorial logic	2	(1) 多级组合逻辑设计方法 (1) Design methods of multilevel combinational logic
6.3	卡诺图与逻辑简化 Karnaugh map and logic minimization		(1) 利用卡诺图进行逻辑简化 (1) Logic minimization with Karnaugh map
6.4	组合逻辑模块与时序 Combinational building block and timing	2	(1) 乘法器 (2) 译码器 (3) 组合逻辑中的传播延迟 (4) 组合逻辑中的时序毛刺问题 (1) Multiplexer (2) Decoder (3) Propagation and contamination delay (4) Glitche

第七章 时序逻辑设计（Sequential Logic Design）

章节序号 Chapter Number	章节名称 Chapters	课时 Class Hour	知　识　点 Key Points
7.1	锁存器与触发器 Latche and flip-flop	2	（1）SR 锁存器 （2）D 锁存器 （3）D 触发器 （4）寄存器 （5）带使能触发器 （6）带复位触发器 （7）晶体管级锁存器和触发器设计 （1）SR latch （2）D latch （3）D flip-flop （4）Register （5）Enabled flip-flop （6）Resettable flip-flop （7）Transistor-level latch and flip-flop design
7.2	同步逻辑设计 Synchronous logic design		（1）同步时序逻辑电路 （2）同步与异步电路 （1）Synchronous sequential circuit （2）Synchronous and asynchronous circuit
7.3	有限状态机 Finite State Machine（FSM）	2	（1）FSM 设计举例 （2）状态编码设计 （3）摩尔与米利有限状态机设计 （4）从原理图推导有限状态机 （1）FSM design example （2）State encoding （3）Moore and Mealy FSM （4）Deriving a FSM from a schematic

续表

章节序号 Chapter Number	章节名称 Chapters	课时 Class Hour	知识点 Key Points
7.4	时序逻辑的时序 Timing of sequential logic	2	（1）建立时间约束 （2）保持时间约束 （3）时钟抖动 （4）亚稳态 （5）同步器 （1）Setup time constraint （2）Hold time constraint （3）Clock skew （4）Metastability （5）Synchronizer

第八章 SystemVerilog 硬件描述语言（SystemVerilog Hardware Description Language）

章节序号 Chapter Number	章节名称 Chapters	课时 Class Hour	知识点 Key Points
8.1	语法介绍及 结构化模块 Introduction and structured module	2	（1）SystemVerilog 语言起源 （2）SystemVerilog 仿真与综合方法 （3）模块化设计方法 （1）SystemVerilog language origin （2）SystemVerilog simulation and synthesis （3）Modules design
8.2	组合逻辑的 SystemVerilog 实现 SystemVerilog implementation of combinatorial logic	2	（1）按位运算设计，注释与空格，条件赋值，内部变量表示，优先级策略，数值表示及延迟问题 （1）Bitwise operator, comment and white space, conditional assignment, internal variable, precedence, number and delay

续表

章节序号 Chapter Number	章节名称 Chapters	课时 Class Hour	知识点 Key Points
8.3	时序逻辑的 SystemVerilog 实现 SystemVerilog implementation of sequential logic	2	（1）时序逻辑中的寄存器、可复位寄存器、带使能寄存器、多级寄存器、锁存器的设计与实现 （1）Design and implementation of register, resettable register, enabled register, multiple register and latch for sequential logic
8.4	状态机 SystemVerilog 实现及 SystemVerilog 模块设计 SystemVerilog implementation of FSM and SystemVerilog module design		（1）状态机、数据类型、参数化模块、测试用例的设计与实现 （1）Design and implementation of FSM, data type, parameterized module and test bench

第九章 数字系统模块（Digital System Module）

章节序号 Chapter Number	章节名称 Chapters	课时 Class Hour	知识点 Key Points
9.1	算术电路 Arithmetic circuit	2	（1）加法器、减法器、比较器、算术逻辑单元的设计 （2）移位器、乘法器及除法器的设计 （1）Design of addition, subtraction, comparator, ALU （2）Design of shifter and rotator, multiplication, division
9.2	数值系统 Numerical system	1	（1）定点数系统与浮点数系统简介 （1）Introduction to fixed-point number system and floating-point number system
9.3	时序电路模块 Sequential circuit module	1	（1）计数器 （2）移位寄存器 （1）Counter （2）Shift register

续表

章节序号 Chapter Number	章节名称 Chapters	课时 Class Hour	知 识 点 Key Points
9.4	存储阵列 Storage array	2	(1) 动态随机存储器 (2) 静态随机存储器 (3) 寄存器文件 (4) 只读存储器 (5) 逻辑的存储阵列实现 (1) DRAM (2) SRAM (3) Register file (4) ROM (5) Logic using memory array
9.5	逻辑阵列 Logic array		(1) 可编程逻辑阵列 (2) 现场可编程门阵列 (3) 阵列实现 (1) Programmable Logic Array（PLA） (2) Field Programmable Gate Array（FPGA） (3) Array implementation

第三部分 计算机体系结构基础(Computer Architecture Basics)

第十章 架构(Architecture)

章节序号 Chapter Number	章节名称 Chapters	课时 Class Hour	知 识 点 Key Points
10.1	指令集与汇编语言 Instruction set and assembly language	2	(1) 操作数：寄存器、内存和常量 (2) 数据处理指令 (3) 数据传送指令 (4) 跳转指令 (5) 寻址方式 (6) 循环设计 (7) 函数调用 (1) Operands：registers，memory，and constants (2) Data-processing instructions (3) Load-store instructions (4) Branching instructions (5) Addressing modes (6) Loop (7) Function calls

续表

章节序号 Chapter Number	章节名称 Chapters	课时 Class Hour	知　识　点 Key Points
10.2	编译、汇编与加载 Compile, assemble, and load	2	（1）编译、汇编与程序加载流程介绍 （1）Introduction to compiling, assembling, and loading
10.3	典型处理器架构介绍 Introduction to x86 architecture		（1）典型处理器架构的寄存器、操作数、状态标志寄存器、指令类型、指令编码等 （1）Register, operand, status flag, instruction and instruction encoding of x86 architecture

第十一章　微体系结构（Microarchitecture）

章节序号 Chapter Number	章节名称 Chapters	课时 Class Hour	知　识　点 Key Points
11.1	性能分析 Performance analysis		（1）计算程序执行时间 （2）计算单位指令时钟周期数 （1）Calculating execution time of a program （2）Calculating Cycle-Per-Instruction (CPI)
11.2	单周期处理器与多周期处理器 Single-cycle processor and multi-cycle processor	4	（1）单周期或多周期的数据通道 （2）单周期或多周期的控制 （3）单周期或多周期的处理器性能分析 （1）Single/multi-cycle datapath （2）Single/multi-cycle control （3）Performance analysis of single/multi-cycle processor
11.3	流水线处理器 Pipelined processor	4	（1）流水线数据通道 （2）流水线冲突 （3）流水线处理器性能分析 （1）Pipelined data path （2）Hazards （3）Performance analysis of pipelined processor

续表

章节序号 Chapter Number	章节名称 Chapters	课时 Class Hour	知 识 点 Key Points
11.4	高级微结构 Advanced microarchitecture	4	（1）深流水线 （2）微操作 （3）分支预测 （4）超标量处理器 （5）乱序处理器 （6）寄存器重命名 （7）多线程 （8）多处理器 (1) Deep pipelines (2) Micro-operations (3) Branch prediction (4) Superscalar processor (5) Out-of-order processor (6) Register renaming (7) Multithreading (8) Multiprocessors

第十二章 存储系统（Memory Hierarchy）

章节序号 Chapter Number	章节名称 Chapters	课时 Class Hour	知 识 点 Key Points
12.1	存储系统性能分析 Performance analysis of memory system	1	（1）计算存储的未命中率、命中率和平均存储时间 （2）改善存取时间的策略 (1) Calculating miss rate, hit rate and average memory access time (2) Methods to improve access time
12.2	高速缓存 Cache	4	（1）存储的层次化设计 （2）高速缓存的数据映射方式 （3）高速缓存的替换策略 （4）多级高速缓存设计 (1) Hierarchical design of memory system (2) Data mapping methods for cache (3) Data replacement strategies for cache (4) Multiple-level cache design

续表

章节序号 Chapter Number	章节名称 Chapters	课时 Class Hour	知识点 Key Points
12.3	虚拟存储器 Virtual memory	2	（1）地址转换 （2）页表 （3）转换后备缓冲 （4）存储器保护 （5）替换策略 （6）多级页表 (1) Address Translation (2) The Page Table (3) The Translation Lookaside Buffer (4) Memory Protection (5) Replacement Policies (6) Multilevel Page Tables

4.2.5 实验环节（Experiments）

序号 Num.	实验内容 Experiment Content	课时 Class Hour	知识点 Key Points
1	数字电路实验：算术加法器与乘法器设计 Digital circuit design of arithmetic adder and multiplier	4	（1）硬件描述语言 SystemVerilog 的基础语法 （2）同步电路中组合逻辑与时序逻辑设计 （3）FPGA 的基础知识与软件开发环境 （4）超前进位链算术加法器的设计方法 （5）基 4 的布斯乘法器的设计方法 (1) The basic conception of the Hardware Description Language（HDL）SystemVerilog (2) The combinational and sequential logic in synchronous circuits (3) The basic knowledge of FPGA and the software development environment (4) The design method for the carry-lookahead adder (5) The design method of the Base-4 Booth multiplier

续表

序号 Num.	实验内容 Experiment Content	课时 Class Hour	知 识 点 Key Points
2	数字电路实验：基于 ARM 处理器的片上系统设计 Digital circuit design of the System on Chip (SoC) based on ARM processor	4	(1) 以 ARM Cortex M0 为核心的片上系统 (2) 常见的 ARM 总线协议与使用方法 (3) Xilinx FPGA 开发板和 Vivado 开发环境 (4) 常见的片上外设,包括串口、LED、KEY 等 (5) 在 ARM 开发系统中扩展片上外设的方法 (1) On-chip systems based on ARM Cortex M0 core (2) Common ARM bus protocols and their usage (3) Xilinx FPGA development boards and the Vivado development environment. (4) Common on-chip peripherals, including UART, LEDs, keys, etc. (5) The method of extending on-chip peripherals in ARM development systems
3	嵌入式系统实验：基于树莓派的信号发生器设计 Embedded system design of digital signal generator with Raspberry Pi	4	(1) 树莓派的硬件资源与软件开发环境 (2) 基于树莓派的命令行软件编程方法 (3) SPI 总线协议并可以使用 GPIO 模拟总线协议 (4) 树莓派模数转换扩展板的硬件资源 (5) 模数转换器设计简易信号发生器的方法 (1) The hardware resources and software development environment of Raspberry Pi (2) Command-line software programming methods on Raspberry Pi (3) The SPI bus protocol and be able to simulate the bus protocol using GPIO of the Raspberry Pi (4) The hardware resources of Raspberry Pi's Analog-to-Digital Converter (ADC) expansion board (5) The method of designing a simple digital signal generator using ADC

续表

序号 Num.	实验内容 Experiment Content	课时 Class Hour	知 识 点 Key Points
4	嵌入式系统实验：基于树莓派的简易示波器设计 Embedded system design of simple oscilloscope with Raspberry Pi	4	（1）基于树莓派的图形化软件编程方法 （2）专业示波器的使用方法与常见的参数设置 （3）树莓派数模转换扩展版的硬件资源 （4）使用数模转换器设计简易示波器的方法 (1) Graphical User Interface (GUI) software programming methods based on Raspberry Pi (2) The usage and common parameter settings of professional oscilloscopes. (3) The hardware resources of Raspberry Pi's Digital-to-Analog Converter (DAC) expansion board (4) The method of designing a simple oscilloscope using DAC

大纲指导者：郑南宁教授（西安交通大学人工智能学院）

大纲制定者：孙宏滨教授（西安交通大学人工智能学院）、杨建国教授（西安交通大学电气工程学院）、梅魁志教授（西安交通大学人工智能学院）、葛晨阳副教授（西安交通大学人工智能学院）、刘龙军副教授（西安交通大学人工智能学院）

大纲审定：西安交通大学人工智能学院本科专业知识体系建设与课程设置第二版修订工作组

4.3 "现代控制工程"课程大纲

课程名称：现代控制工程

Course：Modern Control Engineering

先修课程：工科数学分析、线性代数与解析几何、复变函数与积分变换、电子技术与系统

Prerequisites：Mathematical Analysis for Engineering, Linear Algebra and Analytic Geometry, Complex Analysis and Integral Transformation, Electronic

Technology and System

学分：3

Credits：3

4.3.1　课程目的和基本内容（Course Objectives and Basic Content）

本课程是人工智能学院本科专业必修课。

This course is a compulsory course for undergraduates in College of Artificial Intelligence.

课程主要以线性控制系统分析与设计为主线，对自动控制的基本理论和应用展开全面讨论，包括自动控制的基本概念、控制系统的数学模型以及控制系统的分析与设计方法，并通过应用举例阐述控制理论在现代重大工程中的作用。第一章介绍了自动控制的发展历史以及自动控制系统的组成和分类；第二章结合典型的物理系统介绍了连续控制系统的各种数学模型以及各类模型之间的关系；第三章到第六章主要讨论连续控制系统基于时域、根轨迹、频率响应和状态空间的分析与设计方法，这部分内容是本课程的核心；第七章进一步讨论离散控制系统分析与设计方法以及连续系统与离散系统的关联与区别，第八章介绍了自动控制理论的工程应用。

课程以课堂讲解为主要教学形式，并辅之以研究性实验等教学手段，使学生深入理解自动控制的基本原理，熟练掌握线性控制系统的分析与设计方法，训练学生运用基本理论和方法分析解决实际问题的能力。课程通过若干真实控制系统的设计与实验，培养学生从实际问题中提炼控制问题并实现自动控制的思维能力，加强学生独立分析问题、解决问题的能力，为今后的学习和工作打下良好的基础。

The course focuses on the analysis and design of linear control system, and comprehensively discusses the basic theory and application of automatic control, including the basic concept of automatic control, the mathematical model of control system and the analysis and design methods of control system, and expounds the role of control theory in modern major projects through application examples. Chapter 1 introduces the development history of automatic control and the composition and classification of automatic control system. Chapter 2 introduces various mathematical models of continuous control system and the relationship between various models in combination with typical physical systems. Chapter 3 to Chapter 6 mainly discuss the analysis and design methods of continuous control system based on time domain, root locus, frequency response and state space. This part is the core of this course. Chapter

7 further discusses the analysis and design methods of discrete control system and the connection and difference between continuous system and discrete system. Chapter 8 introduces the engineering application of automatic control theory.

The course takes classroom explanation as the main teaching form, supplemented by research experiments and other teaching methods, in order to facilitate students to deeply understand the basic principles of automatic control, master the analysis and design methods of linear control system, and to be trained to use basic theories and methods to analyze and solve practical problems. Through several design experiments of real control systems, this course cultivates students' thinking ability to extract control problems from actual problems and realize automatic control, strengthens students' ability to analyze and solve problems independently, and lays a good foundation for future study and work.

4.3.2 课程基本情况（Course Arrangements）

课程名称	现代控制工程 Modern Control Engineering								
开课时间	一年级		二年级		三年级		四年级		科学与工程
	秋	春	秋	春	秋	春	秋	春	
课程定位	本科生"科学与工程"课程群必修课								必修（学分）：现代物理与人工智能(7)；电子技术与系统(5)；现代控制工程(3)；数字信号处理(3)
学　分	3 学分								
总学时	56 学时（授课 48 学时、实验 8 学时）								选修（学分）：/
授课学时分配	课堂讲授(48 学时)								
先修课程	工科数学分析、线性代数与解析几何、复变函数与积分变换、电子技术与系统								
后续课程	机器人学基础								
教学方式	课堂教学、大作业与实验、小组讨论								
考核方式	课程结束笔试成绩占 70%，平时成绩占 15%，实验成绩占 10%，考勤占 5%								
参考教材	胡寿松.自动控制原理[M].7 版.北京：科学出版社，2019.								
参考资料	1. 张爱民.自动控制原理[M].2 版.北京：清华大学出版社，2019. 2. Ogata K.现代控制工程[M].5 版.卢伯英，佟明安，译.北京：电子工业出版社，2017.								
其他信息									

4.3.3 教学目的和基本要求(Teaching Objectives and Basic Requirements)

(1) 掌握自动控制的基本思想和概念以及自动控制系统的基本组成和分类,建立系统化思维方式;

(2) 熟练掌握连续控制系统描述的基本方法,运用恰当的方法建立其数学模型(包括微分方程、传递函数以及状态空间等数学模型),并熟悉各种数学模型之间的相互转化方法;

(3) 熟练掌握暂态性能、稳态性能和稳定性等控制系统性能指标的理论和物理含义,熟练运用时域法、根轨迹法和频域法分析连续控制系统的暂态性能、稳态性能和稳定性以及改善这些性能的思路;

(4) 熟练掌握基于状态空间模型的控制系统的稳定性、能控性和能观性分析方法以及状态反馈控制器和状态观测器的设计思路;

(5) 初步掌握离散控制系统的基本分析与设计方法;

(6) 熟练掌握及运用自动控制系统的分析与设计工具。

4.3.4 教学内容及安排(Syllabus and Arrangements)

第一章 绪论(Introduction)

章节序号 Chapter Number	章节名称 Chapters	课时 Class Hour	知 识 点 Key Points
1.1	绪论 Introduction	2	(1) 控制理论发展概况 (2) 自动控制系统的结构 (3) 自动控制系统分类 (4) 自动控制系统的基本要求 (1) Overview of control theory development (2) Typical structure of control system (3) Classification of control system (4) Basic requirements on control system

第二章　连续控制系统的数学模型（Mathematical Model of Continuous Control System）

章节序号 Chapter Number	章节名称 Chapters	课时 Class Hour	知　识　点 Key Points
2.1	控制系统数学模型基本概念 Basic concepts of mathematical model of control system	2	（1）物理系统的微分方程 （2）传递函数 （3）系统方框图 （4）信号流图与梅逊公式
2.2	控制系统的微分方程 Differential equation of control system		（1）Differential equation of physical system （2）Transfer function （3）System block diagram （4）Signal flow diagram and Mason's rule
2.3	控制系统的传递函数 Transfer function of control system		
2.4	控制系统的状态空间模型 State-space model of control system	1	（1）状态空间的基本概念 （2）状态空间模型的建立 （3）状态空间模型的结构图和状态变量图 （1）Basic concept of state-space （2）Establishment of state-space model （3）Structure diagram and state variable diagram of state-space model
2.5	各种数学模型之间的关系 Relationship between various mathematical model	1	（1）由微分方程或传递函数转化为状态空间模型 （2）由状态空间模型求传递函数（矩阵） （3）由方框图求状态空间模型 （1）Calculating the state-space model from differential equation or transfer function （2）Calculating the transfer function (matrix) from the state-space model （3）Calculating the state-space model from system block diagram

第三章 连续控制系统的时域分析与设计（Time-domain Analysis and Design of Continuous Control System）

章节序号 Chapter Number	章节名称 Chapters	课时 Class Hour	知 识 点 Key Points
3.1	连续系统暂态性能分析 Transient performance analysis of continuous system	2	(1) 典型输入信号 (2) 暂态性能指标 (3) 一阶系统的暂态性能分析 (4) 典型二阶系统的暂态性能分析 (5) 高阶系统暂态性能近似分析 (1) Typical input signal (2) Transient performance index (3) Transient performance analysis of first-order system (4) Transient performance analysis of typical second-order system (5) Approximate analysis of transient performance of high order system
3.2	连续系统稳态性能分析 Steady-state performance analysis of continuous system	2	(1) 控制系统误差与稳态误差的定义 (2) 系统类型 (3) 终值定理法 (4) 误差系数法 (5) 扰动作用下的稳态误差分析 (1) Definition of control system error and steady-state error (2) System type (3) Terminal value theorem (4) Error coefficient method (5) Steady-state error analysis under disturbance
3.3	连续系统状态方程的求解与分析 Solution and analysis of state equation of continuous system	2	(1) 定常齐次状态方程的解 (2) 状态转移矩阵 (3) 定常系统的状态响应及输出响应 (1) Solutions of the steady homogeneous state equation (2) State transition matrix (3) State response and output response of steady systems

续表

章节序号 Chapter Number	章节名称 Chapters	课时 Class Hour	知 识 点 Key Points
3.4	连续控制系统的稳定性分析 Stability analysis of continuous control system	2	(1) 稳定性基本概念 (2) 劳斯稳定判据 (3) 赫尔维茨稳定判据 (4) 李雅普诺夫第一法 (5) 稳定性控制 (1) Basic concept of stability (2) Routh stability criterion (3) Herwitz stability criterion (4) Lyapunov's first method (5) Stability control
3.5	连续控制系统的时域设计 Time-domain design of continuous control system		

第四章 连续控制系统的根轨迹法（Root-Locus Method for Continuous Control System）

章节序号 Chapter Number	章节名称 Chapters	课时 Class Hour	知 识 点 Key Points
4.1	根轨迹的基本概念 Basic concepts of Root-Locus method	1	(1) 根轨迹定义 (2) 闭环零、极点和开环零、极点之间关系 (3) 根轨迹方程 (1) Definition of Root-Locus (2) Relationship between closed loop pole-zero and open loop pole-zero (3) Root-Locus equation
4.2	根轨迹的绘制方法 Plotting method of Root-Locus	2	(1) 常规根轨迹作图规则 (2) 参数根轨迹 (3) 零度根轨迹 (4) 纯滞后系统的根轨迹
4.3	广义根轨迹绘制 Plotting of generalized Root-Locus		(1) General rules for constructing Root-Locus (2) Parameter Root-Locus (3) Zero-degree Root-Locus (4) Root-Locus of pure delay system

续表

章节序号 Chapter Number	章节名称 Chapters	课时 Class Hour	知识点 Key Points
4.4	基于根轨迹的系统性能分析 System performance analysis based on root-locus	2	（1）开环点对系统性能的影响 （2）开环零点对系统性能的影响 （3）增益 K 的选取 (1) Influence of open-loop point on system performance (2) Influence of open-loop zero point on system performance (3) Selection of gain K
4.5	基于根轨迹的系统补偿器设计 Design of system compensator based on root-locus	3	（1）超前补偿器的设计 （2）滞后补偿器的设计 （3）PID 控制器的设计 (1) Design of lead compensator (2) Design of lag compensator (3) Design of PID compensator

第五章 连续控制系统的频率法（Frequency Method for Continuous Control System）

章节序号 Chapter Number	章节名称 Chapters	课时 Class Hour	知识点 Key Points
5.1	频率特性 Frequency characteristic	1	（1）频率特性的定义 （2）频率响应 （3）频率特性的几何表示 (1) Definition of Frequency characteristic (2) Frequency response (3) Geometric representation of frequency characteristic
5.2	奈奎斯特图 Nyquist diagram	1	（1）典型环节的奈奎斯特图 （2）开环系统的奈奎斯特图 (1) Nyquist diagram of typical unit (2) Nyquist diagram of open-loop system

续表

章节序号 Chapter Number	章节名称 Chapters	课时 Class Hour	知　识　点 Key Points
5.3	伯德图 Bode diagram	1	(1) 典型环节的伯德图 (2) 开环系统的伯德图 (3) 由伯德图确定传递函数 (1) Bode diagram of typical unit (2) Bode diagram of open-loop system (3) Calculating the transfer function from the Bode diagram
5.4	奈奎斯特稳定判据 Nyquist stability criterion	1	(1) 辐角原理 (2) 奈奎斯特稳定判据 (3) 奈奎斯特稳定判据的应用 (1) Argument principle (2) Nyquist stability criterion (3) Application of Nyquist stability criterion
5.5	控制系统相对稳定性分析 Relative stability analysis of control system	1	(1) 用奈奎斯特图表示相位裕量和幅值裕量 (2) 用伯德图表示相位裕量和幅值裕量 (1) Phase margin and amplitude margin used by Nyquist diagram (2) Phase margin and amplitude margin used by Bode diagram
5.6	基于频率响应的补偿器设计 Design of system compensator based on frequency response	3	(1) 频率指标与时域指标的关系 (2) 超前补偿器的设计 (3) 滞后补偿器的设计 (4) PID 控制器的设计 (1) Relationship between frequency index and time index (2) Design of lead compensator (3) Design of lag compensator (4) Design of PID compensator

第六章　连续控制系统的状态空间法（State-space Method for Continuous Control System）

章节序号 Chapter Number	章节名称 Chapters	课时 Class Hour	知　识　点 Key Points
6.1	连续系统的能控性与能观性分析 Controllability and observability analysis of continuous system	2	（1）系统能控性和能观性的直观示例 （2）连续系统能控性及其判据 （3）连续系统能观性及其判据 (1) Visual examples of controllability and observability (2) Controllability of continuous system and its criteria (3) Observability of continuous system and its criteria
6.2	连续系统的线性变换与结构分解 Linear transformation and structural decomposition of continuous system	2	（1）非奇异线性变换 （2）状态空间表达式的几种标准形式 （3）结构分解 （4）能控性、能观性与传递函数（矩阵）的关系 (1) Non-singular linear transformation (2) Several standard forms of state-space expression (3) Structural decomposition (4) Relationship among controllability, observability and transfer function (matrix)
6.3	连续系统的状态反馈控制 State feedback control of continuous system	2	（1）状态反馈控制系统组成 （2）状态反馈控制系统极点任意配置的条件 （3）单输入系统的极点配置算法 （4）状态反馈对系统能控性和能观性的影响 (1) Composition of state feedback control system (2) Conditions for arbitrary pole assignment of state feedback control system (3) Pole assignment algorithm for single input system (4) Influence of state feedback on controllability and observability

续表

章节序号 Chapter Number	章节名称 Chapters	课时 Class Hour	知 识 点 Key Points
6.4	连续系统的状态观测器设计 State observer design for continuous system	2	(1) 状态重构与全维状态观测器 (2) 降维状态观测器 (3) 引入观测器的状态反馈控制系统 (1) State reconstruction and full-dimensional state observer (2) Reduced-dimension state observer (3) State feedback control system with observer

第七章 离散控制系统分析与综合（Analysis and Synthesis of Discrete Control System）

章节序号 Chapter Number	章节名称 Chapters	课时 Class Hour	知 识 点 Key Points
7.1	离散控制系统基础知识 Basic knowledge of discrete control system	1	(1) 采样过程与采样定理 (2) Z变换 (3) 离散控制系统的数学描述及求解 (1) Sampling process and sampling theorem (2) Z-transform (3) Mathematical description and solution of discrete control system
7.2	离散控制系统的频域与复频域分析与设计 Analysis and design of discrete control system in frequency domain and complex frequency domain	2	(1) 离散系统的稳定性分析 (2) 基于频域特性的分析与设计 (3) 基于Z域的分析与设计 (1) Stability analysis of discrete system (2) Analysis and design based on frequency domain characteristics (3) Analysis and design based on Z domain

章节序号 Chapter Number	章节名称 Chapters	课时 Class Hour	知识点 Key Points
7.3	离散控制系统的状态空间分析与设计 Analysis and design of discrete control system in state-space	2	（1）离散系统的时域分析 （2）离散系统的能控性与能观性分析 （3）离散系统的状态反馈控制 （1）Time domain analysis of discrete system （2）Controllability and observability analysis of discrete system （3）State feedback control of discrete system
7.4	连续系统与离散系统的关联与区别 Correlation and difference between continuous system and discrete system	1	（1）连续系统离散化的稳定性 （2）连续系统离散化的能控性和能观性 （3）数字控制系统分析与设计 （1）Stability of discretized continuous system （2）Controllability and observability of discretized continuous system （3）Analysis and design of digital control system

第八章 自动控制理论的应用举例（Application Examples of Automatic Control Theory）

章节序号 Chapter Number	章节名称 Chapters	课时 Class Hour	知识点 Key Points
8.1	飞行控制系统的分析与设计 Analysis and design of flight control system	1	（1）副翼偏转对应滚转角系统反馈控制设计（根轨迹法） （2）方向舵对应偏航角系统反馈控制设计（根轨迹法） （1）Feedback control design of roll angle system corresponding to aileron deflection (root-locus method) （2）Feedback control design of yaw angle system corresponding to rudder (root-locus method)

续表

章节序号 Chapter Number	章节名称 Chapters	课时 Class Hour	知　识　点 Key Points
8.2	水面无人艇运动控制系统建模与控制系统设计 Modeling and design of motion control system for surface unmanned vehicle	1	（1）无人艇运动模型的建立 （2）无人艇运动模型的简化 （3）无人艇 PID 控制及运动仿真 （1）Establishment of motion model of unmanned vehicle （2）Simplification of motion model of unmanned vehicle （3）PID control and motion simulation of unmanned vehicle
8.3	全自主双轮平衡车数学建模与控制系统设计 Mathematical modeling and design of control system for fully autonomous two-wheel balance vehicle	1	（1）全自主双轮平衡车的状态空间模型建立 （2）系统的能控性和能观性判别 （3）系统的状态反馈控制器和状态观测器设计 （1）Establishment of state-space model for fully autonomous two-wheel balance vehicle （2）Controllability and observability of system （3）Design of state feedback controller and state observer for system
8.4	倒立摆离散控制系统设计 Design of discrete control system for inverted pendulum	1	（1）倒立摆离散状态空间模型 （2）倒立摆系统的稳定性 （3）倒立摆状态反馈控制器 （1）Discrete state-space model of inverted pendulum （2）Stability of inverted pendulum system （3）Inverted pendulum state feedback controller

4.3.5 实验环节(Experiments)

序号 Num.	实验内容 Experiment Content	课时 Class Hour	知　识　点 Key Points
1	液位控制系统的设计与仿真 Design and simulation of the tank water control system	2	(1) 闭环控制系统组成 (2) 时域分析法 (3) MATLAB/Simulink 的使用 (1) Composition of closed-loop control system (2) Time domain analysis method (3) Use of MATLAB/Simulink
2	硬盘读/写碰头组件控制设计与仿真 Design and simulation of control of the read/write head assembly of a hard disk	2	(1) 传递函数模型 (2) 伯德图与带宽 (3) 超前补偿器设计 (1) Transfer function model (2) Bode diagram and bandwidth (3) Design of lead compensator
3	智能小车速度与方向控制器设计 Speed and direction control method for intelligent min-car	2	(1) 环境传感器 (2) PID 控制器 (3) 智能小车速度控制 (4) 智能小车方向控制 (1) Environment sensors (2) PID controller (3) Speed control method for intelligent min-car (4) Direction control method for intelligent min-car

序号 Num.	实验内容 Experiment Content	课时 Class Hour	知 识 点 Key Points
4	二级倒立摆控制系统的设计与仿真 Design and simulation of two-stage inverted pendulum control system	2	（1）状态空间模型 （2）系统的稳定性 （3）状态响应 （4）状态反馈控制器 （5）状态观测器 (1) State-space model (2) System stability (3) Status response (4) State feedback controller (5) State observer

大纲制定者：刘妹琴教授（西安交通大学人工智能学院）、张雪涛副教授（西安交通大学人工智能学院）

大纲审定：西安交通大学人工智能学院本科专业知识体系建设与课程设置第二版修订工作组

4.4 "数字信号处理"课程大纲

课程名称：数字信号处理
Course：Digital Signal Processing
先修课程：工科数学分析、概率统计与随机过程、复变函数与积分变换、电子技术与系统
Prerequisites：Mathematical Analysis for Engineering, Probability Theory and Stochastic Process, Complex Analysis and Integral Transformation, Electronic Technology and System
学分：3
Credits：3

4.4.1 课程目的和基本内容（Course Objectives and Basic Content）

本课程是人工智能学院本科专业必修课。

This course is a compulsory course for undergraduates in College of Artificial Intelligence.

　　课程以变换分析为主线,对采样信号表示、频谱分析、离散傅里叶变换和数字滤波器的基本理论与设计方法展开讨论,同时介绍实时滤波以及离散随机信号分析的基本方法。第一章到第五章分别讨论信号的傅里叶分析与采样信号、离散时间序列与系统的基本分析方法、Z变换、离散傅里叶变换和快速傅里叶变换算法,这部分内容的重点是数字信号的产生及其在时域和频域的表示方法以及离散时间系统的基本性质和分析方法。第六章到第九章主要讨论数字滤波器的基本原理、设计方法和实时滤波,其中专门介绍利用 ROM 查表法的实时滤波方法。第十章讨论离散时间随机信号分析的基本方法。

　　课程通过对基本理论、设计方法和应用技术的学习,帮助学生建立关于数字信号处理基本原理和应用设计方面的知识框架。课程采用小组学习模式,并辅之以研究性实验、课堂测验、小组讨论及综述报告等教学手段,训练学生用基本理论和方法分析解决实际问题的能力,掌握数字信号处理应用设计所必须的基本知识和技能。课程通过语音信号的采集处理及识别系统设计实验使学生巩固和加深数字信号处理的理论知识,通过实践进一步加强学生独立分析问题、解决问题的能力,培养综合设计及创新能力,为今后的工作打下良好的基础。

　　随着计算机和超大规模集成电路技术的发展,数字信号处理不仅在信息技术领域扮演着十分重要的角色,而且其基本原理和方法几乎应用在所有的物理系统和社会计算中,成为一种重要的数值分析、处理与计算的工具。因此,理解和掌握好数字信号处理的基本概念、基本原理和方法,在遇到实际问题时,能激发学生去寻找新的理论和技术,也能使学生利用一种熟悉的工具进入一个生疏的研究领域。

　　The course focuses on transform analysis, and discusses the basic theories and the design methods of sampling signal representation, spectrum analysis, discrete Fourier transform and digital filter. Moreover, it introduces the basic methods of real-time filtering and discrete random signal analysis. Chapter 1 to Chapter 5 discuss the Fourier analysis and sampling signals, the basic analysis methods of discrete time series and systems, the Z transform, the discrete Fourier transform and the fast Fourier transform algorithms, respectively. The focus of this part is on the generation of signals and their representation in the time and frequency domains, as well as the basic properties and the analysis methods of discrete time systems. Chapter 6 to Chapter 9 mainly discuss the basic principles and the design methods of digital filters, and real-time filtering. The real-time filtering method using the ROM look-up table is specifically introduced. Chapter 10 discusses the basic methods of discrete-time random signal analysis.

The course helps students build a knowledge framework for the basic principles of digital signal processing and application design through the study of basic theories, design methods, and applied techniques. The course adopts group study method, supplemented by experiments, in-class tests, discussions and reports, in order to train students the ability to solve practical problems with basic theories and methods and master the basic knowledge and skills for digital signal processing application design. The course includes several experiments on speech signal collection and recognition system design in order to consolidate students' theoretical knowledge of digital signal processing, further strengthen their ability to analyze and solve problems independently, and develop their comprehensive abilities on system design and innovations.

With the development of computer and VLSI technologies, digital signal processing not only plays a very important role in the field of information technology, but its basic principles and methods are applied to almost all physical systems and social computing, becoming an important tool for numerical analysis, processing, and calculation. Therefore, understanding the basic concepts, principles and methods of digital signal processing can stimulate students to find new theories and techniques when they encounter practical problems. Moreover, it also helps students to enter a strange research area with familiar tools.

4.4.2 课程基本情况（Course Arrangements）

课程名称	数字信号处理 Digital Signal Processing							
开课时间	一年级		二年级		三年级		四年级	
	秋	春	秋	春	秋	春	秋	春
课程定位	本科生"科学与工程"课程群必修课							
学　　分	3 学分							
总 学 时	56 学时 （授课 48 学时、实验 8 学时）							
授课学时分配	课堂讲授（46 学时）， 大作业讨论（2 学时）							
先修课程	工科数学分析、概率统计与随机过程、复变函数与积分变换、电子技术与系统							
后续课程	人工智能芯片设计导论							

科学与工程	
必修 （学分）	现代物理与人工智能（7）
	电子技术与系统（5）
	现代控制工程（3）
	数字信号处理（3）
选修 （学分）	/

续表

教学方式	课堂教学、大作业与实验、小组讨论、综述报告
考核方式	课程结束笔试成绩占 60%，平时成绩占 15%，实验成绩占 10%，综述报告占 10%，考勤占 5%
参考教材	郑南宁.数字信号处理简明教程[M].2版.西安：西安交通大学出版社，2019.
参考资料	1. 郑南宁，杨勐，刘剑毅.数字信号处理简明教程习题解析[M].2版.西安：西安交通大学出版社，2022. 2. 郑南宁，张元林，杨勐，等.数字信号处理实验指导书[M]，2018.
其他信息	

4.4.3 教学目的和基本要求（Teaching Objectives and Basic Requirements）

（1）理解离散时间信号与系统的基本概念，掌握其分析的基本工具和方法，了解数字信号处理的基本应用；

（2）熟悉离散傅里叶变换的原理及其算法；

（3）理解时域和频域采样定理；

（4）理解快速傅里叶变换的原理，掌握其算法的实现；

（5）熟悉数字滤波器的基本结构，掌握 FIR 和 IIR 数字滤波器的常用设计方法；

（6）了解数字信号处理中的实时滤波方法及其有限字长效应；

（7）了解离散随机信号的基本分析方法；

（8）熟练使用 C 语言和 MATLAB 实现数字信号处理算法。

4.4.4 教学内容及安排（Syllabus and Arrangements）

绪论 （Introduction）

章节序号 Chapter Number	章节名称 Chapters	课时 Class Hour	知　识　点 Key Points
0.1	绪论 Introduction	2	（1）信号与系统的基本术语 （2）数字信号处理的一般原理 （3）数字信号处理的变换分析方法 (1) Basic terminologies of signal and system (2) General principles of digital signal processing (3) Transform analysis in digital signal processing

第一章 傅里叶分析与采样信号(Fourier Analysis and Sampling Signal)

章节序号 Chapter Number	章节名称 Chapters	课时 Class Hour	知　识　点 Key Points
1.1	连续时间周期信号的傅里叶级数表示 Fourier series representation of continuous-time periodical signal	2	（1）三角函数型傅里叶级数表示 （2）指数型傅里叶级数表示 （3）傅里叶级数的波形分解 （1）Triangular Fourier Series (FS) （2）Exponential FS （3）Waveform decomposition of FS
1.2	非周期信号的连续时间傅里叶变换表示 Continuous-time Fourier transform representation of aperiodic signal		（1）非周期信号的连续时间傅里叶变换表示的推导 （2）傅里叶变换存在的条件：狄利克雷条件 （1）Derivation of Continuous-Time Fourier Transform (CTFT) representation of aperiodic signals （2）Conditions for the Fourier transform: Dirichlet conditions
1.3	连续时间傅里叶变换的性质 Properties of continuous-time Fourier transform	2	（1）连续时间傅里叶变换的性质：线性、对偶性、时间尺度变化、频率尺度变化、时间移位、频率移位、奇偶性、微分、积分等 （1）Properties of CTFT: linearity, duality, time-scaling, frequency-scaling, time-shift, frequency-shift, parity, differentiation, Integration, and so on
1.4	卷积与相关 Convolution and correlation		（1）卷积积分、卷积定理和频域卷积定理 （2）相关和相关定理 （1）Convolution integration, convolution theorem, and convolution theorem in the frequency domain （2）Correlation and correlation theorem

续表

章节序号 Chapter Number	章节名称 Chapters	课时 Class Hour	知识点 Key Points
1.5	连续时间信号的采样 Sampling of continuous-time signal	2	(1) 采样过程与采样函数 (2) 离散时间傅里叶变换的推导 (1) Sampling and sampling function (2) Derivation of Discrete-Time Fourier Transform (DTFT)
1.6	用信号样本表示连续时间信号：采样定理 Representing continuous-time signal with sample：sampling theorem		(1) 采样定理 (1) Sampling theorem
1.7	利用内插由样本重建信号 Reconstructing signal from sample using interpolation		(1) 如何由样本重建信号 (1) How to reconstruct a continuous-time signal from sample
1.8	A/D 转换的量化误差分析 Quantization error analysis of A/D conversion		(1) 采样信号的量化过程 (2) 量化误差的计算 (1) Quantization process of sampled signal (2) Calculation of quantization error

第二章 离散时间序列与系统（Discrete Time Sequence and System）

章节序号 Chapter Number	章节名称 Chapters	课时 Class Hour	知识点 Key Points
2.1	离散时间信号：序列 Discrete time signal：sequence	2	(1) 基本序列及序列的基本运算 (2) 序列的稳定性和因果性 (1) Elementary sequences and basic operations of sequences (2) The stability and causality of sequence

续表

章节序号 Chapter Number	章节名称 Chapters	课时 Class Hour	知　识　点 Key Points
2.2	序列的离散时间傅里叶变换表示 Discrete time Fourier transform representation of a sequence		（1）序列的离散时间傅里叶变换表示的推导 （1）Derivation of the DTFT of sequence
2.3	离散时间傅里叶变换的性质 Properties of discrete time Fourier transform	2	（1）离散时间傅里叶变换的性质：周期性、对称性、线性、时间移位、频率移位、共轭、反转、卷积、相乘、频域微分、帕塞瓦定理等 （1）Properties of DTFT：periodicity, symmetry, linearity, time-shift, frequency-shift, conjugate symmetry, time-reversal, convolution, multiplication, differentiation in the frequency domain, Parseval's theorem, and so on
2.4	离散时间系统 Discrete time system	2	（1）离散线性时不变系统的因果性和稳定性 （2）离散时间系统的差分方程表示 （1）The causality and stability of discrete Linear Time-Invariant (LTI) system （2）Describing the LTI system with difference equation
2.5	离散时间系统的频率响应 Frequency response of discrete time system		（1）复指数序列的频率响应 （2）任意序列的频域响应 （3）由差分方程求频率响应函数 （1）Frequency response of complex exponential sequence （2）Frequency response of general sequence （3）Calculating the frequency response from difference equation

第三章　Z 变换(Z-Transform)

章节序号 Chapter Number	章节名称 Chapters	课时 Class Hour	知　识　点 Key Points
3.1	Z 变换的定义及收敛域 Definition of Z-transform and its convergence domain	2	(1) Z 变换的定义和收敛域 (2) Z 变换与 DTFT 的关系 (1) Definition of Z-transform and its convergence domain (2) Relationship between Z-transform and DTFT
3.2	Z 反变换 Inverse Z-transform		(1) Z 反变换的计算方法：围线积分法、部分分式展开法、幂级数展开法 (1) Calculation of inverse Z-transform: contour integration, partial fractionation, power series expansion
3.3	Z 变换的性质 Properties of Z-transform	2	(1) Z 变换的性质：线性、移位、Z 域微分、指数加权、初值定理、终值定理、卷积定理、复卷积定理、帕塞瓦定理等 (1) Properties of Z transform: linearity, time-shift, differentiation in the Z-transform domain, exponential weighting, initial value theorem, final value theorem, convolution theorem, complex convolution theorem, Parseval's theorem, and so on
3.4	Z 变换域中离散时间系统的描述 Description of discrete time system in the Z-transform domain	2	(1) 由线性常系数微分方程导出系统函数 (2) 系统函数的频域分析 (1) Deriving system functions from linear constant coefficient difference equations (2) Analyzing the system function in the frequency domain
3.5	单边 Z 变换 Unilateral Z-transform		(1) 单边 Z 变换的定义和性质 (1) The definition and properties of the unilateral Z-transform

续表

章节序号 Chapter Number	章节名称 Chapters	课时 Class Hour	知识点 Key Points
3.6	用单边Z变换求解线性差分方程 Solving linear difference equations with unilateral Z-transform		(1) 零输入响应和零状态响应 (1) Zero input response and zero state response

第四章 离散傅里叶变换(Discrete Fourier Transform)

章节序号 Chapter Number	章节名称 Chapters	课时 Class Hour	知识点 Key Points
4.1	离散傅里叶级数 Discrete Fourier series	2	(1) 离散傅里叶级数的性质：线性、移位、对偶性、对称性、周期卷积等 (1) Properties of Discrete Fourier Series (DFS): linearity, time-shift property, duality, symmetry, periodic convolution, and so on
4.2	离散傅里叶变换 Discrete Fourier transform	2	(1) 离散傅里叶变换的定义 (2) 离散傅里叶变换的性质：线性、对偶性、共轭对称性、反转、循环移位、循环卷积等 (3) 利用循环卷积计算线性卷积 (4) Z域频率采样 (1) Definition of Discrete Fourier Transform (DFT) (2) Properties of DFT: linearity, duality, conjugate symmetry, time-reversal, cyclic shift, cyclic convolution, and so on (3) Computing cyclic convolution by cyclic convolution (4) Sampling in the Z-transform domain

146

续表

章节序号 Chapter Number	章节名称 Chapters	课时 Class Hour	知识点 Key Points
4.3	离散傅里叶变换应用中的问题与参数选择 Problems and parameter selection in discrete Fourier transform applications	2	（1）DFT 中的频谱混叠、频谱泄漏、栅栏效应 （2）离散傅里叶变换的物理分辨率、频率分辨率与计算长度 (1) The spectrum aliasing, spectrum leakage, picket fence effect of DFT (2) Physical resolution and frequency resolution of DFT

第五章 快速傅里叶变换（Fast Fourier Transform）

章节序号 Chapter Number	章节名称 Chapters	课时 Class Hour	知识点 Key Points
5.1	FFT 算法的基本原理 The basic principle of the FFT algorithm	2	（1）FFT 的基本原理 (1) Basic principle of Fast Fourier Transform（FFT）
5.2	按时间抽取的 FFT 算法 Decimation-in-time FFT algorithm		（1）按时间抽取 FFT 算法 （2）FFT 中的蝶形运算和码位倒序 (1) The decimation-in-time FFT algorithm (2) The basic butterfly operation and bit-reversed order in FFT
5.3	按频域抽取的 FFT 算法 Decimation-in-frequency FFT algorithm		（1）按频域抽取 FFT 算法 (1) The decimation-in-frequency FFT algorithm
5.4	任意基数的 FFT 算法 Radix-X FFT		（1）任意基数的 FFT 算法 （2）混合基 FFT 算法 (1) Radix-X FFT algorithm (2) Split-radix FFT algorithm

续表

章节序号 Chapter Number	章节名称 Chapters	课时 Class Hour	知识点 Key Points
5.5	IDFT 的快速运算方法 The fast algorithm of IDFT		（1）快速 IDFT 算法 （1）Inverse FFT algorithm
5.6	实数序列的 FFT 运算方法 Computation of the FFT of real sequence		（1）同时运算两个实序列的 FFT （2）用 N 点变换计算 $2N$ 点实序列的 FFT （1）Computation of the FFT of two real sequence （2）Computation of the FFT of a $2N$-point sequence
5.7	FFT 的软件实现 Implementation of FFT	2	（1）FFT 权函数的计算 （1）Determination of the weight term in FFT
5.8	Chirp-Z 变换 Chirp-Z transform		（1）Chirp-Z 变换的定义 （2）Chirp-Z 的算法实现 （1）Definition of the Chirp-Z transform （2）Implementation of the Chirp-Z transform
5.9	FFT 算法中有限寄存器长度量化效应分析 Analysis of quantization effect of finite register length in FFT		（1）直接法计算 DFT 的舍入量化误差 （2）定点 FFT 运算的量化误差 （3）浮点 FFT 运算的量化误差 （4）FFT 运算的系数量化误差 （1）Direct method to calculate the rounding quantization error of DFT （2）Quantization error of fixed-point FFT operation （3）Quantization error of floating point FFT operation （4）Coefficient quantization error of FFT operation

第六章　数字滤波器的基本原理与特性（Basic Principles and Characteristics of Digital Filters）

章节序号 Chapter Number	章节名称 Chapters	课时 Class Hour	知　识　点 Key Points
6.1	数字滤波器 的基本原理 Basic principles of digital filter	2	（1）数字滤波器的基本指标、基本方程、分类、系统函数、冲激响应 （1）Basic characteristics, equation, type, system function, and impulse response of digital filter
6.2	数字滤波器的 基本特性 Basic characteristics of digital filter		（1）FIR 滤波器的基本特性 （2）IIR 的基本特性 （3）FIR 和 IIR 滤波器的比较 （1）Basic characteristics of FIR filter （2）Basic characteristics of IIR filter （3）Difference between FIR and IIR filter

第七章　FIR 数字滤波器设计（FIR Digital Filter Design）

章节序号 Chapter Number	章节名称 Chapters	课时 Class Hour	知　识　点 Key Points
7.1	傅里叶级数展开法 Fourier series expansion	2	（1）傅里叶级数展开法 （2）吉布斯现象 （1）Fourier series expansion method （2）Gibbs phenomenon
7.2	窗函数设计法 Design of FIR using windows		（1）窗函数设计法 （1）Design of FIR using windows
7.3	FIR 滤波器的 计算机辅助设计 FIR design by computer		（1）频率采样法 （2）切比雪夫逼近设计法 （1）Frequency-sampling （2）Chebyshev approximation method

续表

章节序号 Chapter Number	章节名称 Chapters	课时 Class Hour	知识点 Key Points
7.4	FIR滤波器的实现结构 Structures of FIR filter	2	(1) FIR滤波器的实现结构：直接型、级联型、FFT变换型、频率采样型 (1) Structures of FIR filter: direct structure, series-connected structure, FFT transform structure, frequency sampling structure
7.5	非递归型FIR滤波器量化误差分析 Quantization error analysis of non-recursive FIR filter		(1) 系数量化误差 (2) 运算量化误差 (1) Quantization errors of coefficients (2) Quantization errors of operations

第八章 IIR数字滤波器设计（IIR Digital Filter Design）

章节序号 Chapter Number	章节名称 Chapters	课时 Class Hour	知识点 Key Points
8.1	S-Z变换设计法 S-Z transformation	2	(1) 冲激响应不变法 (2) 双线性变换法 (3) 匹配Z变换法 (1) IIR design method by impulse invariance (2) IIR design method by the bilinear transformation (3) The matched-Z transformation
8.2	频率变换设计法 Frequency transformations method		(1) 频率变换设计法 (1) Frequency transformation method
附录B	模拟滤波器 Analog filter	2	(1) 巴特沃斯滤波器 (2) 切比雪夫滤波器 (1) Butterworth filter (2) Chebyshev filter

续表

章节序号 Chapter Number	章节名称 Chapters	课时 Class Hour	知 识 点 Key Points
8.3	IIR 数字滤波器的计算机辅助设计 Computer aided design of IIR filter	2	(1) 最小平方逆滤波设计法 (2) 频率最小均方误差设计法 (3) 时域设计法 (1) FIR least-square inverse filters method (2) Design method of IIR in the frequency domain (3) Design of IIR in the time domain
8.4	IIR 数字滤波器的实现结构 Structure of IIR filter		(1) IIR 数字滤波器的实现结构：直接型、级联型、并联型、梯形结构 (1) Structures of IIR filter: direct structure, series-connected structure, parallel structure, and trapezoidal structure
8.5	递归型 IIR 滤波器量化误差分析 Quantization error analysis of recursive IIR filter		(1) 系数量化误差 (2) 定点运算量化误差 (3) 浮点运算量化误差 (1) Coefficient quantization error (2) Fixed point quantization error (3) Floating point quantization error

第九章　实时滤波(Real-time Filtering)

章节序号 Chapter Number	章节名称 Chapters	课时 Class Hour	知 识 点 Key Points
9.1	ROM 查表式乘法 Multiply operation with ROM table	2	(1) ROM 查表式乘法 (1) Multiply operation method with ROM table
9.2	滤波器的定点运算实现 Fixed-point operations of digital filter		(1) 滤波器的定点运算实现过程 (1) The process of fixed-point operations of digital filter

续表

章节序号 Chapter Number	章节名称 Chapters	课时 Class Hour	知识点 Key Points
9.3	IIR滤波器的查表法实现 IIR filter design by look-up table		（1）一阶 IIR 滤波器的查表法实现 （2）二阶 IIR 滤波器的查表法实现 （3）压缩比例因子的选择 (1) First-order IIR filter design by look-up table (2) Second-order IIR filter design by look-up table (3) Scaling factor selection
9.4	噪声滤除 Noise removal		（1）加性噪声滤除 （2）乘性噪声滤除 （3）同态系统 (1) Additive noise removal (2) Multiplicative noise removal (3) Homomorphic systems

第十章 离散随机信号的统计分析基础（Statistical Analysis Basis of Discrete Random Signal）

章节序号 Chapter Number	章节名称 Chapters	课时 Class Hour	知识点 Key Points
10.1	随机过程的定义 Definition of random process		（1）随机过程的定义 (1) Definition of random process
10.2	离散随机过程的时域统计描述 Statistical descriptions of discrete random process in the time domain	2	（1）离散随机过程的概率分布函数和概率密度函数 （2）平稳随机过程的定义 （3）概率分布特性的特征量 （4）相关序列与协方差序列的性质 （5）各态历经性与时间平均 (1) Probability distribution function and probability density function of discrete random process (2) Definition of stationary stochastic process (3) Characteristic quantities of probability distribution (4) The properties of correlation sequence and covariance sequence (5) Ergodicity and time average

续表

章节序号 Chapter Number	章节名称 Chapters	课时 Class Hour	知 识 点 Key Points
10.3	离散随机过程的频域统计描述 Statistical description of discrete random process in the frequency domain		(1) 功率谱密度 (2) 互功率谱密度 (1) Power Spectrum Density (PSD) (2) Cross-power spectrum density
10.4	离散线性系统对随机信号的响应 Response of random signal in discrete linear system		(1) 系统的稳态响应 (2) LTI 系统的输入输出互相关定理 (3) 互功率谱与系统频率响应的关系 (1) Steady state response of the system (2) Input-output cross-correlation theorem (3) Relationship between cross power spectrum and frequency response

4.4.5 实验环节(Experiments)

序号 Num.	实验内容 Experiment Content	课时 Class Hour	知 识 点 Key Points
1	基于时域分析技术语音识别 Speech recognition based on time-domain analysis	2	(1) 语音信号的格式与采集方法 (2) 语音信号的预处理方法 (3) 语音信号的分帧与加窗处理 (4) 基于双门限法的端点检测 (5) 基于时域分析的孤立字语音识别方法 (1) Format and acquisition methods of speech signal (2) Preprocessing methods of speech signal (3) Speech signal framing and windowing (4) Endpoint detection based on double threshold method (5) Isolated word recognition based on time domain analysis

续表

序号 Num.	实验内容 Experiment Content	课时 Class Hour	知 识 点 Key Points
2	语音信号的频域特征分析 Characteristics analysis of speech signal in the frequency domain	2	（1）短时傅里叶变换的原理 （2）梅尔频率倒谱系数的提取 （1）Principle of short time Fourier transform （2）Calculation of Mel Frequency Cepstrum Coefficients（MFCC）
3	基于动态时间规整（DTW）的孤立字语音识别 Isolated word recognition based on dynamic time warping (DTW)	2	（1）模板匹配法 （2）动态时间规整技术 （3）基于动态时间规整的阿拉伯数字识别 （1）Template matching method （2）Dynamic Time Warping (DTW) （3）Arabic numerals recognition based on DTW
4	独立于内容的说话人识别 Content-independent speaker recognition	2	（1）独立于内容的说话人识别方法 （1）Principle of content independent speaker recognition

大纲指导者：郑南宁教授（西安交通大学人工智能学院）

大纲制定者：郑南宁教授（西安交通大学人工智能学院）、张元林副教授（西安交通大学人工智能学院）、杨勐副教授（西安交通大学人工智能学院）

大纲审定：西安交通大学人工智能学院本科专业知识体系建设与课程设置第二版修订工作组

第 5 章

"计算机科学与技术"课程群

5.1 "计算机程序设计"课程大纲

课程名称：计算机程序设计
Course：Computer Programming
先修课程：无
Prerequisites：None
学分：2
Credits：2

5.1.1 课程目的和基本内容（Course Objectives and Basic Content）

本课程是人工智能学院本科专业必修课。

This course is a compulsory course for undergraduates in College of Artificial Intelligence.

课程以计算机程序设计为主线，结合 C/C++/Python 语言，对程序设计的基本理论、面向过程和面向对象程序设计方法展开讨论，同时介绍排序等基本算法。第一章到第三章讨论 C++ 程序设计的数据结构，包括整型、浮点、字符、数组、指针、结构体等。第四和第五章主要讨论 C++ 的控制结构，包括顺序、分支和循环结构。第六和第七章讨论函数的声明、定义、调用、重载以及模板。第八章简要介绍内存模型与名称空间。第九和第十章讨论 C++ 面向对象编程的基本概念、思想和方法，包括类和对象的定义、类的构造函数和析构函数以及类运算符的重载。

课程通过对基本理论、编程方法的学习,帮助学生建立计算机程序设计方面的知识框架。课程采用小组学习模式,并辅之以研究性实验、课堂测验、小组讨论及综述报告等教学手段,训练学生用基本理论和方法分析解决实际问题的能力,掌握计算机程序设计所必须的基本知识和技能。课程通过专门的设计实验使学生巩固和加深理论知识,通过实践进一步加强学生独立分析问题、解决问题的能力,培养综合设计及创新能力,为今后的工作打下良好的基础。

本课程实验环节包括面向 C/C++ 的专门实验以及扩展的 Python 编程实验,学生通过课堂学习已经掌握了程序设计的基础,在实验环节将进一步通过实验扩展学习 Python 基础编程,并进行专门的实验,加强学生的编程拓展学习能力。

The course focuses on computer program design. Based on C/C++ programming language, it discusses the basic theories and the design methods of programming design, Procedure-Oriented and Object-Oriented programming methods, along with the Sort and other basic algorithms. Chapter 1 to Chapter 3 discuss the data structures of C++ programming, including Integer, Float, Character type, Array, Pointer and Structure. Chapter 4 to Chapter 5 mainly discuss the basic control structures of C++, including sequence, branch and loop. Chapter 6 to Chapter 7 discuss the declaration, definition, invoke, overload and template of Functions. Chapter 8 provides a brief introduction to the memory and namespaces. Chapter 9 to Chapter 10 discuss the basic concepts, ideas and methods of Object Oriented Programming including the concept of Class, Object, Constructor, Destructor as well as Operators overloading.

This course helps students build a solid knowledge for the basic principles, methods of programming. The course adopts group study method, supplemented by experiments, in-class tests, discussions and reports to train students the ability of solving practical problems with basic theories and methods and mastering the basic knowledge and skills for computer programming. The course includes several experiments in order to consolidate students' theoretical knowledge of computer programming, further strengthen their ability to analyze and solve problems independently, and develop their comprehensive abilities on system design and innovations.

The experiments of this course include experiments for C/C++ and extended Python programming experiments. Students have studied the basis of programming design through classroom learning，and students will further learn basic Python programming through more experiments. The special experiments will further strengthen students' ability of programming.

5.1.2　课程基本情况（Course Arrangements）

课程名称	计算机程序设计 Computer Programming							
开课时间	一年级		二年级		三年级		四年级	计算机科学与技术
^	秋	春	秋	春	秋	春	秋　春	^
课程定位	本科生"计算机科学与技术"课程群必修课							必修 （学分）
学　分	2 学分							^
总 学 时	52 学时 （授课 32 学时、实验 20 学时）							^
授课学时 分配	课堂讲授（32 学时）							选修 （学分）
先修课程	无							
后续课程	数据结构与算法							
教学方式	课堂教学、编程实验							
考核方式	课程结束机试成绩占 50％，机试成绩占 30％，实验成绩占 15％，考勤占 5％							
参考教材	Prata S. C++ Primer Plus[M]. 6 版. 北京：人民邮电出版社,2015.							
参考资料	1. Prata S. C++ Primer Plus[M]. 张海龙,译.6 版. 北京：人民邮电出版社,2012. 2. Lippman S. C++ Primer[M]. 5 版. 北京：电子工业出版社,2013. 3. Matthes E. Python 编程：从入门到实践[M]. 3 版. 袁国中,译. 北京：人民邮电出版社,2023. 4. Shaw Z A. 笨办法学 Python 3[M]. 王巍巍,译. 北京：人民邮电出版社,2018.							
其他信息								

计算机科学与技术必修（学分）：
- 计算机程序设计(2)
- 数据结构与算法(3)
- 计算机体系结构(3)
- 理论计算机科学的重要思想(1)

选修（学分）：
- 3D 计算机图形学(2)
- 智能感知与移动计算(2)

5.1.3 教学目的和基本要求（Teaching Objectives and Basic Requirements）

（1）熟悉基本数据结构，能够基于基本数据结构定义和使用变量；

（2）系统掌握结构化程序设计方法的特点，初步建立程序设计的概念。熟练掌握程序的三种基本结构，深刻理解顺序、选择、循环三种逻辑在程序设计中的作用；

（3）建立数据顺序存储的概念，深刻理解数据顺序存储的意义、作用，掌握数组的定义和使用，掌握数组编程技巧；

（4）了解函数的声明、定义和函数调用；

（5）掌握指针的概念和使用，认识指针的作用和意义，弄清指针与数组的关系，了解使用指针指向数组在程序设计所带来的方便；

（6）了解结构体的定义、引用和结构体数组的定义和引用；

（7）理解面向对象的基本概念，掌握类和对象的定义和使用；

（8）熟悉构造函数、析构函数和 this 指针的基本原理和使用方法；

（9）掌握类中运算符重载的原理和方法，理解函数和类模板的概念。

（10）拓展程序设计语言 Python 的基本概念、语法。

（11）掌握 Python 编程实验的基本方法、基本程序设计及相关库。

5.1.4 教学内容及安排（Syllabus and Arrangements）

绪论 （Introduction）

章节序号 Chapter Number	章节名称 Chapters	课时 Class Hour	知识点 Key Points
0.1	绪论 Introduction	1	（1）计算机编程语言的发展 （2）面向过程和面向对象编程的比较 （3）C++程序开发工具 (1) The development of computer programming language (2) The comparison of Procedure-Oriented and OO programming (3) The tools for C++ programming

第一章 C++初步（C++ Initials）

章节序号 Chapter Number	章节名称 Chapters	课时 Class Hour	知　识　点 Key Points
1.1	C++基本知识 C++ initials		（1）程序运行的起点-main函数 （2）C++注释和源代码的格式化 （3）C++预处理器和iostream文件 （4）头文件和名字空间 （5）Cout进行输出 (1) The main function (2) C++ comments and source code formatting (3) The preprocessor and the iostream file (4) Header file and name space (5) Output with Cout
1.2	C++语句 C++ statement	2	（1）声明语句和变量 （2）赋值语句 （3）Cin进行输入 （4）Cin和Cout-初次使用类 (1) Declaration statement and variable (2) Assignment statement (3) Using Cin (4) Cin and Cout: a touch of class
1.3	函数 Function		（1）带返回值的函数 （2）函数变体 （3）用户定义函数 （4）用户定义有返回值的函数 (1) A function with a return value (2) Function variation (3) User-defined function (4) Using a user-defined function that has a return value

第二章　处理数据(Dealing with Data)

章节序号 Chapter Number	章节名称 Chapters	课时 Class Hour	知　识　点 Key Points
2.1	简单变量 Simple variable	2	(1) 变量名 (2) 整型 (3) 整型 short、int、long 和 long long (4) 无符号类型 (5) 选择整型类型 (6) 常量类型 (7) char 类型 (8) bool 类型 (9) const 限定符 (1) Name for variable (2) Integer type (3) The short，int，long，and long long integer types (4) Unsigned type (5) Choosing an integer type (6) Integer literal (7) The char type (8) The bool type (9) The const qualifier
2.2	浮点数 Floating-point number		(1) 浮点数 (2) 浮点类型 (3) 浮点常量 (1) Floating number (2) Floating-point type (3) Floating-point constant
2.3	C++算术运算符 C++ arithmetic operator		(1) 运算符的优先级和结合性 (2) 除法运算符 (3) 取模 (4) 类型转换 (1) Order of operation：operator precedence and associativity (2) Division diversion (3) The modulus operator (4) Type conversion

第三章　复合类型(Compound Type)

章节序号 Chapter Number	章节名称 Chapters	课时 Class Hour	知　识　点 Key Points
3.1	数组 Introducing array		(1) 数组介绍 (2) 数组初始化规则 (1) Introduction of array (2) Initialization rules for array
3.2	字符串 String	2	(1) 拼接字符串常量 (2) 数组中使用字符串 (3) 字符串输入 (4) 每次读取一行字符串输入 (5) 混合输入字符串和数字 (1) Concatenating string literal (2) Using string in an array (3) Adventure in string input (4) Reading string input a line at a time (5) Mixing string and numeric input
3.3	string 类 Introducing the string class		(1) string 类 (2) 赋值、拼接和附加以及其他操作 (3) string 类 I/O (1) String class (2) Assignment, concatenation, and appending (3) I/O of string
3.4	结构简介 Introducing structure		(1) 在程序中使用结构 (2) 结构属性 (3) 结构数组 (1) Using a structure in a program (2) Properties of structure (3) Array of structure
3.5	指针和自由存储空间 Pointers and the free store	2	(1) 声明和初始化指针 (2) 指针和数字 (3) 使用 new (4) 使用 delete (1) Declaring and initializing pointer (2) Pointer and number (3) Allocating memory using new (4) Freeing memory with delete

续表

章节序号 Chapter Number	章节名称 Chapters	课时 Class Hour	知识点 Key Points
3.6	指针、数组和指针算术 Pointer, array, and pointer arithmetic		(1) 指针小结 (2) 指针和字符串 (3) 使用 new 创建动态结构 (4) 自动存储、静态存储和动态存储 (1) Summarizing pointer (2) Pointer and string (3) Using new to create dynamic structure (4) Automatic storage, static storage, and dynamic storage

第四章 循环和关系表达式（Loops and Relational Expression）

章节序号 Chapter Number	章节名称 Chapters	课时 Class Hour	知识点 Key Points
4.1	for 语句 Introducing for loop	2	(1) for 循环的组成部分 (2) 修改步长 (3) 使用 for 访问字符串 (4) ++和―― (5) 复合赋值运算符 (6) 复合语句 (7) 关系表达式 (8) 赋值、比较 (9) C-风格字符串的比较 (10) 比较 string 类字符串 (11) 冒泡排序 (1) Parts of a for loop (2) Changing the step size (3) Inside strings with the for loop (4) ++ and ―― (5) Combination assignment operators (6) Compound statement, or block (7) Relational expression (8) Assignment and comparison (9) Comparing C-style string (10) Comparing string class string (11) Bubble sort algorithm

续表

章节序号 Chapter Number	章节名称 Chapters	课时 Class Hour	知识点 Key Points
4.2	while 循环和 do while 循环 The while loop and do while loop	1	（1）for 与 while （2）编写延时循环 （3）do while 循环 （4）循环和文本输入 （5）嵌套循环和二维数组 (1) The for and while (2) Building a time-delay loop (3) The loop of do while (4) Loops and text input (5) Nested loop and tow-dimensional array

第五章　分支语句和逻辑运算符(Branching Statement and Logical Operation)

章节序号 Chapter Number	章节名称 Chapters	课时 Class Hour	知识点 Key Points
5.1	if 语句 The if statement		（1）if else 语句 （2）格式化 if else 语句 （3）if else if else 语句 (1) The if else statement (2) Formatting if else statement (3) The if else if else construction
5.2	逻辑表达式 Logical expressions	2	（1）逻辑 OR 运算符 || （2）逻辑 AND 运算符 && （3）用 && 来设置取值范围 （4）逻辑 NOT 运算符 ! （5）逻辑运算符细节 （6）字符函数库 cctype （7）?：运算符 （8）switch 语句 （9）break 和 continue 语句 (1) The logical OR operator || (2) The logical AND operator && (3) Setting up range with &&

续表

章节序号 Chapter Number	章节名称 Chapters	课时 Class Hour	知 识 点 Key Points
5.2	逻辑表达式 Logical expression		(4) The logical NOT operator ！ (5) Logical operator fact (6) The cctype library of character Function (7) The ？：operator (8) The switch statement (9) The break and continue statement

第六章 函数——C++的编程模块（Functions：C++'s Programming Module）

章节序号 Chapter Number	章节名称 Chapters	课时 Class Hour	知 识 点 Key Points
6.1	函数的基本知识 Function review	2	(1) 定义函数 (2) 函数原型和函数调用 (1) Defining a function (2) Prototyping and calling a function
6.2	函数参数和按值传递 Function argument and passing by value		(1) 多个参数 (2) 接受两个参数的函数 (1) Multiple argument (2) Another two-argument function
6.3	函数和数组 Function and array	2	(1) 函数如何使用指针来处理数组 (2) 数组作为参数 (3) 数组函数实例 (4) 使用数组区间的函数 (5) 指针和const (6) 函数和二维数组 (1) How pointer enable array-processing function (2) The implications of using array as argument (3) More array function example (4) Function using array range (5) Pointer and const (6) Function and two-dimensional array

续表

章节序号 Chapter Number	章节名称 Chapters	课时 Class Hour	知 识 点 Key Points
6.4	函数和 C-风格字符串 Function and C-Style string		（1）C-风格字符串作为参数 （2）返回 C-风格字符串 （1）Functions with C-style string argument （2）Functions that return C-style string
6.5	函数和结构、 string 和 array 对象 Function and structure，string and array object		（1）传递和返回结构 （2）传递结构的地址 （3）函数和 string 对象 （4）函数和 array 对象 （1）Passing and returning structures （2）Passing structure addresses （3）Functions and string class objects （4）Functions and array objects
6.6	递归 Recursion	2	（1）单一递归调用 （2）多重递归调用 （1）Recursion and a single recursive call （2）Recursion with multiple recursive call

第七章 函数探幽（Adventure in Function）

章节序号 Chapter Number	章节名称 Chapters	课时 Class Hour	知 识 点 Key Points
7.1	C++ 内联函数 C++ inline function	2	（1）内联函数使用 （1）Using inline function

续表

章节序号 Chapter Number	章节名称 Chapters	课时 Class Hour	知　识　点 Key Points
7.2	引用变量 Reference variable		(1) 创建引用变量 (2) 将引用作为函数参数 (3) 引用的属性 (4) 引用和结构 (5) 引用和类 (6) 对象、继承和引用 (7) 何时使用引用 (1) Creating a reference variable (2) Reference as function parameter (3) Reference properties (4) Using reference with a structure (5) Using reference with a class object (6) Objects, inheritance and reference (7) When to use reference argument
7.3	函数重载 Function overloading		(1) 函数重载 (2) 何时使用函数重载 (1) An overloading example (2) When to use function overloading
7.4	函数模板 Function template		(1) 重载的模板 (2) 模板的局限性 (3) 显式具体化 (4) 实例化和具体化 (1) Overloaded template (2) Template limitation (3) Explicit specialization (4) Instantiations and specialization

第八章　内存模型和名称空间（Memory Model and Namespace）

章节序号 Chapter Number	章节名称 Chapters	课时 Class Hour	知　识　点 Key Points
8.1	单独编译 Separate compilation		（1）单独编译 （1）Separate compilation
8.2	存储持续性、 作用域和连接性 Storage duration, scope, and linkage	2	（1）作用域和链接 （2）自动存储持续性 （3）静态持续变量 （4）静态持续性、外部链接性 （5）静态持续性、内部链接性 （6）静态存储持续性、无链接性 （7）说明符和限定符 （8）函数和链接性 （9）语言链接性 （10）存储方案和动态分配 （1）Scope and linkage （2）Automatic storage duration （3）Static duration variables （4）Static duration, external linkage （5）Static duration, internal linkage （6）Static storage duration, no linkage （7）Specifier and qualifier （8）Function and linkage （9）Language linking （10）Storage schemes and dynamic allocation
8.3	名称空间 Name space	2	（1）传统的C++名字空间 （2）新的名字空间特性 （3）实例 （1）Traditional C++ namespace （2）New namespace features （3）A namespace example

第九章 对象和类(Objects and Class)

章节序号 Chapter Number	章节名称 Chapters	课时 Class Hour	知　识　点 Key Points
9.1	过程性编程和OO编程 Procedural and OO Programming		(1) 过程性编程和OO编程 (1) Procedural and OO Programming
9.2	抽象和类 Abstraction and Class	2	(1) 类型 (2) C++中的类 (3) 实现类成员函数 (4) 使用类 (5) 修改实现 (1) Type (2) Class in C++ (3) Implementing class member function (4) Using class (5) Changing the implementation
9.3	类的构造函数和析构函数 Class constructor and destructor	2	(1) 声明和定义构造函数 (2) 使用构造函数 (3) 默认构造函数 (4) 析构函数 (5) 改进stock类 (6) 构造函数和析构函数小结 (1) Declaring and defining constructor (2) Using constructor (3) Default constructor (4) Destructor (5) Improving the stock class (6) Conclusion of constructor and destructor function
9.4	this指针 The this pointer		(1) this指针的概念 (2) this指针的使用场景 (1) The definition of this pointer (2) The using scenarios for this pointer

第十章 使用类(Working with Class)

章节序号 Chapter Number	章节名称 Chapters	课时 Class Hour	知 识 点 Key Points
10.1	运算符重载 Operator overloading	2	(1) 运算符重载 (1) Operator overloading
10.2	运算符重载实例 Developing an operator overloading example		(1) 加法运算符 (2) 重载限制 (3) 重载其他运算符 (1) Adding an addition operator (2) Overloading restriction (3) More overload operator

5.1.5 实验环节(Experiments)

序号 Num.	实验内容 Experiment Content	课时 Class Hour	知 识 点 Key Points
1	C++编程训练1： 输入输出格式化控制、数据类型、表达式、程序控制结构 C++ programming training 1: I/O,data type,expression, loop,branch	2	(1) 标准化输入与输出 (2) 基本数据类型 (3) 表达式 (4) if-else 语句与 switch 语句 (5) while 与 do-while 循环结构 (1) Standard input and output (2) Basic data type (3) Expression (4) if-else and switch statements (5) while loop and do-while loop

续表

序号 Num.	实验内容 Experiment Content	课时 Class Hour	知 识 点 Key Points
2	C++编程训练2： 指针、数组、字符串、函数、结构体 C++ programming training 2：pointer,array,string,function,structure	2	（1）一维数组、二维数组 （2）字符串操作 （3）指针地址传送、指针内容获取 （4）函数的定义和调用 （5）结构体的定义和使用 （1）One-dimensional array and two-dimensional array （2）String operation （3）The address and content transfer for pointer operation （4）The definition and callingof function （5）The definition and use of structure
3	C++编程训练3： 类与对象、类的构造与析构函数、继承与派生、友元与多态、重载 C++ programming training 3：class and object,constructor and destructor,inheritance and derivation,friend and polymorphism,overload	2	（1）类和对象的定义 （2）类的构造函数和析构函数 （3）类的继承和派生 （4）友元、多态和重载 （1）The definition of classand object （2）Constructor and destructor function of class （3）Class inheritance and derivation （4）Friend,polymorphism and overload
4	C++编程训练4： 完整的学籍管理信息系统的程序训练 C++ programming training 4：student status management program training	4	（1）学籍信息系统需求分析 （2）学籍信息系统概要设计 （3）学籍信息系统详细设计 （4）文件打开、关闭、读取、存储 （1）Requirement analysis of the information system （2）Outline design of the information system （3）Detailed design of the information system （4）The operations of open,close,read and write for file system

续表

序号 Num.	实验内容 Experiment Content	课时 Class Hour	知　识　点 Key Points
5	Python 编程拓展 1： Python 基本语法、数据类型 Python Programming training 1： Python basic grammar and data type	2	（1）Python 编程环境配置和程序运行调试 （2）输入输出格式化控制 （3）变量、字符串、数字、注释 （1）Python programming environment configuration and debugging method （2）Input and output （3）Variable，string，number，comment
6	Python 编程拓展 2： 程序控制结构、组合数据类型 Python Programming training 2： branch，loop，exception and composite data type	2	（1）分支结构 （2）循环结构 （3）程序异常处理 （4）列表 （5）元组 （6）字典 （7）集合 （1）Branch （2）Loop （3）Program exception handling （4）List （5）Tuple （6）Dictionary （7）Set
7	Python 编程拓展 3： 函数、类、模块 Python Programming training 3： function，class，module	2	（1）函数概念和使用 （2）类的创建和使用 （3）模块安装和使用 （1）Concept and usage of function （2）Creation and usage of class （3）Installation and usage of module
8	Python 编程拓展 4： 数据组织和文件读写 Python Programming training 4： data organization and file operation	2	（1）文件数据读取 （2）文件数据写入 （3）文件异常处理 （4）Txt、csv、json、xlsx、jpg 文件读写实例 （1）File data reading （2）File data writing （3）File exception handling （4）Txt，csv，json，xlsx，jpg file operation practice

序号 Num.	实验内容 Experiment Content	课时 Class Hour	知 识 点 Key Points
9	Python 编程拓展 5： 标准库和第三方库 Python Programming training 5: standard and third-party libraries	2	（1）第三方库安装方法 （2）Numpy、SciPy、Pandas、Matplotlib （3）OpenCV、Scikit-learn、Pytorch （4）Pybind11 (1) Third-party library installation method (2) Numpy、SciPy、Pandas、Matplotlib (3) OpenCV、Scikit-learn、Pytorch (4) Pybind11

大纲制定者：唐亚哲副教授（西安交通大学计算机科学与技术学院）、李昊副教授（西安交通大学计算机科学与技术学院）、刘龙军副教授（西安交通大学人工智能学院）、张玥工程师（西安交通大学人工智能学院）

大纲审定：西安交通大学人工智能学院本科专业知识体系建设与课程设置第二版修订工作组

5.2 "数据结构与算法"课程大纲

课程名称：数据结构与算法
Course：Data Structure and Algorithm
先修课程：计算机程序设计
Prerequisites：Computer Programming
学分：3
Credits：3

5.2.1 课程目的和基本内容（Course Objectives and Basic Content）

本课程是人工智能学院本科专业必修课。

This course is a compulsory course for undergraduates in College of Artificial Intelligence.

本课程培养学生的数据抽象能力,学会分析研究计算机加工的数据结构的特性,以便为应用涉及的数据选择适当的逻辑结构、存储结构及实现应用的相应算法,初步掌握分析算法的时间和空间复杂度的技术,以及算法设计方法。本课程的内容注重数据结构基础知识、算法设计的核心思想。第一章到第三章主要介绍了本课程基本概念、算法评估的时空复杂的方法、线性表及受限的线性表基本数据结构及操作。第四章扩展线性表为自学内容。第五章和第六章主要讨论高级的数据结构如树结构与图结构的概念及操作。第七章和第八章主要以查找和排序为基本例子介绍了算法设计的概念、方法及步骤等。

通过本课程的学习,使学生了解和掌握数据结构和算法的基本思想,学习分析、设计和实现解决实际问题的策略;使学生了解和基本掌握典型的数据结构类型及其应用;结合实际问题分析,加深对所学知识的理解,并为后续课程和未来的工程实践打下良好的基础。

随着计算机编程语言的发展与丰富,数据结构与算法设计在计算机编程中扮演着重要的角色,掌握好数据结构与算法设计对编程及软件设计等起着非常重要的作用。因此,掌握好本课程的基本概念、基本原理和方法,对学生今后用计算机程序解决实际问题将更加容易。

This course trains students' ability of data abstraction and analyzing the characteristics of data structure processed by computer, so as to select appropriate logical structure, storage structure and corresponding algorithm for data in practical applications. The course also enable students master the analysis methods of time and space complexity, as well as the design method of computer basic algorithms. This course focuses on the basic knowledge of data structure and the core idea of algorithm design. Chapter 1 to Chapter 3 mainly introduce the basic concepts of the course, the time and space complex analysis methods, the basic data structure and operation for the linear table and the restricted linear table. Chapter 4 expands linear list to self-study content. Chapter 5 to Chapter 6 discuss the concepts and operations of advanced data structures such as tree structures and graph structures. Chapter 7 to Chapter 8 mainly introduce the concept, methods and steps for algorithm design based on the basic examples of search and sorting.

Through the study of this course, students can understand and master the basic ideas and common knowledge of data structure and algorithms, learn to analyze, design and solve practical problems. The course enables students to understand and master the typical data structure and their applications, to understand practical programing problem based on the knowledge of this course, and lay a good foundation for follow-up courses and future engineering practice.

With the development and enrichment of computer programming languages, data structure and algorithm design play an important role in computer programming. Mastering data structure and algorithm design are very important for programming and software design. Therefore, mastering the basic concepts, basic principles and methods of this course will make it easier for students to solve practical problems of computer programs in the future.

5.2.2 课程基本情况(Course Arrangements)

课程名称	数据结构与算法 Data Structure and Algorithm							
开课时间	一年级		二年级		三年级		四年级		计算机科学与技术
	秋	春	秋	春	秋	春	秋	春
课程定位	本科生"计算机科学与技术"课程群必修课								必修 (学分)
学 分	3 学分							
总 学 时	56 学时 (授课 48 学时,实验 8 学时)							
授课学时 分配	课堂讲授(48 学时)								选修 (学分)
先修课程	计算机程序设计							
后续课程	人工智能概论、计算机视觉与模式识别							
教学方式	课堂教学							
考核方式	闭卷考试成绩占 70%,平时成绩占 30%							
参考教材	赵仲孟.数据结构与算法[M].北京:高等教育出版社,2016.							
参考资料	Shaffer C A.数据结构与算法分析[M].New York:Dover Publications,2013.							
其他信息								

必修(学分):
- 计算机程序设计(2)
- 数据结构与算法(3)
- 计算机体系结构(3)
- 理论计算机科学的重要思想(1)

选修(学分):
- 3D 计算机图形学(2)
- 智能感知与移动计算(2)

5.2.3 教学目的和基本要求(Teaching Objectives and Basic Requirements)

(1) 具备分析掌握基本数据结构及其算法的能力;

(2) 具备学习分析、设计和实现解决实际问题的能力;

(3) 掌握基本数据结构概念,理解线性表的结构及操作,包括顺序表、链表、栈、队列的增、删、改、查等;

(4) 掌握高级数据结构类型的结构及操作，理解树与图的建立与遍历等；

(5) 理解排序与查找算法，了解其他基本的算法，如贪婪算法、分治算法、回溯算法、动态规划等；

(6) 具备上机编程解决一般应用问题的能力。

5.2.4　教学内容及安排(Syllabus and Arrangements)

第一章　绪论(Introduction)

章节序号 Chapter Number	章节名称 Chapters	课时 Class Hour	知　识　点 Key Points
1.1	数据结构的基本概念 Basic concept of data structure	2	(1) 数据、数据元素、数据对象、数据结构定义、数据的存储方式 (1) Data, data elements, data objects, data structure, data storage
1.2	抽象数据类型 Abstract data type (ADT)		(1) ADT 的表示和实现 (1) Representation and implementation of ADT
1.3	问题、算法和程序介绍 Introduction of problem, algorithm and program		(1) 问题、算法和程序的定义，算法的特性 (1) Problem, algorithm and program, characteristics of the algorithm
1.4	算法分析概述 Algorithm analysis overview	2	(1) 渐近算法分析，渐近时间复杂度，算法增长率 (1) Asymptotic algorithm analysis, asymptotic time complexity, algorithm growth rate
1.5	时间复杂度 Time complexity		(1) 时间复杂度分析规则 (1) Time complexity analysis rules
1.6	渐近分析 Asymptotic analysis		(1) 上限表示法，下限表示法，Θ表示法，化简法则 (1) Upper limit representation, lower limit representation, Θ representation, simplification rule
1.7	空间复杂度 Space complexity		(1) 空间复杂度分析方法 (1) Analysis method for spatial complexity

第二章 线性表（Linear Table）

章节序号 Chapter Number	章节名称 Chapters	课时 Class Hour	知　识　点 Key Points
2.1	线性表的定义 A linear table	2	(1) 线性表的定义 (1) A linear table
2.2	线性表的顺序存储结构 Sequential storage structure of linear tables	2	(1) 顺序存储结构，顺序存储结构的实现 (1) Sequential storage structure，and implementation of sequential storage structure
2.3	线性表的链式存储结构 Linked storage structure of linear table	2	(1) 单链表，双向链表，循环链表 (1) Single linked list，doubly linked list，circularly linked list
2.4	线性表的应用举例 Application examples for linear table	2	(1) 一元多项式的表示，商品链更新 (1) Representation of unary polynomial，commodity chain update

第三章　受限线性表——栈、队列及串（Restricted Linear Tables-Stack，Queue，and String）

章节序号 Chapter Number	章节名称 Chapters	课时 Class Hour	知　识　点 Key Points
3.1	操作受限线性表——栈 Operational restricted linear table-stack	2	(1) 栈的定义，栈的抽象数据类型定义 (1) Stack，abstract data type
3.2	栈的存储结构 Storage structure of stack		(1) 顺序栈，链栈 (1) The sequence stack and chain stack
3.3	栈的应用 Applications of stack	2	(1) 括号匹配检验，栈与递归 (1) Bracket matching test，stack and recursion
3.4	操作受限线性表-队列 Operational restricted linear table-queue		(1) 队列的定义，队列的抽象数据类型定义 (1) Queue，abstract data type definition of queue

章节序号 Chapter Number	章节名称 Chapters	课时 Class Hour	知 识 点 Key Points
3.5	队列的存储结构及实现 Storage structure and implementation of queue		(1) 顺序队列的概念及实现,队列的链式存储结构及实现 (1) The concept and implementation of the sequence queue and chain queue
3.6	队列的应用 Applications of queue		(1) 杨辉三角形,火车车厢重排 (1) Yang Hui triangle, train compartment rearrangement

第四章 扩展线性表—数组与广义表（Extended Linear Tables-Array and Generalized Table）

章节序号 Chapter Number	章节名称 Chapters	课时 Class Hour	知 识 点 Key Points
4.1	数组与广义表 Arrays and generalized table	0 （自学）	(1) 数组与广义表的概念与操作 (1) The concept and operations of array and generalized table

第五章 树和二叉树（Tree and Binary Tree）

章节序号 Chapter Number	章节名称 Chapters	课时 Class Hour	知 识 点 Key Points
5.1	树的概念 The concept of tree	2	(1) 树的概念,相关的基本术语 (1) The concept of tree and basic terminologies
5.2	二叉树 Binary tree		(1) 二叉树的概念,二叉树的主要性质,二叉树的存储结构 (1) The concept of a binary tree, the main property of a binary tree and the storage structure of a binary tree

续表

章节序号 Chapter Number	章节名称 Chapters	课时 Class Hour	知　识　点 Key Points
5.3	二叉树的遍历 Traversal of binary tree	2	（1）二叉树的先序遍历 （2）二叉树的中序遍历 （3）二叉树的后序遍历 （1）Preorder traversal of binary tree （2）In-order traversal of binary tree （3）The post-order traversal of the binary tree
5.4	二叉树的应用 1： 哈夫曼树 Application of binary tree 1：Huffman tree	2	（1）哈夫曼树的构造 （2）哈夫曼编码 （1）Huffman tree construction （2）Huffman coding
5.5	二叉树的应用 2： 二叉查找树 Binary tree application 2：binary search tree		（1）二叉查找树的概念 （2）二叉查找树的查找 （3）二叉查找树的插入 （4）二叉查找树的删除 （1）The concept of binary search tree （2）Search of binary search tree （3）Insertion of binary search tree （4）Deletion of binary search tree
5.6	二叉树的应用 3： 平衡二叉查找树 Binary tree application 3：balanced binary search tree	2	（1）平衡二叉树的定义 （2）平衡化旋转 （3）平衡二叉查找树的插入 （4）平衡二叉查找树的删除 （1）The concept of balanced binary tree （2）Balance rotation （3）The insertion operation in a balance binary search tree （4）The deletion operation of a balanced binary search tree

续表

章节序号 Chapter Number	章节名称 Chapters	课时 Class Hour	知 识 点 Key Points
5.7	二叉树的应用4： 堆与优先队列 Binary tree application 4： heap and priority queue	2	(1) 堆与优先队列的概念及实现 (2) 堆的插入和堆顶删除 (1) Heap and priority queue concept and implementation (2) Heap insertion and heap top deletion
5.8	树与森林 Tree and forest		(1) 树的存储结构 (2) 树、森林与二叉树的转换 (3) 树与森林的遍历 (1) Tree storage structure (2) Tree, forest and binary tree transformation (3) Tree and forest traversal

第六章　图（Graphics）

章节序号 Chapter Number	章节名称 Chapters	课时 Class Hour	知 识 点 Key Points
6.1	图的概念 The concept of graphics	2	(1) 图的定点、边，无向图，有向图，带权图，无环图，连通图 (1) Fixed point, edge, undirected graph, directed graph, weighted graph, acyclic graph, connected graph
6.2	图的存储结构 Storage structure of graphics		(1) 邻接矩阵存储方法 (2) 邻接表存储方法 (1) Adjacency matrix storage method (2) Adjacency table storage method

续表

章节序号 Chapter Number	章节名称 Chapters	课时 Class Hour	知识点 Key Points
6.3	图的遍历 Traversal of graphs	2	(1) 深度优先搜索 (2) 广度优先搜索 (1) Depth-first search (2) Breadth-first search
6.4	图的应用1：拓扑排序 Application of graphs 1: topological sorting		(1) 图谱排序的概念 (2) 拓扑排序算法 (1) The concept of map ordering (2) Topological sorting algorithm
6.5	图的应用2：关键路径 Application of graphs 2: critical Path	2	(1) AOE网 (2) 关键路径算法 (1) AOE network (2) Critical path algorithm
6.6	图的应用3：最短路径 Application of graphs 3: shortest path		(1) 单源点最短路径问题 (2) 任意对顶点之间的最短路径 (1) Single source point shortest path problem (2) The shortest path between any pair of vertices
6.7	图的应用4：图的最小生成树 Application of graphs 4: minimum spanning tree	2	(1) 普里姆算法 (2) 克鲁斯卡尔算法 (1) Prim algorithm (2) Kruskal algorithm

第七章 排序算法（Sorting Algorithm）

章节序号 Chapter Number	章节名称 Chapters	课时 Class Hour	知　识　点 Key Points
7.1	排序的基本概念 Basic concept of sorting	2	（1）排序的含义 （2）排序算法的稳定性含义 （3）排序算法的两个因素 (1) The meaning of sorting (2) The meaning of the stability of the sorting algorithm (3) Two factors of the sorting algorithm
7.2	简单排序 Simple sort		（1）简单插入排序 （2）冒泡排序 （3）简单选择排序 (1) Simple insert sorting (2) Bubble sorting (3) Simple sorting
7.3	高级排序 Advanced sorting	4	（1）希尔排序 （2）快速排序 （3）归并排序 （4）树形选择排序1：锦标赛排序 （5）树形选择排序2：堆排序 (1) Hill sorting (2) Quick sorting (3) Merging and sorting (4) Tree selection sorting 1: tournament sorting (5) Tree selection sorting 2: heap sorting

续表

章节序号 Chapter Number	章节名称 Chapters	课时 Class Hour	知 识 点 Key Points
7.4	关键字比较排序下界问题 Keyword comparison sorting lower bound problem	2	(1) 关键字比较排序下界分析 (1) Keyword comparison sorting lower bound analysis
7.5	非关键字比较的排序 Non-keyword comparison sorting		(1) 基数排序 (2) 多关键字排序 (1) Cardinal sorting (2) Multi-keyword sorting
7.6	各种排序算法的比较 Comparison of various sorting algorithms		(1) 各种排序算法的事件复杂度 (2) 存储和稳定性分析比较 (1) Comparison of event complexity (2) Storage and stability analysis of various sorting algorithms

第八章 查找算法（Search Algorithm）

章节序号 Chapter Number	章节名称 Chapters	课时 Class Hour	知 识 点 Key Points
8.1	查找的基本概念 Basic concept of search algorithm	2	(1) 查找操作的概念 (2) 查找表 (1) The concept of search operation (2) Lookup table
8.2	静态查找表 Static lookup table		(1) 顺序表的查找 (2) 折半查找 (1) Lookup of the sequence table (2) Half-interval search

续表

章节序号 Chapter Number	章节名称 Chapters	课时 Class Hour	知识点 Key Points
8.3	哈希列表 Hash list	2	(1) 哈希函数的常用构建方法 (2) 解决冲突的方法 (3) 哈希表的实现 (4) 哈希表的分析 (1) Common construction methods for hash functions (2) Methods for resolving conflicts (3) Implementation of hash tables (4) Analysis of hash tables
8.4	线性索引,树形索引: 2-3 树 Linear index, tree index: 2-3 trees	2	(1) 线性索引的概念,分块索引的定义和实现,2-3 树 (1) The concept of linear index, the definition and implementation of block index, 2-3 tree
8.5	树形索引:B 树,B+树 Tree index: B tree, B+ tree	2	(1) B 树 (2) B+树 (1) B tree (2) B+ tree

第九章 其他算法设计(Other algorithms design)

章节序号 Chapter Number	章节名称 Chapters	课时 Class Hour	知识点 Key Points
9.1	其他算法设计 Other algorithms design	0 (自学)	(1) 贪婪算法 (2) 分治算法 (3) 回溯算法 (4) 动态规划 (5) 随机化算法 (1) Greedy algorithm (2) Divide and conquer algorithm (3) Backtracking algorithm (4) Dynamic algorithm (5) Randomization algorithm

5.2.5 实验环节（Experiments）

序号 Num.	实验内容 Experiment Content	课时 Class Hour	知 识 点 Key Points
1	线性表操作实验 Experiments of linear table	2	（1）顺序表、链表、栈、队列的增、删、改、查操作 (1) Add, delete, change, and search operations of sequence table, linked list, stack, and queue
2	二叉树结构实验 Experiments of Tree structure	2	（1）树的实现和遍历操作 (1) Implementation and traversal operations of binary tree
3	图结构实验 Experiments of graph structure	2	（1）图的实现和遍历操作 (1) Implementation and traversal operations of graph
4	排序算法实验 Experiments of sorting algorithm	2	（1）简单插入排序、冒泡排序、简单选择排序、希尔排序、快速排序、归并排序 (1) Algorithms of insertion sort, bubble sort, selection sort, shell sort, quick sort and merge sort

大纲制定者：朱晓燕副教授（西安交通大学计算机科学与技术学院）、刘龙军副教授（西安交通大学人工智能学院）

大纲审定：西安交通大学人工智能学院本科专业知识体系建设与课程设置第二版修订工作组

5.3 "计算机体系结构"课程大纲

课程名称：计算机体系结构
Course：Computer Architecture
先修课程：数据结构与算法、电子技术与系统
Prerequisites：Data Structure and Algorithm, Electronic Technology and System
学分：3
Credits：3

5.3.1 课程目的和基本内容（Course Objectives and Basic Content）

本课程是人工智能学院本科专业必修课。

This course is a compulsory course for undergraduates in College of Artificial Intelligence.

本课程在电子技术与系统先修课程的基础上,特别是结合计算机体系结构最新的发展趋势和技术特点,系统地介绍了高级计算机系统的设计基础、RISC-V 指令集系统结构、高级流水线、超标量处理器、乱序执行等指令集并行技术、层次化存储系统与存储设备、向量处理器、单指令多数据以及 GPGPU 等数据并行技术、线程并行技术、多核处理器、片上互连网络以及面向领域应用的计算架构,特别是面向以深度学习为代表的人工智能计算应用的加速器设计。

本课程向学生提供了当前计算体系结构的最新知识,使他们能够洞悉体系结构,特别是摩尔定律接近尾声时,面向领域应用的软硬件结合的系统结构设计方法。

The course systematically introduces the design basis of advanced computer system and key techniques including, RISC-V instruction set architecture, advanced pipeline technique, SuperScalar Processor, Out-of-Order Execution, instruction-level parallelism, memory hierarchy design, data-level parallelism in Vector, SIMD and GPGPU architectures, thread-level parallelism, Many/Multi-core, On-Chip Network and domain-specific architecture—especially for accelerator designs that aimed for artificial intelligence applications, like deep neural network.

This course provides students with up-to-date information on current computing platforms, giving them insight into the architecture. Especially when the Moore's Law is nearing the end, the system structure design method needs combining software and hardware efforts for domain-oriented applications.

5.3.2 课程基本情况(Course Arrangements)

课程名称	计算机体系结构 Computer Architecture							
开课时间	一年级		二年级		三年级		四年级		计算机科学与技术
	秋	春	秋	春	秋	春	秋	春
课程定位	本科生"计算机科学与技术"课程群必修课								必修(学分)	计算机程序设计(2)
学 分	3 学分									数据结构与算法(3)
总 学 时	48 学时 (授课 48 学时、实验 0 学时)									计算机体系结构(3)
		理论计算机科学的重要思想(1)
授课学时分配	课堂讲授(46 学时),小组讨论(2 学时)								选修(学分)	3D 计算机图形学(2)
		智能感知与移动计算(2)

续表

先修课程	数据结构与算法、电子技术与系统
后续课程	人工智能芯片设计导论
教学方式	课堂教学、大作业与实验、小组讨论、综述报告
考核方式	课程结束笔试成绩占 70%，平时成绩占 20%，考勤占 10%
参考教材	Hennessy J,Patterson D. 计算机体系结构：量化研究方法[M]. 6 版. 贾洪峰，译. 北京：机械工业出版社,2019.
参考资料	Hennessy J,Patterson D. 计算机组成与设计[M]. 5 版. 王党辉，康继昌，安建峰，译. 北京：机械工业出版社,2019.
其他信息	

5.3.3 教学目的和基本要求（Teaching Objectives and Basic Requirements）

（1）了解高级计算机体系结构的挑战和发展趋势；

（2）熟悉计算机体系结构的量化分析方法并有效地指导系统设计；

（3）掌握 RISC-V 指令集系统结构；

（4）掌握高级流水线和指令集并行技术；

（5）掌握层次化存储系统与存储设备；

（6）了解向量处理器、单指令多数据以及 GPGPU 等数据并行技术；

（7）熟悉线程并行技术；

（8）熟悉面向领域应用的计算架构——特别是面向以深度学习为代表的人工智能计算应用的加速器设计。

5.3.4 教学内容及安排（Syllabus and Arrangements）

第一章 量化设计与分析基础（Fundamentals of Quantitative Design and Analysis）

章节序号 Chapter Number	章节名称 Chapters	课时 Class Hour	知 识 点 Key Points
1.1	计算机的分类 Classes of computers		（1）物联网和嵌入式计算机、个人移动终端、桌面计算、服务器、集群/仓库级计算机、并行度与并行体系结构的分类 （1）Internet of Things/embedded computer, personal mobile device, desktop computing, server, cluster/warehouse-scale computer, classes of parallelism and parallel architectures

续表

章节序号 Chapter Number	章节名称 Chapters	课时 Class Hour	知　识　点 Key Points
1.2	计算机体系结构定义 Defining computer architecture	2	(1) 指令集体系结构：近距离审视真正的计算机体系结构 (2) 设计满足目标和功能需求的组成和硬件 (1) Instruction set architecture: the myopic view of computer architecture (2) Designing the organization and hardware to meet goals and functional requirements
1.3	技术趋势 Trends in technology		(1) 带宽胜过延迟、晶体管性能与物理连线的发展 (1) Bandwidth over latency, scaling of transistor performance and wires
1.4	集成电路中的功率和能耗趋势 Trends in power and energy in integrated circuits		(1) 微处理器内部的能耗和功率 (1) Energy and power within a microprocessor
1.5	成本趋势 Trends in cost	2	(1) 时间、产量和量产的影响、集成电路的成本、成本与价格、制造成本与运行成本 (1) The impact of time, volume, and commoditization, cost versus price, cost of manufacturing versus cost of operation
1.6	计算机可靠性 Dependability		(1) 计算机可靠性 (1) Dependability
1.7	性能的测量、报告和汇总 Measuring, reporting, and summarizing performance		(1) 基准测试、报告性能测试结果、性能结果汇总 (1) Benchmarks, reporting performance results, summarizing performance results

续表

章节序号 Chapter Number	章节名称 Chapters	课时 Class Hour	知 识 点 Key Points
1.8	计算机设计的量化原理 Quantitative principles of computer design		(1) 充分利用并行、局域性原理、重点关注常见情形、阿姆达尔定律、处理器性能公式 (1) Take advantage of parallelism, principle of locality, focus on the common case, Amdahl's law, the processor performance equation
1.9	融会贯通：性能、价格和功耗 Putting it All together: performance, price, and power		(1) 性能、价格和功耗的关系 (1) Performance, price, and power

第二章 层次化存储设计（Memory Hierarchy Design）

章节序号 Chapter Number	章节名称 Chapters	课时 Class Hour	知 识 点 Key Points
2.1	缓存基本概念 Introduction		基本概念 Introduction
2.2	存储器技术和优化方法 Memory technology and optimizations	2	(1) SRAM 技术、DRAM 技术 (2) 提高 DRAM 芯片内部的存储器性能 (3) 降低 SDRAM、图形数据 RAM 和闪存中的功耗，提高存储器系统的可靠性 (1) SRAM, DRAM (2) Improving memory performance inside a DRAM chip (3) Reducing power consumption in SDRAM, graphics data RAM, flash memory, enhancing dependability in memory system

188

续表

章节序号 Chapter Number	章节名称 Chapters	课时 Class Hour	知 识 点 Key Points
2.3	缓存性能的10种高级优化方法 Ten advanced optimizations of cache performance	4	（1）小而简单的第一级缓存,用以缩短命中时间、降低功率 （2）采用路预测以缩短命中时间 （3）实现缓存访问的流水化,以提高缓存带宽 （4）采用无阻塞缓存,以提高缓存带宽 （5）关键字优先和提前重启动以降低缺失代价 （6）合并写缓冲区以降低缺失代价 （7）采用编译器优化以降低缺失率 （8）对指令和数据进行硬件预取,以降低缺失代价或缺失率 （9）用编译器控制预取,以降低缺失代价或缺失率 （10）采用HBM技术增加存储层次 (1) Small and simple first-level caches to reduce hit time and power (2) Way prediction to reduce hit Time (3) Pipelined access and multi-banked caches to increase bandwidth (4) Nonblocking caches to increase cache bandwidth (5) Critical word first and early restart to reduce miss penalty (6) Merging write buffer to reduce miss penalty (7) Compiler optimizations to reduce miss rate (8) Hardware prefetching of instructions and data to reduce miss penalty or miss rate (9) Compiler-controlled prefetching to reduce miss penalty or miss rate (10) Using HBM to extend the memory hierarchy

续表

章节序号 Chapter Number	章节名称 Chapters	课时 Class Hour	知 识 点 Key Points
2.4	虚拟存储器和虚拟机 Virtual memory and virtual machine	4	（1）通过虚拟存储器提供保护，对虚拟机监视器的要求，虚拟机（缺少）的指令集体系结构支持 （2）虚拟机对虚拟存储器和 I/O 的影响 （3）VMM 实例：Xen 虚拟机 （1）Protection via virtual memory, protection via virtual machine, requirements of a virtual machine monitor, instruction set architecture support for virtual machine （2）Impact of virtual machine on virtual memory and I/O （3）The Xen virtual machine
2.5	存储器层次结构的设计 The design of memory hierarchies		（1）保护和指令集体系结构，缓存数据的一致性 （1）Protection, virtualization, and instruction set architecture
2.6	融会贯通：ARM Cortex-A5 和 Intel Core i7-6700 中的存储器层次结构 Memory hierarchies in the ARM Cortex-A53 and Intel Core i7 6700	2	（1）ARM Cortex-A5 和 Intel Core i7-6700 中的存储器层次结构 （1）Memory hierarchies in the ARM Cortex-A53 and Intel Core i7-6700

第三章 指令级并行及其开发(Instruction-Level Parallelism and Its Exploitation)

章节序号 Chapter Number	章节名称 Chapters	课时 Class Hour	知 识 点 Key Points
3.1	指令级并行：概念与挑战 Instruction-level parallelism: concepts and challenges	2	(1) 数据相关与冒险 (2) 数据相关，控制相关 (1) Data dependences and hazards (2) Data dependences, control dependences
3.2	揭示 ILP 的基本编译器技术 Basic compiler techniques for exposing ILP		(1) 基本流水线调度和循环展开 (1) Basic pipeline scheduling and loop unrolling
3.3	用高级分支预测降低分支成本 Reducing branch costs with advanced branch prediction		(1) 相关分支预测器，竞争预测器：局部预测器与全局预测器的自适应联合，标记混合预测器 (2) Intel Core i7 分支预测器 (1) Correlating branch predictors, tournament predictors: adaptively combining local and global predictors, tagged hybrid predictors (2) Intel Core i7 branch predictor
3.4	用动态调度克服数据冒险 Overcoming data hazards with dynamic scheduling	2	(1) 动态调度、使用托马苏洛算法进行动态调度 (1) Dynamic scheduling and using Tomasulo's algorithm for dynamic scheduling
3.5	动态调度：示例和算法 Dynamic scheduling: examples and the algorithm		(1) 动态调度：示例和算法 (1) Dynamic scheduling: examples and the algorithm
3.6	基于硬件的推测 Hardware-based speculation		(1) 基于硬件的推测 (1) Hardware-based speculation
3.7	以多发射和静态调度来开发 ILP Exploiting ILP using multiple issue and static scheduling		(1) 以多发射和静态调度来开发 ILP (1) Exploiting ILP using multiple issue and static scheduling

续表

章节序号 Chapter Number	章节名称 Chapters	课时 Class Hour	知 识 点 Key Points
3.8	以动态调度、多发射和推测来开发ILP Exploiting ILP using dynamic scheduling, multiple issue, and speculation	2	(1) 以动态调度、多发射和推测来开发ILP (1) Exploiting ILP using dynamic scheduling, multiple issue,and speculation
3.9	用于指令传送和推测的高级技术 Advanced techniques for instruction delivery and speculation		(1) 指令传送和推测的高级技术 (1) Advanced techniques for instruction delivery and speculation
3.10	ILP方法与存储器系统 Cross-cutting issues		(1) 可实现处理器上ILP的局限性、硬件推测与软件推测 (1) Hardware versus software speculation
3.11	多线程：开发线程级并行提高单处理器吞吐 Multithreading: exploiting thread-level parallelism to improve uniprocessor throughput	2	(1) 同步多线程技术对超标量处理器的作用 (1) Effectiveness of simultaneous multithreading on superscalar processors
3.12	融会贯通：Intel Core i7 6700和ARM Cortex-A53 Putting it All together: The Intel Core i7 6700 and ARM Cortex-A53		(1) Intel Core i7 6700 和 ARM Cortex-A53 (1) The Intel Core i7 6700 and ARM Cortex-A53

第四章 数据级并行——向量、SIMD 和 GPU 体系结构（Data-Level Parallelism in Vector,SIMD,and GPU Architectures）

章节序号 Chapter Number	章节名称 Chapters	课时 Class Hour	知识点 Key Points
4.1	数据级并行简介 Introduction of data-level parallelism		(1) 数据级并行 (1) Data-level parallelism
4.2	向量处理器体系结构 Vector architecture	2	(1) 向量处理器如何工作、向量执行时间、多赛道 (2) 每个时钟周期超过一个元素、向量长度寄存器、向量屏蔽寄存器、内存组、处理向量体系结构中的多维数组、在向量体系结构中处理稀疏矩阵 (3) 向量体系结构编程 (1) Vector execution time, multiple lanes (2) Beyond one element per clock cycle, vector-length registers: handling loops, supplying bandwidth for vector load/store units, handling multidimensional arrays in vector architectures (3) Vector architectures programming
4.3	单指令多数据指令（SIMD）对于多媒体的扩展 SIMD instruction set extensions for multimedia	2	(1) 多媒体 SIMD 体系结构编程 (2) Roofline 可视性能模型 (1) Programming multimedia SIMD architecture (2) The roofline visual performance model

续表

章节序号 Chapter Number	章节名称 Chapters	课时 Class Hour	知 识 点 Key Points
4.4	图形处理器 Graphics Processing Units(GPU)	2	(1) GPU 编程 (2) NVIDIA GPU 计算结构 (3) NVIDA GPU 指令集体系结构 (4) GPU 中的条件分支 (5) GPU 存储器结构 (6) 帕斯卡 GPU 体系结构中的创新 (7) 向量体系结构与 GPU 的相似与不同 (8) 多媒体 SIMD 计算机与 GPU 之间的相似与不同 (1) Programming the GPU (2) NVIDIA GPU computational structures (3) NVIDA GPU instruction set architecture (4) Conditional branching in GPU (5) NVIDIA GPU memory structures (6) Innovations in the Pascal GPU architecture (7) Similarities and differences between vector architecture and GPU (8) Similarities and differences between multimedia SIMD computer and GPU
4.5	检测和增强循环内并行度 Detecting and enhancing loop-level parallelism		(1) 查找相关性、消除相关性计算 (1) Finding dependences, eliminating dependent computations
4.6	交叉影响 Cross-cutting issues	2	(1) 能耗与 DLP (2) 慢而宽与快而窄、分组存储器和图形存储器、步幅访问和 TLB 缺失 (1) Energy and DLP (2) Slow and wide versus fast and narrow, banked memory and graphics memory, strided accesses and TLB misses
4.7	嵌入式和服务器 GPU Putting it All together: embedded versus server GPU and tesla versus Core i7		(1) 移动与服务器 GPU、Tesla 与 Intel Core i7 (1) Embedded versus Server GPU and Tesla Versus Intel Core i7

第五章 线程级并行(Thread-Level Parallelism)

章节序号 Chapter Number	章节名称 Chapters	课时 Class Hour	知 识 点 Key Points
5.1	线程级并行简介 Introduction of thread-level parallelism		(1) 多处理器体系结构、并行处理的挑战 (1) Multiprocessor architecture: issues and approach, challenges of parallel processing
5.2	集中式共享存储器体系结构 Centralized shared-memory architecture	2	(1) 多处理器缓存一致性 (2) 一致性的基本实现方案 (3) 监听一致性协议 (4) 基本一致性协议的扩展 (5) 对称共享存储器多处理器与监听协议的局限性 (6) 多处理器和监听协议 (7) 监听缓存一致性的实施 (1) Multiprocessor cache coherence (2) Basic schemes for enforcing coherence (3) Snooping coherence protocol (4) Extensions to the basic coherence protocol (5) Limitations in symmetric shared-memory (6) Multiprocessors and snooping protocol (7) Implementing snooping cache coherence
5.3	对称共享存储器多处理器的性能 Performance of symmetric shared-memory multiprocessor	2	(1) 工作负载的性能测量 (2) 多重编程和操作系统工作负载 (3) 多重编程和操作系统工作负载的性能 (1) A multiprogramming (2) OS workload (3) Performance of the multiprogramming and OS workload

续表

章节序号 Chapter Number	章节名称 Chapters	课时 Class Hour	知 识 点 Key Points
5.4	分布式共享存储器和目录式一致性 Distributed shared-memory and directory-based coherence	2	（1）目录式缓存一致性协议：基础知识和举例 （1）Directory-based cache coherence protocols: the basics
5.5	同步 Synchronization		（1）基本硬件原语、使用一致性实现锁 （1）Basic hardware primitives, implementing locks using coherence
5.6	存储器一致性简介 Models of memory consistency		（1）程序员的观点、宽松连贯性模型 （1）The programmer's view, relaxed consistency models: the basics and release consistency
5.7	交叉问题 Cross-cutting issues	2	（1）编译器优化与连贯性模型 （2）利用推测来隐藏严格连贯性模型中的延迟 （3）多重处理和多线程的性能增益 （1）Compiler optimization and the consistency model （2）Using speculation to hide latency in strict consistency models （3）Performance gains from multiprocessing and multithreading
5.8	多核处理器及其性能 Multicore processors and their performance		（1）多核处理器及其性能 （1）Performance of multicore-based multiprocessors on a multiprogrammed workload

第六章 通用图形处理器(General Purpose Graph Processing Unit,GPGPU)

章节序号 Chapter Number	章节名称 Chapters	课时 Class Hour	知　识　点 Key Points
6.1	简介 Introduction of GPGPU		(1) 通用图形处理器简介 (1) Introduction of GPGPU
6.2	GPGPU的工作方式和编程模型 Programming mode of GPGPU		(1) 异构协处理器 (2) 块和线程的组织方式 (3) 显存的管理及Kernel执行 (1) Heterogeneous CPU-GPU system (2) Block and arrays of parallel threads (3) Device memory management and kernel execution
6.3	GPGPU的存储结构 Memory structure of GPGPU	2	(1) 局部存储、共享存储、全局存储 (1) Local Memory, shared memory and global memory
6.4	GPGPU计算架构及编程示例 GPGPU architecture and programming examples		(1) GPGPU的硬件执行和调度模式 (2) Warp的工作模式 (3) 分支散度以及动态合并 (4) 存储散度以及地址合并 (1) Hardware execution and scheduling mode of GPGPU (2) Warp grouping and dispatch (3) Branch divergence and dynamic warp Formation (4) Memory divergence and address coalescing

第七章 面向领域应用的计算机体系结构（Domain-Specific Architecture）

章节序号 Chapter Number	章节名称 Chapters	课时 Class Hour	知　识　点 Key Points
7.1	简介 Introduction of domain-specific architecture		（1）面向领域应用的计算机简介 （1）Introduction of domain-specific architecture
7.2	深度神经网络计算介绍 Deep neural network	2	（1）DNN 神经元 （2）训练与推理 （3）多层感知器 （4）卷积神经网络 （5）递归神经网络 （6）批量，量化 （1）The Neurons of DNN （2）Training versus inference （3）Multilayer perceptron （4）Convolutional neural network （5）Recurrent neural network （6）Batches, quantization
7.3	谷歌 TPU Google TPU		（1）TPU 体系结构、指令集、微体系机构、实现、软件和改进方法 （1）TPU architecture, TPU instruction set architecture, TPU microarchitecture, TPU implementation, TPU software, improving the TPU
7.4	微软 Catapult Microsoft Catapult	2	（1）微软 Catapult 灵活的数据中心加速，结构、软件、卷积神经网络运行、搜索加速 （1）Microsoft Catapult, a flexible data center accelerator, Catapult implementation and architecture, Catapult software, CNN on Catapult, search acceleration on Catapult
7.5	英特尔 Crest Intel Crest		（1）用于加速训练的英特尔 Crest 数据中心加速器 （1）Intel Crest, a data center accelerator for training

续表

章节序号 Chapter Number	章节名称 Chapters	课时 Class Hour	知　识　点 Key Points
7.6	个人移动计算图像处理 A personal mobile device image processing unit	2	（1）个人移动计算图像处理器和指令集 （1）Personal mobile device image processing unit, pixel visual core instruction set architecture
7.7	GPGPU、CPU、DNN 加速器比较 CPU versus GPGPU versus DNN accelerator		（1）GPGPU、CPU、DNN 加速器性能、执行时间、吞吐率、能效比等 （1）GPGPU versus DNN accelerators, performance: rooflines, response time, and throughput, cost-performance, TCO, and performance/watt

大纲指导者：郑南宁教授（西安交通大学人工智能学院）

大纲制定者：任鹏举教授（西安交通大学人工智能学院）、赵文哲助理教授（西安交通大学人工智能学院）、夏天副教授（西安交通大学人工智能学院）

大纲审定：西安交通大学人工智能学院本科专业知识体系建设与课程设置第二版修订工作组

5.4　"理论计算机科学的重要思想"课程大纲

课程名称：理论计算机科学的重要思想

Course：Great Ideas in Theoretical Computer Science

先修课程：无

Prerequisites：None

学分：1

Credits：1

5.4.1　课程目的和基本内容（Course Objectives and Basic Content）

本课程是人工智能学院本科专业必修课。

This course is a compulsory course for undergraduates in College of Artificial Intelligence.

课程以理论计算机科学的核心思想为主线，对确定型算法、随机化算法、可计算性理论、密码学、博弈论、数论、数值线性代数等展开讨论，着重介绍其中使用的严格的数学论证方法。除绪论之外，第一章到第七章在以上每个领域分别选择一个具体课题进行讨论，包括图灵机停机问题、卡拉楚巴算法、拉斯维加斯算法与蒙特卡洛算法、零知识证明、纳什均衡、连分数与无理数的逼近、条件数与病态矩阵等等。

本课程通过对基本理论和解题方法的学习，帮助学生建立关于理论计算机科学的初步印象。课程采用师生互动的课堂讲授模式，鼓励学生积极参与到论证和推导之中，予以学生充分的问题解决训练，力求在课程完成后学生能够初步掌握理论计算机科学的思维模式和工作方法。通过解题训练，进一步加强学生的问题求解能力、综合理解能力和书面表达能力，培养严谨科学的作风，为今后从学习走向研究打下扎实的基础。

随着人类进入信息时代，理论计算机科学不仅继续在信息技术领域做出重要贡献，而且将其影响扩展到社会的方方面面。因此，学习好理论计算机科学的重要思想和解题方法对于下一代科学技术人才有着极其深远的意义。在学习完本课程之后，每当面临新的技术挑战时，学生应当知晓理论计算机科学可以提供给自己的强有力的工具和方法。通过运用这些工具和方法，并将其与实践结合，有望取得更大、更多的创新成果。

This course focuses on the key idea of theoretical computer science and covers topics like deterministic algorithms, randomized algorithms, computability theory, cryptography, game theory, discrete mathematics, and numerical linear algebra, especially introducing the rigorous mathematical proofs. Chapter 1 to Chapter 7 discuss one specific problem from one of the above topics, respectively, including Halting Problem of Turing Machine, Karatsuba Algorithm, Las Vegas Algorithm and Monte Carlo Algorithm, Zero-Knowledge Proofs, Nash Equilibrium, Continued Fractions and Approximation of Irrationals, Condition Numbers of Matrices and Ill-Conditioned Matrices, etc.

This course helps students have a preliminary understanding of theoretical computer science through the study of basic theories and problem-solving methods. The course adopts an interactive teaching strategy, encourages students to actively participate in demonstrating and deducing problems, and gives students sufficient training on solving problem, and strives to master the basic idea and method of theoretical computer science after the course is completed. Through problem-solving training, students' problem solving ability, comprehensive understanding ability and writing skills will be further strengthened, and a rigorous and scientific style will be cultivated, laying a solid foundation from learning to research in the future.

As humans enter the information age, theoretical computer science not only continues to make important contributions in the field of information technology, but also extends its influence to all aspects of society. Therefore, learning the important ideas and solving methods of theoretical computer science has far-reaching significance for the next generation of scientific and technological talents. After completing this course, students should be aware of the powerful tools and methods that theoretical computer science can provide to them whenever they face new technical challenges. By applying these tools and methods, and combining them with practice, it is expected to achieve greater and more innovative results.

5.4.2 课程基本情况（Course Arrangements）

课程名称	理论计算机科学的重要思想 Great Ideas in Theoretical Computer Science								
开课时间	一年级	二年级		三年级		四年级		计算机科学与技术	
	秋 春	秋 春	小学期	秋	春	秋	春		
课程定位	本科生"计算机科学与技术"课程群必修课							必修 （学分）	计算机程序设计(2)
									数据结构与算法(3)
学　　分	1学分								计算机体系结构(3)
总 学 时	16学时 （授课16学时、实验0学时）								理论计算机科学的重要思想(1)
授课学时 分配	课堂讲授(14学时)， 数学摸底测验(2学时)							选修 （学分）	3D计算机图形学(2)
									智能感知与移动计算(2)
先修课程	无								
后续课程									
教学方式	课堂讲授、习题演练、小组讨论、综述报告								
考核方式	平时成绩（含考勤）占50%，开卷期末考试占50%								
参考教材	自编讲义								
参考资料	1. MIT Open Courseware 6.080/6.089 2. CMU Course 15-251								
其他信息									

5.4.3 教学目的和基本要求（Teaching Objectives and Basic Requirements）

（1）理解图灵机的定义、图灵机停机问题的不可判定性及其证明；
（2）理解高精度乘法的算法挑战性，掌握卡拉楚巴算法；
（3）掌握拉斯维加斯算法和蒙特卡洛算法的定义，理解二者之间的联系和区别；
（4）理解各种零知识证明的定义，初步学会设计零知识证明；
（5）理解纳什均衡的定义、理解纳什定理，了解角谷不动点定理与纳什定理之间的关联；
（6）掌握代数数、超越数等概念，了解无理数的逼近问题及其重要性；
（7）理解条件数与相对误差之间的关系，学会使用条件数来推算相对误差。

5.4.4 教学内容及安排（Syllabus and Arrangements）

绪论（Introduction）

章节序号 Chapter Number	章节名称 Chapters	课时 Class Hour	知识点 Key Points
0.1	数学摸底考试 Mathematical examination	2	（1）1.5小时测验 （2）对测验题进行简单讲解 (1) A preliminary mathematical screening of 1.5 hour (2) A brief presentation of solutions

第一章 可计算性理论选讲—图灵机停机问题（Lectures on Computational theory: Halting Problem of Turing Machines）

章节序号 Chapter Number	章节名称 Chapters	课时 Class Hour	知识点 Key Points
1.1	图灵机 Turing machine	1	（1）图灵机的定义 （2）通用图灵机定理 （3）关于磁带数量的注记 (1) Definition of Turing machine (2) Universal Turing machine theorem (3) Notes on the number of tapes

续表

章节序号 Chapter Number	章节名称 Chapters	课时 Class Hour	知识点 Key Points
1.2	递归可枚举 语言与递归语言 Recursive enumerable language and recursive language		（1）递归可枚举语言的定义 （2）递归语言的定义 （3）递归可枚举语言与停机的关系 （4）递归可枚举语言与递归语言的关系 (1) Definition of recursive enumerable language (2) Definition of recursive language (3) Relationship between recursive enumerable language and halting problem (4) Relationship between recursive enumerable language and recursive language
1.3	图灵机停机问题 Halting problem of Turing machine	1	（1）图灵机停机问题 （2）该问题不可判定性的证明 （3）赖斯定理 (1) Halting problem of Turing machine (2) Undecidability of the halting problem (3) Rice theorem
1.4	图灵机停机问题不可判定性的现实价值 The practical value of the undecidability of the Turing machine shutdown problem		（1）图灵机停机问题不可判定性在现实中的含义与价值 (1) Practical implications of the undecidability of halting problem

第二章 确定型算法选讲—卡拉楚巴算法（Lectures on Deterministic Algorithm: Karatsuba Algorithm）

章节序号 Chapter Number	章节名称 Chapters	课时 Class Hour	知识点 Key Points
2.1	高精度乘法 High precision multiplication	1	（1）高精度乘法及其算法挑战性 （2）幼稚算法及其分析 (1) High precision multiplication and its algorithmic challenge (2) A naïve algorithm with analysis

续表

章节序号 Chapter Number	章节名称 Chapters	课时 Class Hour	知识点 Key Points
2.2	卡拉楚巴算法 Karatsuba algorithm		(1) 卡拉楚巴算法的基本思想 (2) 算法步骤 (3) 应用 Master 定理的效率分析 (1) The main idea behind Karatsuba algorithm (2) Algorithm specification (3) Efficiency analysis through Master theorem
2.3	不对称型类卡拉楚巴公式 Asymmetric Karatsuba formula	1	(1) 若干不对称型的类卡拉楚巴公式 (1) Some asymmetric Karatsuba-like formulas

第三章　随机化算法选讲—拉斯维加斯算法与蒙特卡洛算法（Lectures on Randomization Algorithm: Las Vegas Algorithm and Monte Carlo Algorithm）

章节序号 Chapter Number	章节名称 Chapters	课时 Class Hour	知识点 Key Points
3.1	随机快排算法 Random quick sort	1	(1) 排序问题 (2) 确定型排序算法及其复杂度 (3) 随机快排算法的基本思想 (4) 随机快排算法的步骤以及效率分析 (1) Sorting problem (2) Deterministic algorithms for sorting and their complexities (3) Basic ideas behind random quick sort (4) Algorithm specification and efficiency analysis

续表

章节序号 Chapter Number	章节名称 Chapters	课时 Class Hour	知 识 点 Key Points
3.2	随机最小割算法 Randomized min cut algorithm	1	(1) 最小割问题 (2) 确定型最小割算法及其复杂度 (3) 随机最小割算法的基本思想 (4) 随机最小割算法的步骤以及效率分析 (1) Min cut problem (2) Deterministic algorithms for min cut (3) Basic ideas behind randomized min cut algorithm (4) Algorithm specification and efficiency analysis
3.3	拉斯维加斯算法与蒙特卡洛算法 Las Vegas algorithm and Monte Carlo algorithm		(1) 拉斯维加斯算法的定义 (2) 蒙特卡洛算法的定义 (3) 拉斯维加斯算法与蒙特卡洛算法的联系和区别 (4) 蒙特卡洛算法的分类 (1) Definition of Las Vegas algorithm (2) Definition of Monte Carlo algorithm (3) Connections and differences between Las Vegas algorithm and Monte Carlo algorithm (4) Classification of Monte Carlo algorithms

第四章 密码学选讲—零知识的知识证明(Lectures on Crytograhpy: Zero-Knowledge Proof of Knowledge)

章节序号 Chapter Number	章节名称 Chapters	课时 Class Hour	知 识 点 Key Points
4.1	交互式证明 Interactive proof	1	(1) 交互式证明的定义、完备性、合理性 (1) Definition of interactive proof system, completeness, soundness

章节序号 Chapter Number	章节名称 Chapters	课时 Class Hour	知识点 Key Points
4.2	零知识证明 Zero-knowledge proof	1	（1）完美零知识证明 （2）统计零知识证明 （3）计算零知识证明 （4）零知识证明实例 (1) Perfect zero-knowledge proof (2) Statistic zero-knowledge proof (3) Computational zero-knowledge proof (4) Examples of zero-knowledge proofs
4.3	零知识的知识证明 Zero-knowledge proof of knowledge	1	（1）知识证明（知识抽取器） （2）零知识的知识证明 （3）零知识与知识证明的悖论 (1) Proof of knowledge (knowledge extractor) (2) Zero-knowledge proof of knowledge (3) Paradox of zero knowledge and proof of knowledge

第五章　博弈论选讲—纳什均衡（Lectures on Game Theory: Nash Equilibrium）

章节序号 Chapter Number	章节名称 Chapters	课时 Class Hour	知识点 Key Points
5.1	策略博弈 Strategic game	1	（1）策略博弈的定义与实例 （2）纯策略与混合策略 (1) Definition and examples of strategic game (2) Pure strategy and mixed strategy
5.2	纳什均衡与纳什定理 Nash equilibrium and Nash theorem		（1）纳什均衡与最优反馈函数 （2）角谷不动点定理 （3）纳什定理及其证明 (1) Nash equilibrium and best response function (2) Kakutani fixed point theorem (3) Nash theorem and its proof

续表

章节序号 Chapter Number	章节名称 Chapters	课时 Class Hour	知识点 Key Points
5.3	纳什均衡与其他解概念之间的关系 Relationship between Nash equilibrium and other equilibrium	1	(1) 纳什均衡与优势策略均衡 (2) 可理性化动作 (3) 帕累托最优之间的关系 (1) Relationship between Nash equilibrium and dominant strategy equilibrium (2) Rationalizable actions (3) Pareto optimality

第六章 数论选讲——连分数与无理数的逼近（Lectures on Number Theory: Continued Fraction and Theory of Approximation of Irrationals by Rationals）

章节序号 Chapter Number	章节名称 Chapters	课时 Class Hour	知识点 Key Points
6.1	有限连分数 Finite continued fraction		(1) 有限连分数、收敛子、正商连分数、简单连分数、有理数的连分数表示 (1) Finite continued fraction, convergent, positive-quotient continued fraction, simple continued fraction, representing rational by continued fraction
6.2	无限连分数 Infinite continued fraction	1	(1) 无限简单连分数 (2) 无理数的连分数表示 (3) 基于收敛子的近似表示 (1) Infinite simple fraction (2) Representing irrational by continued fraction (3) Approximation by convergent
6.3	无理数的逼近理论 Theory of approximation of irrational by rational		(1) 无理数的逼近问题 (2) 狄利克雷定理 (3) 代数数与超越数 (4) 刘维尔定理 (1) Problem of approximation of irrational by rational (2) Dirichlet argument (3) Algebraic number and transcendental number (4) Liouville theorem

第七章 数值线性代数选讲——条件数和病态矩阵（Lectures on Numerical Linear Algebra: Condition Number and Ill-Conditioned Matrix）

章节序号 Chapter Number	章节名称 Chapters	课时 Class Hour	知　识　点 Key Points
7.1	条件数 Condition number		（1）矩阵条件数的定义 （2）条件数与相对误差的关系 (1) Definition of condition number (2) Relationship between the condition number of a matrix and the relative error of the solution
7.2	奇异值分解与条件数 Singular value of matrix and condition number	1	（1）矩阵的奇异值 （2）奇异值分解 （3）欧几里得范数与奇异值之间的关系 （4）条件数与奇异值之间的关系 （5）利用奇异值分解求解条件数 (1) Singular value of matrix (2) Singular value decomposition (3) Relationship between the Euclidean norm and the singular values (4) Relationship between the condition number and the singular values (5) Computing the condition number through singular value decomposition
7.3	条件数的性质 与病态矩阵 Properties of condition number and ill-conditioned matrix	1	（1）条件数的性质 （2）利用条件数推算相对误差 （3）病态矩阵 (1) Properties of condition number (2) Computing relative error from condition number (3) Ill-conditioned matrix

大纲制定者：仲盛教授（南京大学计算机科学与技术系）

大纲审定：西安交通大学人工智能学院本科专业知识体系建设与课程设置第二版修订工作组

5.5 "3D 计算机图形学"课程大纲

课程名称：3D 计算机图形学
Course：3D Computer Graphics
先修课程：线性代数与解析几何、计算机程序设计、数据结构与算法
Prerequisites：Linear Algebra and Analytic Geometry, Computer Programming, Data Structure and Algorithm
学分：2
Credits：2

5.5.1 课程目的和基本内容（Course Objectives and Basic Content）

本课程是人工智能学院本科专业选修课。

This course is an elective course for undergraduates in College of Artificial Intelligence.

图形是由包含几何和属性信息的点、线、面等基本图元构成的可视对象，计算机图形学是研究用计算机表示、生成、处理和显示图形的原理、算法、方法和技术的一门学科。3D 计算机图形学的目的就是从物体的 3D 几何模型产生物体或场景的真实感图形图像。

本课程主要讲述 3D 计算机图形处理的核心概念：3D 物体的表示、建模与绘制。物体绘制重点讨论多边形网格渲染流水线、局部光照模型、多边形表面光强计算以及光栅化计算的数学原理和算法，真实感图像绘制主要介绍表面纹理映射、阴影生成和全局光照模型。

本课程目的是使学生掌握 3D 计算机图形学的基本方法，了解多边形网格绘制的核心技术。课程要求学生阅读一定的文献资料、完成指定课程实验，培养学生的思考、综合和解决问题能力，奠定计算机图形学领域的基础知识和基本技术能力。

A graphic is a visual object made up of basic primitives such as points, lines, and surfaces that contain geometry and attribute information. Computer graphics is a discipline that studies the principles, methods and techniques for representing, generating, and displaying graphical image with the aid of computers. The purpose of 3D computer graphics is to generate realistic graphical images of objects or scenes from 3D geometric models.

This course focuses on the core concepts of computer graphics processing: representation, modeling and rendering of 3D objects. The rendering of the object mainly introduces polygon mesh rendering pipeline, local illumination model, polygonal face luminance calculation, and mathematical principle and algorithm description of rasterization calculation. Realistic image rendering includes surface texture mapping, shadow generation, and global illumination models.

The purpose of this course is to develop students' ability to think, integrate and solve problems. They need to read certain literature materials and complete specified course experiments, to strengthen students' ability to analyze and solve problem, and develop technical skills in the field of computer graphics.

5.5.2 课程基本情况（Course Arrangements）

课程名称	3D计算机图形学 3D Computer Graphics								
开课时间	一年级		二年级		三年级		四年级	计算机科学与技术	
	秋	春	秋	春	秋	春	秋	春	
课程定位	本科生"计算机科学与技术"课程群选修课							必修 （学分）	计算机程序设计(2)
									数据结构与算法(3)
学　分	2 学分								计算机体系结构(3)
总学时	38 学时 （授课 32 学时、实验 6 学时）								理论计算机科学的重要思想(1)
								选修 （学分）	3D计算机图形学(2)
授课学时分配	课堂讲授（32 学时）								智能感知与移动计算(2)
先修课程	线性代数与解析几何、计算机程序设计、数据结构与算法								
后续课程									
教学方式	课堂教学、大作业、实验综述报告								
考核方式	笔试成绩占 60%，大作业成绩占 20%，实验综述报告占 20%								
参考教材	Marschner S, Shirley P. Fundamentals of Computer Graphics[M]. 4th. New York: CRC Press, 2016.								
参考资料	1. Foley J D. Computer Graphics: Principles and Practice[M]. 3rd. New York: Addison-Wesley, 2013. 2. Edward A. 交互式计算机图形学——基于 OpenGL 的自顶向下方法[M]. 5版. 吴文国, 译. 北京: 清华大学出版社, 2007.								
其他信息									

5.5.3 教学目的和基本要求(Teaching Objectives and Basic Requirements)

(1) 理解 3D 计算机图形学的基本概念与理论,了解其应用技术;
(2) 掌握 3D 物体的表示与建模方法;
(3) 掌握 3D 物体与场景的多边形渲染流水线方法;
(4) 熟悉局部光照模型与明暗处理技术;
(5) 熟悉光栅化计算的基本方法;
(6) 掌握 3D 物体的真实感绘制方法与技术;
(7) 熟悉表面纹理映射技术;
(8) 熟悉全局光照模型的基本方法;
(9) 熟悉使用 OpenGL 和其他图形引擎实现物体建模与绘制方法。

5.5.4 教学内容及安排(Syllabus and Arrangements)

第一章 概述(Introduction)

章节序号 Chapter Number	章节名称 Chapters	课时 Class Hour	知 识 点 Key Points
1.1	计算机图形学简介 An introduction to computer graphics	3	(1) 计算机图形学的基本概念 (2) 计算机图形学的发展历程 (3) 计算机图形学的应用领域 (1) Basic concepts of computer graphics (2) Development history of computer graphics (3) Application fields of computer graphics
1.2	计算机图形学的数学基础 Miscellaneous Math		(1) 向量和矩阵 (2) 参数曲线和参数曲面 (1) Vector and matrix (2) Parametric curves and surfaces

第二章 几何建模(Geometric Modeling)

章节序号 Chapter Number	章节名称 Chapters	课时 Class Hour	知 识 点 Key Points
2.1	几何变换 Geometric transformation	9	(1) 二维几何变换 (2) 三维几何变换：旋转矩阵，四元数 (3) 相机模型、投影变换 (1) 2D geometric transformation (2) 3D geometric transformation：rotation matrix，Quaternion (3) Camera models，projective transformation
2.2	三维物体的几何表征 3D Geometric representation		(1) 显式表达：多边形网格、点云 (2) 隐式表达：距离函数、代数曲面、神经辐射场 (1) Explicit representation：polygonal meshes，point clouds (2) Implicit expressions：distance functions，algebraic surfaces，NeRF
2.3	网格处理 Mesh processing		(1) 网格细分 (2) 网格简化 (3) 网格光顺与去噪 (4) 重网格化 (1) Mesh subdivision (2) Mesh simplification (3) Mesh smoothing and denoising (4) Remeshing
2.4	参数曲线与曲面 Parametric curves and surfaces		(1) 参数曲线：Bezier 曲线、B 样条曲线、均匀非有理 B 样条曲线 (2) 参数曲面：Hermite 曲面、Bezier 曲面、B 样条曲面 (1) Parametric curves：Bezier curves，B-spline curves，NURBS (2) Parametric surfaces：Hermite surfaces，Bezier surfaces，B-spline surfaces

第三章 光栅化与光线追踪(Rasterization and Ray Tracing)

章节序号 Chapter Number	章节名称 Chapters	课时 Class Hour	知 识 点 Key Points
3.1	光栅化 Rasterization	6	(1) 光栅化的基本概念：光栅图像、图元、深度缓存、裁剪 (2) 深度测试与抗锯齿 (1) Basic concepts of rasterization: raster image, primitive, z-buffer, clipping (2) Depth testing and anti-aliasing
3.2	光线追踪 Ray tracing		(1) 光线生成 (2) 光线与物体相交 (3) 局部光照计算 (4) 递归光线追踪 (5) 阴影计算 (6) 全局光照 (1) Ray generation (2) Ray-object intersection (3) Local lighting calculation (4) Recursive ray tracing (5) Shadow calculation (6) Global illumination

第四章 着色(Shading)

章节序号 Chapter Number	章节名称 Chapters	课时 Class Hour	知 识 点 Key Points
4.1	光源模型 Luminaire models	4	(1) 辐射度函数 (2) 直接光与间接光 (3) 点光源、平行光源、矩形面光源 (1) The radiance function (2) Direct and indirect light (3) Spot light, parallel light, rectangular area light

213

续表

章节序号 Chapter Number	章节名称 Chapters	课时 Class Hour	知识点 Key Points
4.2	基本着色模型 Shading models		（1）着色模型：恒定着色、平面着色、Gouraud 着色、Phong 着色、Blinn-Phong 着色 （2）漫反射、镜面反射、环境光 （1）Shading models: constant shading, plane shading, Gouraud shading, Phong shading, Blinn-Phong shading （2）Diffuse reflection, specular reflection, ambient light
4.3	纹理映射 Texture mapping		（1）纹理坐标、漫反射纹理、镜面反射纹理、法线纹理和高光纹理 （2）Mipmap 技术 （1）Texture coordinates, diffuse texture, specular texture, normal texture and specular texture （2）Mipmap technology

第五章　真实感绘制（Photorealistic Rendering）

章节序号 Chapter Number	章节名称 Chapters	课时 Class Hour	知识点 Key Points
5.1	材质与外观 Materials and appearances		（1）双向反射分布函数 （2）微平面理论 （1）BRDF （2）Microfacet theory
5.2	高级光线传播与复杂外观建模 Advanced light transport and appearance modeling	5	（1）光线传播：无偏光线传播，有偏光线传播，即时辐射度 （2）外观建模：无曲面模型，曲面模型，程序化生成的外观 （1）Light Transport: Unbiased and biased light transport methods, instant radiosity （2）Appearance Modeling: Non-surface and surface models, procedural appearance

章节序号 Chapter Number	章节名称 Chapters	课时 Class Hour	知识点 Key Points
5.3	颜色 Color		(1) 颜色模型、颜色空间、色彩感知、色彩量化与抖动、Gamma 校正、高动态范围图像 (1) Color model, color space, color perception, color quantization and dithering, Gamma correction, HDR image

第六章 动画与模拟(Animation and Simulation)

章节序号 Chapter Number	章节名称 Chapters	课时 Class Hour	知识点 Key Points
6.1	动画的基本概念 Basic concepts of animation		(1) 关键帧动画、层级动画、骨骼动画 (1) Keyframe animation, hierarchical animation, skeletal animation
6.2	物理模拟 Physical simulation	5	(1) 质心弹簧系统 (2) 粒子系统 (3) 前向动力学与方向动力学 (1) Mass spring system (2) Particle system (3) Forward kinematics and inverse kinematics
6.3	刚体与流体 Rigid bodies and fluids		(1) 刚体模拟 (2) 流体模拟：基于位置的模拟方法,Lagrangian 质点法与 Eulerian 网格法,物质点法 (1) Rigid body simulation (2) Fluid simulation: position-based method, Lagrangian method and Eulerian method, Material point method (MPM)

5.5.5 实验环节（Experiments）

序号 Num.	实验内容 Experiment Content	课时 Class Hour	知　识　点 Key Points
1	3D多边形网格 模型观察器实现 Viewer for 3D polygon mesh model	2	（1）3D多边形网格的空间变换 （2）3D多边形网格的表面细分 （3）层次细节显示 (1) Space transformation of 3D polygon mesh (2) 3D surface tessellation (3) Level Of Details(LOD)
2	绘制贝塞尔曲线与曲面 Drawing Bézier curve and surface	2	（1）贝塞尔曲线 （2）贝塞尔曲面 （3）德卡斯特奥算法 (1) Bézier curve (2) Bézier surface (3) de Casteljau algorithm
3	简易光线追踪	2	（1）真实感表面纹理映射技术 （2）多边形网格空间变换与渲染流水线 (1) Realistic surface texture mapping (2) Space transformation and rendering pipeline of polygon mesh

　　大纲指导者：郑南宁教授(西安交通大学人工智能学院)

　　大纲制定者：刘跃虎教授(西安交通大学人工智能学院)、蒋才桂教授(西安交通大学人工智能学院)、张驰助理教授(西安交通大学人工智能学院)

　　大纲审定：西安交通大学人工智能学院本科专业知识体系建设与课程设置第二版修订工作组

5.6 "智能感知与移动计算"课程大纲

课程名称：智能感知与移动计算
Course：Smart Sensing and Mobile Computing
先修课程：概率统计与随机过程、人工智能概论、数字信号处理、自然语言处理、计算机视觉与模式识别
Prerequisites：Probability Theory and Stochastic Process,Introduction to Artificial Intelligence,Digital Signal Processing,Natural Language Processing,Computer Vision and Pattern Recognition
学分：2
Credits：2

5.6.1 课程目的和基本内容(Course Objectives and Basic Content)

本课程是人工智能学院本科专业选修课,课程由两个主题组成,包括主题1：多维信息采集与处理和主题2：移动计算与移动智能。

This course is an elective course for undergraduates in College of Artificial Intelligence. It consists of two topics,1. Multi-dimensional Information Collection and Processing,2. Mobile Computing and Mobile Intelligence.

主题1 旨在阐述感知数据的获取是连接物理世界和数字世界的桥梁,是人工智能系统的基础和重要组成部分。在信息种类错综复杂和数量爆发式增长的趋势下,智能感知和泛在互联技术已开始凸显其重要的战略性和基础性,感知技术在智慧城市、智能家居、智慧工厂、智慧医疗等诸多领域中发挥着不可或缺的作用,也是推动智慧型产业结构升级的重要手段。本主题旨在研究多模态感知信息的获取和处理的关键技术,除了传统的视觉感知、听觉感知和触觉感知外,重点对新兴的非传感器感知和群智感知技术进行相应介绍。进一步探讨在感知目标特征微弱、感知对象非合作的情况下,如何通过弱信号特征提取和多维感知数据融合的方式,实现无所不在的智能感知计算。

Topic 1 aims to illustrate that the collection of perceptual data is the bridge between the physical world and the digital world,and is the foundation and important component of artificial intelligence systems. Under the trend of intricate information and explosive growth of information types, smart sensing and Ubiquitous Internet technologies have begun to highlight their important strategic and fundamental technologies. In the smart city, smart home, smart factory, smart medical and many other fields, it plays an indispensable role and is also an important mean to promote the upgrading of smart industrial structure. This topic aims to study the key technologies of multi-modal sensing information collection and processing. In addition to traditional visual perception, auditory perception and tactile perception, the focus is on introducing emerging non-sensor sensing and group sensing technology. Furthermore, under the condition that the perceived target features are weak and the perceived objects are non-cooperative, how to achieve ubiquitous intelligent sensing computing through weak signal feature extraction and multi-dimensional sensing data fusion.

主题 2 旨在阐述感知、计算、传输三者之间的关系，探讨计算模式的发展趋势。随着移动设备计算能力的提升，计算任务逐步由服务器端向移动客户端迁移，这种计算前移的现象是人工智能发展的新趋势。移动计算技术使智能手机或其他智能终端设备在无线环境下实现数据传输及资源共享，极大地改变人们的生活方式和工作方式。移动计算是一个"硬件＋软件"的解决方案，通过在移动网络边缘提供智能服务环境和边缘计算能力，以减少网络操作和服务交付的时延。其技术特征主要包括"邻近性、低时延、高宽带和位置相关"，未来有广阔的应用前景，例如车联网（如无人驾驶）、AR、视频优化加速、监控视频分析等。

Topic 2 aims to explain the relationship between perception, computing, and transmission, as well as to explore the development trend of computing models. As the computing power of mobile devices increases, the computing tasks gradually migrate from the server to the mobile client. This phenomenon of computing advancement is a new trend in the development of artificial intelligence. Mobile computing technology enables smartphones or other information intelligent terminal devices to realize data transmission and resource sharing in a wireless environment, which greatly changes people's lifestyle and working methods. Mobile Computing is a "hardware ＋ software" solution that reduces the latency of network operations and service delivery by providing intelligent service environments and edge computing capabilities at the edge of the mobile network. Its technical features mainly include "proximity, low latency, high bandwidth and location awareness", and there are broad application prospects in the future, such as car networking (such as driverless), AR, video optimization acceleration, surveillance video analysis, and so on.

5.6.2 课程基本情况(Course Arrangements)

课程名称	智能感知与移动计算 Smart Sensing and Mobile Computing								
开课时间	一年级		二年级		三年级		四年级		计算机科学与技术
^^	秋	春	秋	春	秋	春	秋	春	^^
课程定位	本科生"计算机科学与技术"课程群选修课								必修 (学分) · 计算机程序设计(2) 数据结构与算法(3) 计算机体系结构(3) 理论计算机科学的重要思想(1)
学　分	2 学分								
总学时	38 学时 (授课 32 学时、实验 6 学时)								
授课学时分配	课堂讲授(26 学时)、 文献阅读与小组讨论(6 学时)								选修 (学分) · 3D 计算机图形学(2) 智能感知与移动计算(2)
先修课程	概率统计与随机过程、人工智能概论、数字信号处理、自然语言处理、计算机视觉与模式识别								
后续课程									
教学方式	课堂讲授、文献阅读与小组讨论、课外实验、大作业								
考核方式	课程结束笔试成绩占 60%，大作业占 30%，实验占 10%								
参考教材	1. Meijer G. 智能传感器系统：新技术及应用[M]. 靖向萌，译. 北京：机械工业出版社，2018. 2. 赵志为，闵革勇. 边缘计算：原理、技术及实践[M]. 北京：机械工业出版社，2018.								
参考资料	1. Spencer Jr. B F, Ruiz-Sandoval M E, Kurata N. Smart Sensing Technology: Opportunities and Challenges[J]. Structural Control and Health Monitoring, 2010, 11(4): 349-368. 2. Jiang H B, et al. Smart Home based on WiFi Sensing: A Survey[J]. IEEE Access, 2018, 6: 13317-13325. 3. Santos P M, et al. PortoLivingLab: An IoT-Based Sensing Platform for Smart Cities[J]. IEEE Internet of Things Journal, 2018, 5(2): 523-532. 4. Abbas N, et al. Mobile Edge Computing: A Survey[J]. IEEE Internet of Things Journal, 2018, 5(1): 450-465. 5. Chen M, Hao Y X. Task Offloading for Mobile Edge Computing in Software Defined Ultra-Dense Network[J]. IEEE Journal on Selected Areas in Communications, 2018, 36(3): 587-597. 6. Lyu X C, et al. Energy-Efficient Admission of Delay-Sensitive Tasks for Mobile Edge Computing[J]. IEEE Transactions on Communications, 2018, 66(6): 2603-2616.								
其他信息									

5.6.3 教学目的和基本要求（Teaching Objectives and Basic Requirements）

主题 1 多维信息采集与处理（The Collection and Processing of Multi-dimensional Information）
（1）理解感知的基本概念，掌握其属性与分类；
（2）了解智能感知的基本方法、研究对象及其特征；
（3）了解智能感知对人工智能等学科发展的贡献；
（4）掌握非传感器感知的基本原理、主要类型和相应的实现方法；
（5）了解群智感知的核心思想和典型应用。

主题 2 移动计算与移动智能（Mobile Computing and Mobile Intelligence）
（1）理解移动计算的基本概念，了解感知、传输、计算三者间的关系；
（2）了解移动计算中面临的安全和隐私的问题，以及相应的解决方案；
（3）了解移动网络的基础理论和关键技术；
（4）了解移动边缘计算的技术背景、优势和挑战；
（5）了解感知与计算融合的智能系统发展趋势。

5.6.4 教学内容及安排（Syllabus and Arrangements）

主题 1 多维信息采集与处理（The Collection and Processing of Multi-dimensional Information）

章节序号 Chapter Number	章节名称 Chapters	课时 Class Hour	知 识 点 Key Points
1	智能感知的 基本概念与分类 Basic concepts and classifications of smart sensing	0.5	（1）人工智能与智能感知之间的关系 （2）智能感知的概念与分类 （3）智能感知的未来发展 (1) The relationship between artificial intelligence and smart sensing (2) The concepts and classifications of smart sensing (3) The perspective of smart sensing

续表

章节序号 Chapter Number	章节名称 Chapters	课时 Class Hour	知 识 点 Key Points
2	视觉感知 Visual perception	1	(1) 人类视觉的研究进展 (2) 传统视觉感知技术 (3) 基于深度学习的视觉感知技术 (4) 智能视觉感知技术的应用 (1) The advances on the study of human vision (2) The conventional technologies of visual perception (3) Deep-learning based visual perception technology (4) The application of intelligent visual perception technology
3	听觉感知 Auditory perception	1	(1) 听觉感知技术的理论基础 (2) 声音的来源、特征及听觉感知的实例 (3) 人体听觉的生理结构 (4) 声学信号处理技术 (5) 基于深度学习的听觉感知技术 (6) 听觉感知技术的应用场景 (1) The theoretical foundation for auditory perception (2) The origin of human-auditory and its feature, as well as some real-world examples (3) Physiological structure of human auditory (4) The signal processing technology of acoustic signals (5) Deep-learning based auditory perception technology (6) The real-world application of auditory perception
4	触觉感知 Tactile perception	1	(1) 触觉感知技术的理论基础 (2) 触觉感知的分类及性质 (3) 如何在人工智能领域实现触觉感知 (4) 触觉感知数据的采集及处理 (5) 如何利用触觉感知数据进行建模 (6) 触觉感知技术典型应用及未来发展趋势 (1) Fundamentals of the tactile perception (2) The classifications and properties of tactile perception (3) How to achieve tactile perception in AI (4) The data-collection and data-processing of tactile perception data (5) How to establish a model via tactile perception data (6) The applications and future trend of tactile perception

续表

章节序号 Chapter Number	章节名称 Chapters	课时 Class Hour	知　识　点 Key Points
5	非传感器感知 Sensorless perception	5	（1）非传感器感知技术的理论基础 （2）非传感器感知技术主要类型 （3）每类非传感器感知技术（Wi-Fi 感知、RFID 感知、基于大数据感知）的主要原理 （4）相比于传统的传感器感知技术，非传感器感知技术的优势 （5）目前非传感器感知技术主要的应用领域和典型范例 （6）非传感器感知技术与物联网的关系及未来发展趋势 （1）Basic ideas about sensor less perception technology （2）Basic types of sensor less perception technology （3）Basic mechanics of each non-sensor perception technology （4）The advantages of sensor less perception technology compared with the traditional sensor perception （5）The main applications of sensor less perception technology （6）The relationship between non-sensor sensing technology and the Internet of Things and the development trends in the future

续表

章节序号 Chapter Number	章节名称 Chapters	课时 Class Hour	知识点 Key Points
6	多维感知融合与群智感知 Multi-dimensional perceptual fusion and crowdsourcing	4	（1）多维感知融合技术的概念 （2）多维感知融合技术的实现途径 （3）多维感知融合技术的主要体系结构 （4）群智感知技术的概念及性质 （5）群智感知技术的典型系统架构 （6）群智感知技术面临的主要问题 （7）群智感知激励机制的相关研究 （8）群智感知技术的典型应用及进一步研究方向 (1) The concept of multi-dimensional perceptual fusion (2) The major way to achieve multi-dimensional perceptual fusion (3) The main architecture of multi-dimensional perceptual fusion (4) The basic concepts and properties of crowdsourcing (5) The basic architecture of crowdsourcing (6) The major challenges towards crowdsourcing (7) The research about the incentive mechanism of crowdsourcing (8) The applications and future trends of crowdsourcing

主题 2　移动计算与移动智能（Mobile Computing and Mobile Intelligence）

章节序号 Chapter Number	章节名称 Chapters	课时 Class Hour	知识点 Key Points
1	人工智能与移动终端的深度融合 ——未来人工智能的新方向 The deep fusion of AI and mobile terminals —the new trend of AI	1	（1）传统人工智能 （2）人工智能的未来 （3）人工智能与移动终端深度融合技术的应用 (1) Traditional AI (2) The future of AI (3) The application of deep fusion for AI and mobile terminal

续表

章节序号 Chapter Number	章节名称 Chapters	课时 Class Hour	知 识 点 Key Points
2	数据安全与用户隐私保护 Data security and user privacy protection	3.5	(1) 安全与隐私的区别与联系 (2) 大数据下的安全挑战 (3) 数据安全与用户隐私保护关键技术 (4) 数据安全与隐私保护的发展与展望 (1) The differences and connections between security and privacy (2) The challenge towards the security issues at the age of big data (3) The key technologies for data security and user privacy protection (4) The development and prospect of data security and user privacy protection
3	移动网络技术 Mobile network technology	4	(1) 移动网络的演进 (2) 无线网络技术理论 (3) 无线传感器网络 (4) 移动网络技术的发展趋势 (1) The development of mobile network (2) The theory of wireless network (3) Wireless sensor network (4) The trend of mobile network
4	移动端机器学习 Mobile-terminal based machine learning	1.5	(1) 移动端机器学习的兴起 (2) 移动端机器学习技术 (3) 移动端机器学习的应用场景 (4) 移动端机器学习的局限性及未来发展方向 (1) The origin of mobile-terminal based machine learning (2) Mobile-terminal based machine learning technology (3) Application scenarios of mobile-terminal based machine learning (4) Limitations and future developments of mobile-terminal based machine learning

续表

章节序号 Chapter Number	章节名称 Chapters	课时 Class Hour	知识点 Key Points
5	移动边缘计算 Mobile Edge Computing(MEC)	2.5	(1) 移动边缘计算的理论基础 (2) 移动边缘计算与云计算 (3) 移动边缘计算的优势和挑战 (4) 移动边缘计算的关键技术和应用前景 (1) The theoretical foundation of MEC (2) MEC and cloud computing (3) The advantages and challenges for MEC (4) The key technologies and applications of MEC
6	感知计算融合及其应用 Perceptual computing fusion and its applications	1	(1) 感知计算融合的理论基础 (2) 感知计算融合的应用场景 (3) 感知计算融合面临的问题和挑战 (1) The theoretical foundations of perceptual computing fusion (2) The applications of perceptual computing fusion (3) The challenges for perceptual computing fusion

5.6.5 实验环节(Experiments)

序号 Num.	实验内容 Experiment Content	课时 Class Hour	知识点 Key Points
1	波束成型技术在目标感知中的应用 Object sensing using beam forming	2	(1) 波束成型技术 (2) 多用户多输入多输出 (3) 信道状态信息获取 (4) 基于信道状态信息的目标感知与识别方法 (1) Beam forming (2) Multi-user multiple input and muliple output (3) Channel State Information (CSI) (4) Object sensing based on CSI

续表

序号 Num.	实验内容 Experiment Content	课时 Class Hour	知　识　点 Key Points
2	无线传感器网络 感知数据获取 Sensing data collection using wireless sensor network	2	（1）TinyOS 原理 （2）传播和数据树获取协议 （1）Principle of TinyOS （2）Dissemination & Collection Tree Protocol(CTP)
3	基于物理层信息的 可信认证 Authentication using physical layer informaiton	2	（1）物理层信息 （2）设备认证 （3）密钥生成与分发 （1）Physical layer informaiton （2）Device authentication （3）The secret key generation and dissemination

大纲指导者：杨强教授（香港科技大学计算机科学与工程系）

大纲制定者：惠维副教授（西安交通大学计算机科学与技术学院）

大纲审定：西安交通大学人工智能学院本科专业知识体系建设与课程设置第二版修订工作组

第 6 章

"人工智能核心"课程群

6.1 "人工智能概论"课程大纲

课程名称：人工智能概论
Course：Introduction to Artificial Intelligence
先修课程：工科数学分析、线性代数与解析几何、计算机科学与人工智能的数学基础、概率统计与随机过程
Prerequisites：Mathematical Analysis for Engineering, Linear Algebra and Analytic Geometry, Math Foundation of Computer Science and Artificial Intelligence, Probability Theory and Stochastic Process
学分：2
Credits：2

6.1.1 课程目的和基本内容（Course Objectives and Basic Content）

本课程是人工智能学院本科专业必修课。

This course is a compulsory course for undergraduates in College of Artificial Intelligence.

人工智能是以人工的方法用机器（计算机）模拟、延伸和扩展人类智能的学科，具有多学科交叉、高度复杂、强渗透性的学科特点。本课程围绕人工智能的问题表达与求解，在介绍人工智能的基本概念和发展概况的基础上，着重讲授知识表示、搜索策略、推理方法等人工智能经典三大基本技术，涵盖经典知识表示、知识图谱、盲目搜索与启发式搜索、博弈搜索、确定性推理、不确定性推理和多智能体系统等内容。本课程与"机器学习"紧密联系，相互呼应。

本课程的目的是帮助学生了解人工智能的发展和应用，掌握人工智能的基本原理和方法，拓宽科学视野，提高利用人工智能技术解决实际问题的能力，为今后进一步学习人工智能相关课程奠定基础。

Artificial intelligence is a discipline that uses machines (computers) to simulate, extend and expand human intelligence in an artificial way. This course revolves around problem expressions and solutions for artificial intelligence. On the basis of introducing the basic concepts and development overview of artificial intelligence, it focuses on three basic technologies of artificial intelligence, including knowledge representation, search strategies, and reasoning methods, and it covers classical knowledge representations, knowledge graphs, blind and heuristic search strategies, adversarial search, deterministic reasoning, uncertainty reasoning, and multi-agent systems. This course is closely related to and echoes the course "Modern Approaches of Artificial Intelligence: Machine Learning".

This course aims to help students understand the development and application of artificial intelligence, master the basic principles and methods of artificial intelligence, broaden their scientific horizons, improve their ability to use artificial intelligence technology to solve practical problems, and lay a foundation for further study of artificial intelligence related courses in the future.

6.1.2 课程基本情况（Course Arrangements）

课程名称	人工智能概论 Introduction to Artificial Intelligence									
开课时间	一年级		二年级		三年级		四年级		必修（学分）	人工智能核心
^^	秋	春	秋	春	秋	春	秋	春	^^	人工智能概论(2)
^^									^^	机器学习(3)
课程定位	本科生"人工智能核心"课程群必修课									自然语言处理(2)
学　分	2学分									计算机视觉与模式识别(3)
总学时	32学时（授课32学时、实验0学时）									强化学习(2)
^^	^^									生成式AI与大语言模型(2)
授课学时分配	课堂讲授（32学时）								选修（学分）	/

续表

先修课程	工科数学分析、线性代数与解析几何、计算机科学与人工智能的数学基础、概率统计与随机过程
后续课程	机器学习
教学方式	课堂教学、作业、讨论、报告
考核方式	闭卷考试成绩占70%,作业、考勤等占30%
参考教材	Russell S,Norvig P. 人工智能:一种现代的方法[M]. 殷建平,祝恩,刘越,等译. 4版. 北京:人民邮电出版社,2022.
参考资料	1. Lucci S,Kopec D. 人工智能[M]. 林赐,译. 2版. 北京:人民邮电出版社,2018. 2. Luger G F. 人工智能:复杂问题求解的结构与策略[M]. 史忠植,张银奎,赵志崐,等译. 北京:机械工业出版社,2006. 3. 史忠植,王文杰,马慧芳. 人工智能导论[M]. 北京:机械工业出版社,2019. 4. 吴飞. 人工智能导论:模型与算法[M]. 北京:高等教育出版社,2020. 5. 刘若辰,慕彩红,焦李成,等. 人工智能导论[M]. 北京:清华大学出版社,2021.
其他信息	

6.1.3 教学目的和基本要求(Teaching Objectives and Basic Requirements)

(1) 熟悉人工智能的发展简况、主要学派、发展趋势和应用领域;

(2) 熟悉知识表示的相关概念、掌握谓词逻辑表示法、产生式表示法、语义网络表示法、框架表示法和状态空间表示法;

(3) 掌握知识图谱中本体知识表示、知识图谱逻辑结构、知识图谱关键技术和知识图谱典型应用;

(4) 熟悉搜索算法基础,掌握典型的搜索算法:盲目搜索算法、启发式搜索算法和蒙特卡洛树搜索算法等;

(5) 熟悉博弈搜索基本概念,掌握极小极大算法、α-β 搜索算法和纳什均衡;

(6) 熟悉推理基本概念,掌握推理逻辑基础、自然演绎推理、归纳演绎推理和归结演绎推理;

(7) 熟悉不确定性基本概念,掌握概率推理、主观贝叶斯方法、可信度方法、证据理论和模糊推理方法;

(8) 了解掌握智能体的基本概念、结构、环境任务、类型和应用等。

6.1.4 教学内容及安排(Syllabus and Arrangements)

第一章 绪论(Introduction)

章节序号 Chapter Number	章节名称 Chapters	课时 Class Hour	知 识 点 Key Points
1	绪论 Introduction	3	(1) 人工智能发展简况 (2) 人工智能主要学派 (3) 人工智能发展趋势 (4) 人工智能应用领域 (1) Brief history of artificial intelligence (2) Major paradigms of artificial intelligence (3) Development trends of artificial intelligence (4) Application fields of artificial intelligence

第二章 知识表示(Knowledge Representation)

章节序号 Chapter Number	章节名称 Chapters	课时 Class Hour	知 识 点 Key Points
2	知识表示 Knowledge representation	4	(1) 知识的相关概念 (2) 谓词逻辑表示法 (3) 产生式表示法 (4) 语义网络表示法 (5) 框架表示法 (6) 状态空间表示法 (1) Related concepts of knowledge (2) Predicate logic representation (3) Production rule (4) Semantic network representation (5) Frame representation (6) State space representation

第三章 知识图谱(Knowledge Graphs)

章节序号 Chapter Number	章节名称 Chapters	课时 Class Hour	知　识　点 Key Points
3	知识图谱 Knowledge graph	3	(1) 本体知识表示 (2) 知识图谱逻辑结构 (3) 知识图谱关键技术 (4) 知识图谱典型应用 (1) Ontology knowledge representation (2) Logical structure of knowledge graph (3) Key technologies of knowledge graph (4) Typical applications of knowledge graph

第四章 搜索求解(Solving Problems by Searching)

章节序号 Chapter Number	章节名称 Chapters	课时 Class Hour	知　识　点 Key Points
4	搜索问题求解 Solving problems by searching	4	(1) 搜索算法基础 (2) 盲目搜索算法 (3) 启发搜索算法 (4) 蒙特卡洛树搜索 (1) Search algorithm basics (2) Blind search algorithm (3) Heuristic search algorithm (4) Monte Carlo tree search

第五章 博弈搜索(Adversarial Search)

章节序号 Chapter Number	章节名称 Chapters	课时 Class Hour	知　识　点 Key Points
5	博弈搜索 Game search	4	(1) 博弈基本概念 (2) 极大极小搜索 (3) α-β 剪枝 (4) 纳什均衡 (1) Basic concept of game (2) Minimax search (3) α-β pruning (4) Nash equilibrium

第六章 确定性推理(Deterministic Reasoning)

章节序号 Chapter Number	章节名称 Chapters	课时 Class Hour	知 识 点 Key Points
6	确定性推理 Deterministic reasoning	6	(1) 推理基本概念 (2) 推理逻辑基础 (3) 自然演绎推理 (4) 归纳演绎推理 (5) 归结演绎推理 (1) Basic concept of reasoning (2) Fundamentals of logic reasoning (3) Natural deductive reasoning (4) Inductive deductive reasoning (5) Resolution deductive Reasoning

第七章 不确定性推理(Uncertainty Reasoning)

章节序号 Chapter Number	章节名称 Chapters	课时 Class Hour	知 识 点 Key Points
7	不确定性推理 Uncertainty reasoning	6	(1) 不确定性基本概念 (2) 概率推理 (3) 主观贝叶斯方法 (4) 可信度方法 (5) 证据理论 (6) 模糊推理方法 (1) Basic concept of uncertainty (2) Probabilistic reasoning (3) Subjective Bayesian method (4) Certainty method (5) Evidence theory (6) Fuzzy reasoning method

第八章　多智能体系统（Multi-agents System）

章节序号 Chapter Number	章节名称 Chapters	课时 Class Hour	知 识 点 Key Points
8	多智能体系统 Multi-agents system	2	（1）智能体基本概念 （2）智能体的结构 （3）智能体的环境任务 （4）智能体的类型 （5）多智能体的应用 (1) Basic concept for agent (2) Structure of agent (3) Task environments for agent (4) Types of agent (5) Multi-agent applications

大纲指导者：郑南宁教授（西安交通大学人工智能学院）

大纲制定者：辛景民教授（西安交通大学人工智能学院）、武佳懿助理教授（西安交通大学人工智能学院）

大纲审定：西安交通大学人工智能学院本科专业知识体系建设与课程设置第二版修订工作组

6.2　"机器学习"课程大纲

课程名称：机器学习

Course：Machine Learning

先修课程：工科数学分析、线性代数与解析几何、概率统计与随机过程、计算机科学与人工智能的数学基础、人工智能概论

Prerequisites：Mathematical Analysis for Engineering，Linear Algebra and Analytic Geometry，Probability Theory and Stochastic Process，Math Foundation of Computer Science and Artificial Intelligence，Introduction to Artificial Intelligence

学分：3

Credits：3

6.2.1 课程目的和基本内容（Course Objectives and Basic Content）

本课程是人工智能本科专业必修课。

This course is a compulsory course for undergraduates in College of Artificial Intelligence.

机器学习是一门广泛应用于科学、工程、金融等领域的学科，特别是在人工智能和机器人领域发挥着核心和关键作用。该学科的目的是设计和发展相关模型、算法从数据和经验中学习知识，进而应用于问题的分析、预测和解决。本课程将系统介绍机器学习的概念、思想、技术和方法，具体内容包括机器学习的基础概念和理论、概率、熵与信息、概率图模型、核方法与支持向量机、马尔可夫与隐马尔可夫模型、神经网络与深度学习、生成方法、概率采样等。

课程目标旨在帮助学生掌握机器学习的基本概念、思想和研究方法，了解未来的发展趋势，为今后在该领域的深入研究打下基础。通过本课程的学习，使学生初步具备使用机器学习方法解决实际问题的能力，为今后进一步的学习和工作奠定基础。

Machine learning is a discipline widely used in science, engineering, finance, etc., especially in the field of artificial intelligence and robotics. The purpose of the discipline is to design and develop relevant models and algorithms to learn from data and experience, and then apply them to problem analysis, prediction and solving. This course will introduce the concepts, ideas, techniques and methods of machine learning, such as the basic concepts and theories of machine learning, probability, entropy and information, probabilistic graphical model, kernel method and support vector machine, Markov random field and hidden Markov model, neural network and deep learning, generative method, and probability sampling.

The course aims to help students master the basic concepts, ideas, and research methods of machine learning, understand future development trends, and lay the foundation for the further research in this field. Through the study of this course, students will initially master the ability to use machine learning methods to solve practical problems, laying the foundation for the further study and work in the future.

6.2.2 课程基本情况(Course Arrangements)

课程名称	机器学习 Machine Learning							
开课时间	一年级		二年级		三年级		四年级		人工智能核心
^	秋	春	秋	春	秋	春	秋	春
课程定位	本科生"人工智能核心"课程群必修课							必修 (学分)
学分	3学分						
总学时	48学时(授课48学时、实验0学时)						
授课学时分配	课堂讲授(48学时)							选修 (学分)
先修课程	工科数学分析、线性代数与解析几何、概率统计与随机过程、计算机科学与人工智能的数学基础、人工智能概论							
后续课程	自然语言处理、计算机视觉与模式识别、强化学习							
教学方式	课堂教学、大作业与实验、小组讨论、综述报告							
考核方式	闭卷考试成绩占70%,平时成绩(考勤、作业)占30%							
参考教材	周志华.机器学习[M].北京:清华大学出版社,2016.							
参考资料	1. Bishop C M. Pattern Recognition and Machine Learning[M]. Berlin:Springer,2009. 2. Goodfellow I,Bengio Y,Courville A. 深度学习[M].北京:人民邮电出版社,2017. 3. Koller D,Friedman N. 概率图模型——原理与技术[M].北京:清华大学出版社,2015. 4. Barbu A,Zhu S-C. Monte Carlo Methods[M]. Berlin:Springer,2020.							
其他信息								

人工智能核心必修(学分):
- 人工智能概论(2)
- 机器学习(3)
- 自然语言处理(2)
- 计算机视觉与模式识别(3)
- 强化学习(2)
- 生成式AI与大语言模型(2)

选修(学分): /

6.2.3 教学目的和基本要求(Teaching Objectives and Basic Requirements)

(1) 理解机器学习的基本概念,掌握概率、熵与信息基本概念;
(2) 掌握线性回归和线性分类模型;
(3) 掌握随机图模型与贝叶斯分类,理解马尔可夫随机场与隐马尔可夫模型;
(4) 掌握核方法与支撑向量机;
(5) 掌握集成学习与随机森林;
(6) 理解无监督学习与聚类;
(7) 理解稀疏学习与压缩感知;
(8) 理解深度学习与神经网络基本概念;

(9) 掌握卷积神经网络、循环神经网络、长短记忆网络与注意力模型；

(10) 理解生成方法的概念和生成对抗思想；

(11) 掌握概率采样与蒙特卡洛方法。

6.2.4　教学内容及安排(Syllabus and Arrangements)

第一章　机器学习的基础概念和理论(Basic Concepts and Theories of Machine Learning)

章节序号 Chapter Number	章节名称 Chapters	课时 Class Hour	知　识　点 Key Points
1.1	机器学习概述 Overview of machine learning	1	(1) 机器学习的定义和任务 (2) 机器学习的历史和现状 (3) 机器学习的典型应用 (1) The definition and task of machine learning (2) History and current status of machine learning (3) Typical applications of machine learning
1.2	机器学习算法的一般准则 General principles of machine learning algorithms	1	(1) 性能度量和准则 (2) 最大似然估计 (3) 过拟合与欠拟合 (1) Performance evaluation measurement and principle (2) Maximum likelihood estimation (3) Overfitting and underfitting
1.3	概率与信息基本概念 Basic concepts of probability and information	2	(1) 概率及其定理 (2) 信息与熵 (1) Probability and its theorems (2) Information and entropy

第二章　线性模型(Linear Models)

章节序号 Chapter Number	章节名称 Chapters	课时 Class Hour	知　识　点 Key Points
2.1	线性回归模型 Linear regression model	2	(1) 线性回归基本形式 (2) 线性/非线性基函数 (3) 多输出模型 (1) Basic forms of linear regression model (2) Linear / Non-linear basis functions (3) Multiple output model

续表

章节序号 Chapter Number	章节名称 Chapters	课时 Class Hour	知识点 Key Points
2.2	线性分类模型 Linear classification model	3	(1) 判别函数模型 (2) 概率判别模型 (3) 概率生成模型 (1) Discriminantfunction model (2) Probabilistic generative model (3) Probabilistic discriminative model

第三章 核方法与支撑向量机（Kernel Methods and Support Vector Machines）

章节序号 Chapter Number	章节名称 Chapters	课时 Class Hour	知识点 Key Points
3.1	支撑向量机 Support Vector Machine(SVM)	3	(1) 最大间隔分类 (2) 对偶问题 (3) 软间隔与松弛变量 (4) 回归问题支撑向量机 (1) Max-margin classification (2) Dual problem (3) Soft margin and slack variable (4) SVM for regression
3.2	核方法 Kernel methods	1	(1) 特征空间 (2) 核和构建核 (3) 核替换 (1) Feature space (2) Kernel and constructing kernel (3) Kernel substitution

第四章　概率图模型(Probabilistic Graphical Model)

章节序号 Chapter Number	章节名称 Chapters	课时 Class Hour	知　识　点 Key Points
4.1	概率图模型 Probabilistic graphical model	3	(1) 概率图模型的概念 (2) 有向图、无向图的概率表达 (3) 条件独立性及其判断方法 (4) 图模型推理的概念 (1) The concept of probabilistic graphical model (2) Probabilistic representation of directed graph and undirected graph (3) Conditional independence and its judgement (4) Inference in graphical model
4.2	马尔可夫随机场 Markov random field	1	(1) 随机场的概念 (2) 马尔可夫随机场定义 (3) 马尔可夫随机场的概率表达 (1) The concept of random field (2) The definition of Markov random field (3) Probabilistic representation of Markov random field
4.3	隐马尔可夫随机场 Hidden Markov random field	2	(1) 序列与隐状态 (2) 隐马尔可夫模型 (3) 维特比算法 (1) Sequence and hidden state (2) Hidden Markov model (3) Viterbi algorithm

第五章　集成学习(Ensemble learning)

章节序号 Chapter Number	章节名称 Chapters	课时 Class Hour	知　识　点 Key Points
5.1	集成学习的概念 Concepts of ensemble learning	1	(1) 集成学习的概念与特征 (2) 集成学习的框架和分类 (1) The concepts and characteristics of ensemble learning (2) The framework and categorization of esemble learning

章节序号 Chapter Number	章节名称 Chapters	课时 Class Hour	知 识 点 Key Points
5.2	Boosting 方法 Boosting method	2	(1) Boosting 基本框架 (2) AdaBoost 算法 (1) Boosting framework (2) AdaBoost method
5.3	随机森林方法 Random forest method	2	(1) 随机森林建模方法 (2) 随机森林分类器 (1) Random forest modeling method (2) Random forest classifier

第六章　神经网络与深度学习(Neural Networks and Deep Learning)

章节序号 Chapter Number	章节名称 Chapters	课时 Class Hour	知 识 点 Key Points
6.1	深度学习与神经网络基本概念 Basic concepts of deep learning and neural network	1	(1) 深度学习的基本概念 (2) 深度学习的历史与现状 (1) The basic concepts of deep learning (2) History and current status of deep learning
6.2	深度前向网络 Deep forward network	1	(1) 深度前向网络的结构 (2) 前向网络的函数拟合方法与定理 (1) Architecture of deep forward network (2) Function fitting method and theorem of forward network
6.3	卷积神经网络 Convolutional neural network	2	(1) 卷积 (2) 池化 (3) 激励函数 (4) 卷积神经网络计算 (1) Convolution (2) Pooling (3) Activation function (4) Convolutional neural network

续表

章节序号 Chapter Number	章节名称 Chapters	课时 Class Hour	知 识 点 Key Points
6.4	循环神经网络概念 The concepts of recurrent neural network	1	（1）循环神经网络的基本概念 （2）循环神经网络的结构 （1）The basic concepts of recurrent neural network （2）Architecture of recurrent neural network
6.5	长短记忆网络 Long Short-Term Memory(LSTM) network	2	（1）长短记忆网络的结构 （2）长短记忆网络的计算 （3）长短记忆网络建模 （1）Architecture of LSTM network （2）Computation of LSTM network （3）LSTM modeling
6.6	注意力模型 Attention Model	2	（1）注意力的概念 （2）自注意力模型 （3）交叉注意力模型 （1）Concepts of attention （2）Self-attention model （3）Cross-attention model

第七章　无监督学习（Unsupervised Learning）

章节序号 Chapter Number	章节名称 Chapters	课时 Class Hour	知 识 点 Key Points
7.1	无监督方法 Unsupervised method	1	（1）无监督聚类的概念 （2）无监督聚类的基本方法 （1）The concepts of unsupervised clustering （2）Basic methods of unsupervised clustering
7.2	聚类方法 Clustering method	2	（1）K-means 聚类 （2）密度聚类 （3）层次聚类 （1）K-means clustering （2）Density-based clustering （3）Hierarchical clustering

第八章 压缩感知与稀疏性(Compressed Sensing and Sparsity)

章节序号 Chapter Number	章节名称 Chapters	课时 Class Hour	知 识 点 Key Points
8.1	压缩感知 Compressed sensing	1	(1) 压缩感知的概念 (2) 压缩感知的分类 (1) The concept of compressed sensing (2) Categories of compressed sensing
8.2	稀疏模型 Sparse model	2	(1) 稀疏性概念与L_1范数 (2) 稀疏学习方法 (1) The concept of sparsity and the L_1 norm (2) Sparse learning method

第九章 生成方法(Generative Methods)

章节序号 Chapter Number	章节名称 Chapters	课时 Class Hour	知 识 点 Key Points
9.1	生成方法概述 Overview of generative method	1	(1) 生成方法的概念 (2) 生成方法的基本思想 (1) The concept of generative method (2) The basic idea of generative method
9.2	自编码器和自编码模型 Autoencoder and autoencoder model	1	(1) 编码与解码 (2) 自编码神经网络结构 (1) Encoding and decoding (2) Autoencoder neural network
9.3	生成对抗网络 Generative adversarial network	2	(1) 生成对抗思想与概念 (2) 生成对抗网络 (1) Ideas and concepts of generative adversarial (2) Generative adversarial network

第十章 蒙特卡洛方法（Monte Carlo Methods）

章节序号 Chapter Number	章节名称 Chapters	课时 Class Hour	知 识 点 Key Points
10.1	概率采样方法 Probability sampling	1	（1）概率采样的概念 （2）概率采样的基本思想 （3）蒙特卡洛算法基本流程 （1）The concept of probability sampling （2）The basic idea of probability sampling （3）The general process of Monte Carlo algorithm
10.2	基本采样方法 Basic sampling methods	4	（1）均匀分布采样 （2）逆变换采样 （3）拒绝采样 （4）重要性采样 （5）Metropolis-Hasting 方法 （6）吉布斯采样 （1）Uniform distribution sampling （2）Inverse transform sampling （3）Rejection sampling （4）Importance sampling （5）Metropolis-Hasting method （6）Gibbs sampling

大纲指导者：郑南宁教授（西安交通大学人工智能学院）

大纲制定者：魏平教授（西安交通大学人工智能学院）

大纲审定：西安交通大学人工智能学院本科专业知识体系建设与课程设置第二版修订工作组

6.3 "自然语言处理"课程大纲

课程名称：自然语言处理

Course：Natural Language Processing

先修课程：线性代数与解析几何、概率统计与随机过程

Prerequisites：Linear Algebra and Analytic Geometry，Probability Theory and Stochastic Process

学分：2

Credits：2

6.3.1 课程目的和基本内容（Course Objectives and Basic Content）

本课程是人工智能学院本科专业必修课。

This course is a compulsory course for undergraduates in College of Artificial Intelligence.

语言是人类认识世界的手段，也是人类认知的成果。通过分析人类的语言，可以在一定程度上了解人类认知的规律。另外，如何通过计算机科学和统计方法的手段，研究自然语言理解和生成也是 AI 领域的重要挑战之一。本课程以自然语言处理（Natural Language Processing，NLP）的基础知识与技术、认知科学与语言的理论为主线，对语言模型、词法和句法分析的基本概念与关键技术进行介绍，同时讨论机器翻译以及文本分类等的主要 NLP 技术应用方向，最后从哲学、心理学与认知神经学的不同角度，介绍认知与语言的主要理论与发展。第一章到第五章分别讨论文本预处理技术、语言模型、词性标注、句法分析与语义分析，这部分内容的重点是语言模型、词法和句法分析方法。第六章到第九章主要讨论文本分类和聚类、统计机器翻译、信息检索、推荐系统和情感分析的基本 NLP 应用，第十章主要介绍认知科学和语言的理论基础以及相关的认知语言学知识。

通过对自然语言的基本思想和关键技术的学习，帮助学生建立关于自然语言处理、认知与语言的基础知识框架。课程采用小组学习模式，并辅之以研究性实验、课堂测验、小组讨论及实验报告等教学手段，训练学生用基本理论和方法分析解决实际问题的能力，掌握认知与语言、自然语言处理所必需的基本知识和技能。课程通过机器翻译和文本分类等基本 NLP 系统设计实验使学生巩固和加深自然语言处理的理论知识，通过实践进一步加强学生独立分析问题、解决问题的能力，培养综合设计及创新能力，培养实事求是、严肃认真的科学作风和良好的实验习惯，为今后的工作打下良好的基础。

Language is a method to understand the real world，and also a result from human cognition activity. Through analysis of natural language，it is possible to know the cognitive rules of human. Moreover，it is also a challenge to understand and generate natural language using the computer science and statistical method. This course focuses on basic knowledge and techniques of natural language processing and the principles of cognitive science and languages. Moreover, it introduces the language model, lexical and syntax analysis, and discusses the NLP applications such as machine translation, text classification and finally introduces the basic principles of cognitive science and linguistics from philosophical, psychological and cognitive neuroscience perspectives. Chapter 1 to Chapter 5 discuss text preprocessing, language model, Part Of Speech(POS) tagging, syntax analysis and semantic anglysis separately. Chapter 6 to Chapter 9 mainly focus text classification and clustering, statistical machine translation, information retrieval, recommendation system and sentiment analysis. Chapter 10 introduces the basic principles of cognitive science and language, and basic properties of cognitive linguistics.

This course helps students build a knowledge framework for the basic principles of natural language processing, cognitive science and language through the study of basic theories, design methods, and applied techniques. The course adopts group study method, supplemented by experiments, in-class tests, discussions and reports, in order to train students the ability to solve practical problems with basic theories and methods and master the basic knowledge and skills for natural language processing application design. The course includes several experiments on machine translation and text classification in order to consolidate the students'theoretical knowledge of natural language processing, further strengthen their ability to analyze and solve problems independently, and develop their comprehensive abilities on system design and innovations as well as good habits for future work.

6.3.2 课程基本情况（Course Arrangements）

课程名称	自然语言处理 Natural Language Processing									
开课时间	一年级		二年级		三年级		四年级		\multicolumn{3}{c	}{人工智能核心}
^	秋	春	秋	春	秋	春	秋	春	\multicolumn{3}{c	}{}
课程定位	\multicolumn{8}{c	}{本科生"人工智能核心"课程群必修课}	必修 (学分)	人工智能概论(2)
学　　分	\multicolumn{8}{c	}{2 学分}	^	机器学习(3)
^	^	^	^	^	^	^	^	^	^	自然语言处理(2)
总 学 时	\multicolumn{8}{c	}{40 学时 （授课32学时、实验8学时）}	^	计算机视觉与模式识别(3)
^	^	^	^	^	^	^	^	^	^	强化学习(2)
^	^	^	^	^	^	^	^	^	^	生成式AI与大语言模型(2)
授课学时 分配	\multicolumn{8}{c	}{课堂授课(28学时)， 大作业讨论(4学时)}	选修 (学分)	/

续表

先修课程	线性代数与解析几何、概率统计与随机过程
后续课程	强化学习、生成式 AI 与大语言模型
教学方式	课堂教学、大作业与实验、小组讨论
考核方式	课程结束笔试成绩占 60%，平均成绩占 15%，实验成绩占 20%，考勤占 5%
参考教材	1. 宗成庆. 统计自然语言处理[M]. 北京：清华大学出版社，2013. 2. Jurafsky D, Martin J H. 自然语言处理综论[M]. 冯志伟，译. 北京：电子工业出版社，2018.
参考资料	1. 戴维·凯默勒. 语言的认知神经科学[M]. 王穗苹，周晓林，译. 杭州：浙江教育出版社，2017. 2. 列夫·维果茨基. 思维与语言[M]. 李维，译. 北京：北京大学出版社，2010.
其他信息	

6.3.3 教学目的和基本要求（Teaching Objectives and Basic Requirements）

（1）了解自然语言处理的基本概念及其基本工具和方法，掌握自然语言处理的基本技术和方法；

（2）理解经典语言模型、神经网络的语言模型与大语言模型；

（3）熟悉词法分析和序列标注的经典算法；

（4）熟悉基于短语和依存构架的句法分析方法；

（5）了解语义分析的基本概念和方法；

（6）了解文本分类和聚类的基本方法；

（7）了解统计机器翻译的基本方法；

（8）了解信息检索处理方法和推荐系统的经典算法；

（9）了解情感分析的基本处理方法；

（10）熟悉使用 Python 语言或其他高级语言进行自然语言处理算法的实现；

（11）了解认知科学与语言的哲学基础和心理学基础，了解认知神经科学和语言之间的关系，了解认知语言学的相关知识。

6.3.4　教学内容及安排(Syllabus and Arrangements)

绪论(Introduction)

章节序号 Chapter Number	章节名称 Chapters	课时 Class Hour	知　识　点 Key Points
0.1	绪论 Introduction	1	(1) 自然语言处理的基本术语 (2) 自然语言处理的基本概念 (1) Basic terminologies of natural language processing (2) General principles of natural language processing

第一章　文本预处理技术(Text Preprocessing)

章节序号 Chapter Number	章节名称 Chapters	课时 Class Hour	知　识　点 Key Points
1.1	文本正则化 Text regularization		(1) 文本正则化定义 (2) 文本正则化表达 (1) Definition of text regularization (2) Text regular representation
1.2	文本形符化 Text tokenization	1	(1) 文本形符化的定义 (2) 文本形符化的基本方法 (1) Definition of text tokenization (2) Basic methods of text tokenization
1.3	文本编辑距离 Text edit distance		(1) 文本编辑距离的定义 (2) 文本编辑距离的计算方法 (1) Definition of text edit distance (2) Computing method of text edit distance

第二章　语言模型(Language Model)

章节序号 Chapter Number	章节名称 Chapters	课时 Class Hour	知　识　点 Key Points
2.1	概率语言模型 Probabilistic language model		(1) 语言模型的定义 (2) 概率语言模型的定义 (3) 概率语言模型的性质 (1) Definition of language model (2) Definition of probabilistic language model (3) The properties of probabilistic language model
2.2	n-gram 语言模型 n-gram language model	2	(1) n-gram 语言模型的定义 (2) n-gram 语言模型的性质 (3) n-gram 语言模型参数估计 (1) Definition of n-gram language model (2) The properties of n-gram language model (3) Parameter estimation of n-gram language model
2.3	数据稀疏 Data sparsity		(1) 数据稀疏的定义 (2) 数据平滑的基本方法 (1) Definition of data sparsity (2) Basic methods of data smoothing
2.4	基于神经网络的语言模型与大语言模型 Language models based on neural network LLM	3	(1) 词向量表示 (2) 基于前馈神经网络的语言模型 (3) 大语言模型与 RLHF (1) Word vector representation (2) Language models based on feedforward neural network (3) LLM and RLHF

第三章　词法分析(Lexical Analysis)

章节序号 Chapter Number	章节名称 Chapters	课时 Class Hour	知　识　点 Key Points
3.1	词法分析 Lexical analysis	4	（1）形态分析 （2）中文分词 (1) Morphological Analysis (2) Chinese Segmentation
3.2	词性标注 POS tagging		（1）词性的基本定义 （2）词性标注的基本方法 (1) Definition of POS (2) Basic method of POS tagging
3.3	序列标注 Sequence tagging		（1）命名实体识别的标注方法 （2）分词的标注方法 (1) Tagging for named entity recognition (2) Tagging for word segmentation

第四章　句法分析(Syntax Analysis)

章节序号 Chapter Number	章节名称 Chapters	课时 Class Hour	知　识　点 Key Points
4.1	句法分析 Syntax analysis	4	（1）句法分析的概念 （2）基于短语的句法分析的概念和性质 （3）基于依存语法句法分析的概念及性质 (1) Definition of syntax analysis (2) Definition and properties of phrase based syntax analysis (3) Definition and properties of dependency grammar analysis
4.2	句法分析的方法 Basic methods of syntax analysis		（1）形式语法的基本定义 （2）CKY句法分析的方法 （3）概率上下文无法语法的定义 （4）句法树构建的基本方法 (1) Definition of formal grammar (2) The CKY method of syntax analysis (3) Definition of Probabilistic Context Free Grammar(PCFG) (4) The PCFG method of syntax tree building

第五章　语义分析(Semantic Analysis)

章节序号 Chapter Number	章节名称 Chapters	课时 Class Hour	知　识　点 Key Points
5.1	语义理论 Semantic theory	2	(1) 语义计算的基本概念 (2) 语义分析的常用理论 (1) Basic principle of semantic computing (2) General theories of semantic analysis
5.2	语义分析 Semantic analysis		(1) 词的表示 (2) 词义消歧 (3) 语义角色标注 (4) 句义分析 (1) Representation of word (2) Word sense disambiguation (3) Semantic role labeling (4) Sentence analysis

第六章　文本分类与聚类(Text Classification and Clustering)

章节序号 Chapter Number	章节名称 Chapters	课时 Class Hour	知　识　点 Key Points
6.1	文本分类 Text classification	1	(1) 文本分类的基本概念 (2) 文档空间的距离测度 (3) 基于朴素贝叶斯模型的文本分类 (4) 文本分类模型的评价方法 (1) Basic concept of text classification (2) Distance metric of document space (3) Text classification of naive Bayes model (4) Evaluations of text classification model
6.2	文本聚类 Text clustering	1	(1) 文本聚类的基本概念 (2) 文本聚类的基本方法 (1) Basic principle of text clustering (2) Basic methods of text clustering

第七章 统计机器翻译(Statistics Machine Translation)

章节序号 Chapter Number	章节名称 Chapters	课时 Class Hour	知识点 Key Points
7.1	机器翻译概述 An introduction of machine translation	2	(1) 机器翻译的基本思想 (2) 信道噪声模型的基本概念 (3) 基于词汇的机器翻译模型 (4) 基于短语的机器翻译模型 (1) Basic principles of machine translation (2) Basic principle of channel noisy model (3) Machine translation model based on lexical (4) Machine translation model based on phrase
7.2	基于神经网络的机器翻译模型概述 An introduction of machine translation based on neural network model	1	(1) 基于神经网络的机器翻译模型 (2) 基于循环神经网络的机器翻译模型 (1) Machine translation model based on neural network (2) Machine translation model based on Recurrent Neural Network(RNN)

第八章 信息检索与推荐系统(Information Retrieval and Recommendation System)

章节序号 Chapter Number	章节名称 Chapters	课时 Class Hour	知识点 Key Points
8.1	信息检索 Information Retrieval(IR)	2	(1) 信息检索的基本定义 (2) 信息检索的基本方法 (1) Definition of IR (2) Basic methods of IR
8.2	推荐系统 Recommendation System(RS)	2	(1) 推荐系统的基本概念 (2) 推荐系统的基本方法 (3) 基于协同滤波的推荐系统 (4) 基于隐藏语义的推荐系统 (1) Definition of RS (2) Basic methods of RS (3) Collaborative filtering method of RS (4) Latent factor method of RS

第九章 情感分析(Sentiment Analysis)

章节序号 Chapter Number	章节名称 Chapters	课时 Class Hour	知识点 Key Points
9.1	情感分析 Sentiment analysis	2	(1) 情感分析的基本概念 (2) 情感分析的基本性质 (3) 情感词典的定义 (4) 情感词典的构建方法 (5) 情感分析的基本方法 (1) Basic principles of sentiment analysis (2) Basic properties of sentiment analysis (3) Definition of emotion lexicon (4) Construction method of emotion lexicon (5) Basic methods of sentiment analysis

第十章 认知与语言(Cognition and Language)

章节序号 Chapter Number	章节名称 Chapters	课时 Class Hour	知识点 Key Points
10.1	认知科学与 语言基础 An introduction of cognitive science and language	1	(1) 认知科学与语言的理论基础 (2) 认知科学与语言的历史 (3) 认知神经学与语言的理论基础 (1) Basic theories of cognitive science and language (2) Histories of cognitive science and language (3) Basic theories of cognitive neuroscience and language
10.2	认知语言学基础 An introduction of cognitive linguistics	1	(1) 认知语言学的基本概念 (2) 语言习得的基本概念与方法 (1) Basic principles of cognitive linguistics (2) Basic principles and methods of language learning

6.3.5 实验环节(Experiments)

序号 Num.	实验内容 Experiment Content	课时 Class Hour	知 识 点 Key Points
1	基于最小编辑距离的拼写检查 Spell check based on minimum edit distance	2	(1) 文本预处理的方法 (2) 最小编辑距离的计算方法 (3) 英文单词的拼写检查方法 (1) Text preprocessing method (2) Minimum edit distance method (3) Spell check for English
2	文本分类/聚类系统设计与实现 Design and implementation of text classification/clustering system	2	(1) 文本向量空间的表达方法 (2) 文本向量空间距离的计算方法 (3) 基于朴素贝叶斯的分类方法 (4) 基于K均值的文本聚类方法 (5) 文本分类与聚类的评价方法 (1) Representation of text vector space (2) Distance metric of text vector space (3) Classification method of naive Bayes (4) Clustering method of K-means (5) Evaluation methods of classification and clustering
3	基于概率上下文无关语法的短语句法分析 Syntax analysis based on Probabilistic Context Free Grammar(PCFG)	2	(1) 概率上下文无关语法的表示方法 (2) 基于短语结构的句法树 (3) CYK的句法分析方法 (4) 基于PCFG的句法分析方法 (1) Representation of probabilistic context free grammar (2) Phrase based grammar tree (3) CYK method of syntax analysis (4) PCFG method of syntax analysis
4	基于RNN的机器翻译(中英/中日) RNN based machine translation (Chinese2English/Chinese2Japanese)	2	(1) 双语语料预处理方法 (2) RNN的参数选择方法 (3) 机器翻译的评价方法 (1) Preprocessing methods of parallel corpus (2) Parameter selection methods of RNN (3) Evaluation method of machine translation

大纲指导者：宗成庆研究员(中国科学院自动化研究所)、辛景民教授(西安交通大学人工智能学院)

大纲制定者：辛景民教授（西安交通大学人工智能学院）、姜沛林副教授（西安交通大学人工智能学院）

大纲审定：西安交通大学人工智能学院本科专业知识体系建设与课程设置第二版修订工作组

6.4 "计算机视觉与模式识别"课程大纲

课程名称：计算机视觉与模式识别

Course：Computer Vision and Pattern Recognition

先修课程：工科数学分析、线性代数与解析几何、概率统计与随机过程

Prerequisites：Mathematical Analysis for Engineering, Linear Algebra and Analytic Geometry, Probability Theory and Stochastic Process

学分：3

Credits：3

6.4.1 课程目的和基本内容（Course Objectives and Basic Content）

本课程是人工智能学院本科专业必修课。

This course is a compulsory course for undergraduates in College of Artificial Intelligence.

课程涉及视觉认知神经科学、计算机科学、信号处理与数学等多种学科。课程将结合具体的视觉任务与视觉应用系统，讲解视觉模式识别的基本方法。通过课程学习，学生可了解计算机视觉的发展与应用，熟悉计算机视觉与模式识别相关基础理论，掌握相机模型与标定、立体匹配与运动估计、特征检测与匹配、图像分割与视觉识别的基本方法。通过视觉任务实践环节，培养学生对视觉数据的洞察力与解决视觉问题的创新能力。

课程第一章讲解相机模型与标定方法、颜色视觉与图像模型；第二章和第三章论述立体视觉与运动视觉的相关机理，讲解立体匹配与运动估计的基本方法；第四章论述图像结构检测、表示与匹配的基本方法；第五章讲述视觉感知组织的相关机理、感知聚类与图像分割算法；第六章讲述视觉认知机理，讲解视觉模式识别的传统方法与深度学习方法；第七章结合具体应用系统，介绍计算机视觉系统结构与功能设计。

视觉是现代人工智能系统与机器人系统不可或缺的重要组成部分,视觉技术与视觉应用日新月异。如何从人类视觉感知与认知机理中获得灵感,利用现代机器学习与物理实现方法,构建有效的视觉计算模型与视觉系统,是未来人工智能技术发展的重要方向。了解视觉认知机理,理解与掌握计算机视觉与模式识别的基本概念、理论、模型与算法,将有助于学生在计算机视觉领域开展创新性研究,以及应对未来复杂视觉数据与视觉任务带来的挑战。

The course involves visual cognitive neuroscience, computer science, signal processing, mathematics and other disciplines. It will explain the basic methods of visual pattern recognition in combination with specific visual tasks and visual application systems. Through the course study, students can understand the development and application of computer vision, be familiar with the basic theories of computer vision and pattern recognition, and master the basic methods of camera model and calibration, stereo matching and motion estimation, feature detection and matching, image segmentation and visual recognition. Through the practice of visual tasks, cultivate students' insight into visual data and innovative ability to solve visual problems.

Chapter 1 explains camera model and calibration method, color vision and image model. Chapter 2 to Chapter 3 discuss the relevant mechanism of stereo vision and motion vision, and explains the basic methods of stereo matching and motion estimation. Chapter 4 discusses the basic methods of image structure detection, representation and matching. Chapter 5 describes the relevant mechanism of visual perception organization, perception clustering and image segmentation algorithm. Chapter 6 describes the mechanism of visual cognition, explains the traditional methods and deep learning methods of visual pattern recognition. Chapter 7 introduces the structure and function design of computer vision system combined with specific application system.

Vision, as an indispensable component of modern artificial intelligent and robotic system, is evolving quickly in terms of its research and application. Topics on how to scoop inspiration from human's visual perception and cognition, and how to build effective visual computing models based on contemporary machine learning and physics techniques are going to play an important role in the artificial intelligence advancement. Getting to know visual cognitive mechanism, and understanding the fundamental concepts, theories, models and algorithms of computer vision and pattern recognition can help students conduct innovative research in this field, and in the meantime, tackle the future challenges of complex visual data and visual tasks.

6.4.2 课程基本情况(Course Arrangements)

课程名称	计算机视觉与模式识别 Computer Vision and Pattern Recognition			
开课时间	一年级 / 二年级 / 三年级 / 四年级 秋 春 秋 春 **秋** 春 秋 春		人工智能核心	
课程定位	本科生"人工智能核心"课程群必修课	必修 (学分)	人工智能概论(2)	
			机器学习(3)	
			自然语言处理(2)	
学 分	3学分		计算机视觉与模式识别(3)	
总学时	56学时 (授课48学时,实验8学时)		强化学习(2)	
			生成式AI与大语言模型(2)	
授课学时分配	课堂讲授(46学时), 大作业讨论(2学时)	选修 (学分)	/	
先修课程	工科数学分析、线性代数与解析几何、概率统计与随机过程			
后续课程				
教学方式	课堂教学、大作业与实验、小组讨论、综述报告			
考核方式	闭卷考试成绩占50%,平时成绩占15%,实验成绩占15%,调研综述报告占15%,考勤占5%			
参考教材	Forsyth D A, Ponce J. Computer Vision: A Modern Approach[M]. 北京: 清华大学出版社,2004.			
参考资料	1. 郑南宁. 计算机视觉与模式识别[M]. 北京: 国防工业出版社,1998. 2. Marr D. Vision: A Computational Investigation into the Human Representation and Processing of Visual Information[M]. San Francisco: W. H. Freeman and company,1982. 3. Palmer S E. Vision Science: Photons to Phenomenology[M]. Cambridge: MIT Press,1999. 4. Hartley R, Zisserman A. Multiple View Geometry in Computer Vision[M]. Cambridge: Cambridge University Press,2003. 5. Duda R O, Hart P E, Stork D G. Pattern Classification[M]. New York: Wiley-Interscience,2001. 6. Goodfellow I, Bengio Y, Courville A. Deep Learning[M]. 北京: 机械工业出版社,2018.			
其他信息				

6.4.3 教学目的和基本要求(Teaching Objectives and Basic Requirements)

(1) 了解计算机视觉的发展与应用及视觉认知的相关机理；
(2) 掌握相机模型与标定方法；
(3) 熟悉图像的光照模型与颜色模型；
(4) 熟悉立体视觉原理、掌握立体匹配的基本方法；
(5) 了解运动场的基本特点、掌握光流估计的基本方法；
(6) 掌握图像特征点检测与匹配的基本方法；
(7) 了解视觉感知组织机理、掌握图像分割的基本方法；
(8) 理解视觉模式识别的传统方法和深度学习方法；
(9) 了解现代计算机视觉的主要应用场景和相关算法；
(10) 熟悉计算机视觉相关软件工具。

6.4.4 教学内容及安排(Syllabus and Arrangements)

绪论(Introduction)

章节序号 Chapter Number	章节名称 Chapters	课时 Class Hour	知 识 点 Key Points
0.1	绪论 Introduction	2	(1) 计算机视觉发展 (2) 计算机视觉系统构成 (3) 计算机视觉应用 (4) 计算机视觉软件工具 (1) The development of computer vision (2) Computer vision system composition (3) The applications of computer vision (4) Software and tools of computer vision

第一章 图像形成与图像模型(Image Formation and Image Model)

章节序号 Chapter Number	章节名称 Chapters	课时 Class Hour	知　识　点 Key Points
1.1	相机模型 Camera model	2	(1) 图像传感器 (2) 针孔相机 (3) 透视投影与仿射投影 (4) 相机模型与相机参数 (1) Image sensor (2) Pinhole camera (3) Perspective projection and affine projection (4) Camera model and camera parameter
1.2	相机标定 Camera calibration	2	(1) 最小二乘方法 (2) 相机标定的线性方法 (3) 相机畸变参数估计 (4) 奇异值分解 (1) Least-squares method (2) Linear approach to camera calibration (3) Distortion parameter estimation (4) Singular Value Decomposition(SVD)
1.3	图像模型 Image model	2	(1) 颜色视觉(色觉) (2) 光辐射测度 (3) 光源与光照模型 (4) 色彩与颜色模型 (5) 本征图像分解 (1) Color vision (2) Radiometry (3) Light source and shading model (4) Color and color model (5) Intrinsic image decomposition

第二章　立体视觉(Stereo Vision)

章节序号 Chapter Number	章节名称 Chapters	课时 Class Hour	知识点 Key Points
2.1	两视几何 Two-view geometry	2	(1) 深度感知 (2) 对极几何 (3) 本质矩阵与基础矩阵 (4) 弱标定 (1) Perceiving depth (2) Epipolar geometry (3) Essential matrix and foundation matrix (4) Weak calibration
2.2	立体匹配 Stereo matching	2	(1) 立体对应 (2) 视差与视差线索 (3) 极线约束 (4) 立体标定 (5) 立体匹配模型 (1) Stereo corresponding (2) Disparity and disparity cue (3) Epipolar constraint (4) Stereo calibration (5) Model for stereo matching

第三章　运动视觉(Motion Vision)

章节序号 Chapter Number	章节名称 Chapters	课时 Class Hour	知识点 Key Points
3.1	图像运动 Image motion	2	(1) 感知运动 (2) 运动场与光流场 (3) 光流方程与孔径问题 (4) 运动分解 (5) 光流估计 (1) Perceiving motion (2) Motion field and optical flow field (3) Optical flow equation and aperture problem (4) Motion decomposition (5) Optical flow estimation

续表

章节序号 Chapter Number	章节名称 Chapters	课时 Class Hour	知 识 点 Key Points
3.2	目标运动 Object motion	2	（1）刚体运动与非刚体运动 （2）目标表示与目标匹配 （3）运动模型与观测模型 （4）目标关联与目标追踪方法 （1）Rigid motion and non-rigid motion （2）Object representation and object matching （3）Motion model and observation model （4）Object association and object tracking method

第四章 图像结构检测、表示与匹配（Image Structure Detection，Representation and Matching）

章节序号 Chapter Number	章节名称 Chapters	课时 Class Hour	知 识 点 Key Points
4.1	图像滤波与 几何变换 Image filtering and geometric transformation	2	（1）卷积与线性尺度空间 （2）图像金字塔与几何变换 （3）边缘与边缘检测算子 （1）Convolution and linear scale space （2）Image Pyramids and geometric transformation （3）Edge and edge detector
4.2	形状与纹理 Shape and texture	2	（1）滤波器组 （2）纹理基元与纹理表示 （3）形状基元与形状表示 （4）形状匹配 （1）Filter banks （2）Texton and texture representation （3）Shape primitives and shape representation （4）Shape matching

续表

章节序号 Chapter Number	章节名称 Chapters	课时 Class Hour	知识点 Key Points
4.3	特征检测与表示 Feature detection and feature representation	2	(1) 关键点 (2) 特征描述子 (3) 梯度直方图 (4) 尺度不变性特征变换 (5) 视觉词典 (1) Keypoints (2) Feature descriptors (3) Histogram of Gradient(HoG) (4) Scale-Invariant Feature Transform(SIFT) (5) Visual dictionary
4.4	特征匹配 Feature matching	2	(1) 图像表示 (2) 相似性与鲁棒匹配 (3) 哈希算法 (4) 图像检索 (1) Image representation (2) Similarity and robust matching (3) Hash algorithm (4) Image retrieval

第五章　感知聚类与分割方法(Perceptual Grouping and Segmentatioin)

章节序号 Chapter Number	章节名称 Chapters	课时 Class Hour	知识点 Key Points
5.1	感知组织与分割 Perceptual organization and segmentation	2	(1) 感知组织 (2) 组织目标与场景 (3) 格式塔原理 (4) 感知聚类 (5) 视觉内插与视觉完形 (6) 图像分割与视频分割 (1) Perceptual organization (2) Organizing object and scene (3) Gestalt principle (4) Perceptual grouping (5) Visual interpolation and visual completion (6) Image segmentation and video segmentation

章节序号 Chapter Number	章节名称 Chapters	课时 Class Hour	知 识 点 Key Points
5.2	基于聚类的分割方法 Segmentation method based on clustering	2	（1）特征空间 （2）聚类方法 （3）K-means （4）亲合矩阵 （5）谱聚类与归一化切割 (1) Feature space (2) Clustering methods (3) K-means (4) Affinity matrix (5) Spectral Clustering and Normalized Cut
5.3	基于概率模型的分割方法 Segmentation method based on probabilistic model	2	（1）概率模型 （2）混合模型 （3）EM算法 （4）模型选择 (1) Probabilistic model (2) Mixture model (3) EM algorithm (4) Model selection
5.4	随机场与图割方法 Random field and graph cut	2	（1）随机场 （2）马尔可夫随机场 （3）能量模型 （4）信任传播算法 （5）图割 (1) Random field (2) Markov random field (3) Energy-Based Model（EBM） (4) Belief propagation algorithm (5) Graph cut

第六章　视觉认知与识别方法（Vision Cognition and Recognition Method）

章节序号 Chapter Number	章节名称 Chapters	课时 Class Hour	知　识　点 Key Points
6.1	视觉认知 Visual cognition	2	（1）感知目标属性与部件 （2）感知功能与类别 （3）视觉选择性 （4）视觉注意 （5）视觉记忆与想象 （6）分类、检测与鉴别 (1) Perceiving object properties and parts (2) Perceiving function and category (3) Visual selection (4) Visual attention (5) Visual memory and imagery (6) Classification, detection and identification
6.2	集成学习与支撑向量机 Ensemble learning and support vector machine	2	（1）组合分类器 （2）提升算法 （3）支撑向量机 （4）隐支撑向量机 （5）可形变的部件模型 （6）人脸检测 （7）行人检测 (1) Ensemble classifier (2) Boosting algorithm (3) Support Vector Machine (SVM) (4) Latent Support Vector Machine (LSVM) (5) Deformable Parts Model (DPM) (6) Face detection (7) Pedestrian detection

续表

章节序号 Chapter Number	章节名称 Chapters	课时 Class Hour	知识点 Key Points
6.3	视觉识别的非度量方法 Non-metric method of visual recognition	2	(1) 决策树 (2) 决策森林 (3) 随机蕨 (4) 人脸检测与对齐 (1) Decision tree (2) Decision forest (3) Random fern (4) Face detection and alignment
6.4	卷积网络 Convolutional network	2	(1) 前馈 (2) 感受野/知觉场 (3) 分层结构 (4) 分布式表示 (5) 卷积神经网络 (6) 反向传播算法 (7) 图像分类网络 (8) 目标检测网络 (1) Feedforward (2) Perceptional field (3) Hierarchical structure (4) Distributed representation (5) Convolutional neural network (6) Backpropagation algorithm (7) Image classification network (LeNet, AlexNet) (8) Object detection network
6.5	递归网络 Recursive network	1	(1) 隐马尔可夫模型 (2) 递归神经网络 (3) 循环神经网络 (4) 长短时记忆网络 (5) 前向算法与反向算法 (6) 动作识别 (1) Hidden Markov Model (HMM) (2) Recursive neural network (3) Recurrent Neural Network (RNN) (4) Long Short-Term Memory (LSTM) (5) Forward algorithm and backward algorithm (6) Action recognition

章节序号 Chapter Number	章节名称 Chapters	课时 Class Hour	知 识 点 Key Points
6.6	Transformer	1	（1）Transformer，位置编码 （2）注意力的表达 （3）多头注意力 （4）基于 Transformer 的目标检测 （1）Transformer，Position encoding （2）Expression of attention （3）Multi-head attention （4）Thansformer-based object detection

第七章　视觉应用系统（Visual Application System）

章节序号 Chapter Number	章节名称 Chapters	课时 Class Hour	知 识 点 Key Points
7.1	视觉 ADAS Visual Advanced Driver Assistant System（ADAS）	2	（1）视觉 ADAS （2）车道线检测 （3）行人检测 （4）车辆检测 （5）跟踪 （6）交通标志检测 （7）车道偏离报警 （8）前向碰撞报警 （1）Visual ADAS （2）Lane detection （3）Pedestrian detection （4）Vehicle detection （5）Tracking （6）Traffic sign detection （7）Lane Departure Warning（LDW） （8）Forward Collision Warning（FCW）

续表

章节序号 Chapter Number	章节名称 Chapters	课时 Class Hour	知识点 Key Points
7.2	视觉监控系统 Visual surveillance system	2	（1）视觉监控系统 （2）身份认证 （3）目标检测 （4）目标跟踪 （5）异常行为检测 （6）人群计数与密度估计 （1）Visual surveillance system （2）Identity authentication （3）Object detection （4）Object tracking （5）Abnormal action detection （6）Crowd counting and density estimation

6.4.5 实验环节（Experiments）

序号 Num.	实验内容 Experiment Content	课时 Class Hour	知识点 Key Points
1	相机标定 Camera calibration	2	（1）坐标系统与坐标变换 （2）3D-2D 对应 （3）相机内参数与外参数 （4）张氏标定方法 （1）Coordinate system and coordinate transformation （2）3D-2D correspondence （3）Intrinsic parameters and extrinsic parameters （4）Zhang's calibration method
2	图像拼接 Image stitching	2	（1）关键点检测与描述子 （2）描述子匹配 （3）几何变换 （4）多通道图像融合 （1）Keypoint detection and descriptor （2）Descriptor matching （3）Geometric transformation （4）Multi-channel image fusion

续表

序号 Num.	实验内容 Experiment Content	课时 Class Hour	知 识 点 Key Points
3	车道线检测与拟合 Lane detection and fitting	2	(1) 逆透视变换 (2) 边缘检测 (3) 霍夫变换 (4) 鲁棒曲线拟合 (1) Inverse perspective mapping (2) Edge detection (3) Hough transformation (4) Robust curve fitting
4	行人检测与动作识别 Pedestrian detection and action recognition	2	(1) 卷积神经网络与长短时记忆网络 (2) 行人检测网络 (3) 人体骨架序列 (4) 基于骨架的动作识别网络模型 (1) CNN&LSTM (2) CNN for pedestrian detection (3) Human skeleton sequence (4) Network model for skeleton-based action recognition

大纲指导者：郑南宁教授（西安交通大学人工智能学院）

大纲制定者：袁泽剑教授（西安交通大学人工智能学院）、王进军教授（西安交通大学人工智能学院）

大纲审定：西安交通大学人工智能学院本科专业知识体系建设与课程设置第二版修订工作组

6.5 "强化学习"课程大纲

课程名称：强化学习

Course：Reinforcement Learning

先修课程：概率统计与随机过程、人工智能概论、机器学习、现代控制工程

Prerequisites：Probability Theory and Stochastic Process，Introduction to Artificial Intelligence，Machine Learning，Modern Control Engineering

学分：2

Credits：2

6.5.1 课程目的和基本内容（Course Objectives and Basic Content）

本课程是人工智能学院本科专业必修课。

This course is a compulsory course for undergraduates in College of Artificial Intelligence.

本课程主要围绕如何借鉴生物适应环境的目标导向、试错与奖惩、自主学习、进化与演化等机制，研究动态、开放环境下自主智能体动态决策这一核心问题。课程以构建动态决策模型与优化策略为主线，组织和梳理了强化学习领域的知识体系。通过本课程学习，夯实强化学习和自主智能体动态决策的理论基础，提升运用所学知识分析问题、解决问题和面向真实环境设计智能系统的能力，启发和激励学生进一步钻研探索前沿技术。

课程设置围绕强化学习这一主题，按由浅到深的三个层次组织授课内容。第一层次主要聚焦于强化学习的核心思想与基本概念，对应绪论、第一章和第二章，主要内容为马尔可夫决策过程、反馈评估、价值函数、优化和逼近等；第二层次为强化学习的基本方法，对应第三章到第五章，主要包括：①有模型强化学习，包括价策略评估与改进、动态规划、价值迭代、策略迭代等经典算法；②免模型强化学习，包括蒙特卡洛强化学习、时序差分学习。第三层次介绍深度强化学习及前沿问题，对应第六章和第七章，主要包括：①基于值函数的深度强化学习和基于策略梯度深度强化学习的两种主流技术；②模仿与示教学习、注意与记忆等前沿强化学习技术。

课程采用授课、小组讨论、课后大作业和口头报告等多种形式，训练学生综合应用强化学习的方法解决实际问题的能力，并重视培养学生从思想、概念、模型、算法到系统的全链条科研能力。此外，本课程增加了对强化学习中重要科学思想、核心算法的发展过程和未来趋势相关内容，通过小组讨论等环节，启发和引导学生发现问题、分析问题和解决问题的能力。

强化学习通过借鉴生物从与环境主动交互中学习获得适应性这一机制，经由人工智能与控制学科交叉融合发展而来，一直是机器学习、神经网络、最优控制等领域共同关注的重点方向，并被视为发展通用人工智能技术的重要途径。强化学习的思想、方法和技术已广泛应用于人工智能、经济管理和社会学等多个学科。理解和掌握强化学习的基本概念、基本原理和方法对夯实学生的人工智能基础理论、培养学生分析问题和解决问题的能力以及对他们在人工智能领域创新创业均具有十分重要的意义。

The concepts, computational models, and algorithms organized in this course belonging to the line of thinking inspired by goal guided, the reward and punishment mechanism, and evolving process that form basis for biologic systems live in dynamic and open environments. The purpose of this course is for students to lay a solid foundation of Reinforcement Learning for students, improve their capabilities of applying knowledge to solve real decision-making problems and design intelligent systems for real world, and encourage their future research.

The course focuses on Reinforcement Learning (RL), and the knowledge is organized in three levels. The first level consists of introduction and Chapter 1 to Chapter 2, in which basic ideas and fundamental knowledge of RL are presented. Major issues include Markov decision process, feedback evaluation, value function, optimization and approximation. The second level consists of Chapter 3 to Chapter 5. In this level, we first discuss model based RL methods which include policy evaluation and improvement, dynamic programming, value iteration, and policy iteration. We then introduce the model-free RL techniques, which includes Monte Carlo methods and temporal difference learning. The third level consists of Chapter 6 to Chapter 7. We present two categories of deep RL methods, the value-based methods for deep RL and policy gradient methods for deep RL. We also discuss the frontiers of deep RL which cover topics including imitation learning, learning by demonstration, attention and memory, and challenges of applying RL to real world problems.

Integrating lectures, group discussions, after-school assignments, and oral presentations, the course trains students' ability to apply theoretical principles and methods to solve real problems using reinforcement learning and natural computation from multiple perspectives, dimensions and scales. Furthermore, it aims to strengthen students' capabilities ranging from proposing ideas, formulation of concept, designing model and algorithm, and building systems. The course also inspires students by introducing the history, current state, and future development trends of the disciplines through group discussions. It is beneficious to train students to discover problems, analysis problems and solve problems.

Reinforcement learning is a computational approach to learning whereby an agent tries to maximize the total amount of reward it receives when interacting with a dynamic, uncertain environment. It is an interdisciplinary of artificial intelligence and control theory, and has been a long-term common research focus among several fields including machine learning, neural network, and optimal control. It is believed that reinforcement learning is a promising general approach to artificial intelligence adopted in open, dynamic environments. Reinforcement learning has been widely adopted in many disciplines including artificial intelligence, economics and management, and social science. Thus, understanding and mastering knowledge covered in the course is beneficial for students to lay solid theoretical basis, and provides a source of innovation for their successful career in the future.

6.5.2 课程基本情况(Course Arrangements)

课程名称	强化学习 Reinforcement Learning							
开课时间	一年级		二年级		三年级		四年级		人工智能核心
	秋	春	秋	春	秋	春	秋	春
课程定位	本科生"人工智能核心"课程群必修课								必修 (学分)	人工智能概论(2)
		机器学习(3)
		自然语言处理(2)
学 分	2学分								计算机视觉与模式识别(3)
		强化学习(2)
总学时	32学时(授课32学时)								生成式AI与大语言模型(2)
授课学时 分配	课堂讲授(30学时),小组讨论(2学时)								选修 (学分)	/
先修课程	概率统计与随机过程、人工智能概论、机器学习、现代控制工程
后续课程	
教学方式	课堂教学、大作业与实验、小组讨论、综述报告
课时分配	课堂讲授48学时
考核方式	课程结束笔试成绩占60%,平时成绩占15%,实验成绩占10%,综述报告占10%,考勤占5%
参考教材	Sutton R S, Barto A G. Reinforcement Learning: An Introduction[M]. Cambridge: MIT Press, 2015.
参考资料	
其他信息	

6.5.3 教学目的和基本要求(Teaching Objectives and Basic Requirements)

(1) 理解强化学习的基本概念与基本方法;
(2) 理解 MDP、POMDP 及值迭代方法;
(3) 掌握动态规划、蒙特卡洛学习、时序差分学习等基本算法;
(4) 理解预测与控制、规划与学习等基本方法;
(5) 掌握基于价值的深度强化学习方法;
(6) 掌握基于策略梯度的深度强化学习方法;
(7) 初步具备能利用基本算法解决动态决策问题的能力;

6.5.4 教学内容及安排(Syllabus and Arrangements)

绪论(Introduction)

章节序号 Chapter Number	章节名称 Chapters	课时 Class Hour	知 识 点 Key Points
0.1	绪论 Introduction	2	(1) 强化学习的基本术语及概念 (2) 强化学习的基本要素 (3) 理解探索与利用困境 (4) 强化学习的主要应用 (1) Basic terminologies and concepts of RL (2) Elements of RL (3) Understanding exploration/exploitation dilemma (4) Applications of RL

第一章 评估性反馈(Evaluative Feedback)

章节序号 Chapter Number	章节名称 Chapters	课时 Class Hour	知识点 Key Points
1.1	N-摇臂赌博机问题 N-armed bandit problem	1	(1) N-摇臂赌博机中的探索与利用问题 (2) 交互中的三种反馈：直觉性、评估性与指导性 (1) Exploration and exploitation in N-armed bandit problem (2) Three kinds of feedbacks in interaction: intuitive, evaluative, and instructive
1.2	动作-价值方法 Action-value methods	1	(1) 动作价值与动作选择 (1) Value of action and action selection
1.3	Softmax 动作选择 Softmax action selection		(1) 动作选择的 Softmax 函数 (1) Softmax function for action selection
1.4	增量式实现 Incremental implementation		(1) 动作-价值法的增量式实现 (1) Incremental implementation of action-value methods.

第二章 强化学习问题(The Reinforcement Learning Problem)

章节序号 Chapter Number	章节名称 Chapters	课时 Class Hour	知识点 Key Points
2.1	智能体-环境界面 Agent-environment interface	2	(1) 强化学习的基本要素：智能体、环境、状态、奖励和策略 (1) Elements of RL: agent, environment, state, reward and policy
2.2	目标、奖励与回报 Goal, reward and return		(1) 交互学习的目标、奖励和回报 (2) 马尔可夫过程的基本概念与各组成要素 (3) 马尔可夫奖励过程的基本概念与各组成要素 (4) 价值函数与贝尔曼方程 (1) Goal, reward and return in learning from interaction (2) Concept of Markov process and element (3) Concept of Markov reward process and element (4) Value function and Bellman equation

续表

章节序号 Chapter Number	章节名称 Chapters	课时 Class Hour	知　识　点 Key Points
2.3	马尔可夫决策过程 Markov decision process		（1）马尔可夫决策过程的基本要素 （2）策略及给定策略的价值函数 (1) Elements of Markov decision process (2) Policy and Value function under a policy
2.4	最优价值函数 Value function and Optimal value function	2	（1）贝尔曼期望方程 （2）最优价值函数 （3）最优策略 （4）贝尔曼最优性方程 (1) Bellman expectation equation (2) The optimal value function (3) The optimal policy (4) Bellman optimality equation
2.5	最优性及其近似 Optimality and approximation		（1）贝尔曼最优性方程的逼近求解 (1) Solve Bellman optimality equation approximately.

第三章　动态规划(Dynamic Programming)

章节序号 Chapter Number	章节名称 Chapters	课时 Class Hour	知　识　点 Key Points
3.1	动态规划 Dynamic programming	2	（1）动态规划的基本思想与基本方法 （2）预测与控制 (1) Basic idea and method of dynamic programming (2) Prediction and control
3.2	策略评估与策略改善 Policy evaluation and policy improvement		（1）策略评估 （2）策略改善 (1) Policy evaluation (2) Policy improvement

章节序号 Chapter Number	章节名称 Chapters	课时 Class Hour	知 识 点 Key Points
3.3	策略迭代 Policy iteration	1	(1) 最优策略 (2) 策略迭代算法 (1) The optimal policy (2) Policy iteration algorithm
3.4	价值迭代 Value iteration		(1) 最优性原理 (2) 确定性值迭代 (3) 价值迭代算法 (1) Principle of optimality (2) Deterministic value iteration (3) Value iteration
3.5	异步动态规划 Asynchronous dynamic programming	2	(1) 同步动态规划 (2) 异步动态规划 (3) 实时动态规划 (1) Synchronous dynamic programming (2) Asynchronous dynamic programming (3) Real-time dynamic programming
3.6	通用策略迭代 Generalized policy iteration		(1) 通用策略迭代 (1) Generalized policy iteration

第四章　蒙特卡洛方法(Monte Carlo Methods)

章节序号 Chapter Number	章节名称 Chapters	课时 Class Hour	知 识 点 Key Points
4.1	蒙特卡洛策略评估 Monte Carlo policy evaluation	1	(1) 蒙特卡洛方法的基本思想 (2) 理解回合、首次访问蒙特卡洛方法、每次访问蒙特卡洛方法等基本概念 (3) 累进更新平均值、蒙特卡洛累进更新 (1) Idea of Monte Carlo method (2) Understanding basic concepts including episode, first-visit Monte Carlo method, every-visit Monte Carlo method (3) Incremental mean, Monte Carlo incremental mean
4.2	估计动作价值的 蒙特卡洛方法 Monte Carlo estimation of action values		(1) 给定策略下状态-动作价值估计方法 (2) 持续探索问题 (1) Estimation value of state-action under a determined policy (2) The problem of maintaining exploration
4.3	蒙特卡洛控制 Monte Carlo control	2	(1) 免模型控制 (2) 同轨策略学习与离轨策略学习 (3) 蒙特卡洛策略迭代：蒙特卡洛策略评估＋ε-贪婪策略改进 (1) Model-free control (2) On-policy learning and Off-policy learning (3) Monte Carlo policy iteration: Monte Carlo policy evaluation ＋ ε-greedy policy improvement
4.4	同轨策略 蒙特卡洛控制 On-policy Monte Carlo control		(1) 同轨策略的蒙特卡洛控制 (2) GLIE 蒙特卡洛控制 (1) On-policy Monte Carlo control (2) GLIE Monte Carlo control

章节序号 Chapter Number	章节名称 Chapters	课时 Class Hour	知 识 点 Key Points
4.5	离轨策略 蒙特卡洛控制 Off-policy Monte Carlo control	1	（1）重要性抽样 （2）离轨策略的蒙特卡洛控制 （1）Importance sampling （2）Off-policy Monte Carlo control
4.6	增量式实现 Incremental implementation		（1）增量式实现方法 （1）Incremental implementation

第五章 时序差分学习（Temporal Difference Learning）

章节序号 Chapter Number	章节名称 Chapters	课时 Class Hour	知 识 点 Key Points
5.1	时序差分学习 Temporal-Difference (TD) learning	2	（1）时序差分学习的基本思想 （2）对比时序差分学习与蒙特卡洛方法 （1）Ideas of temporal-difference learning （2）Comparisons between TD and Monte Carlo
5.2	时序差分预测 TD prediction		（1）引导、抽样等基本概念 （2）n 步预测与 n 步回报 （3）前向视角的 TD、反向视角的 TD （1）Understanding concepts of bootstrapping and sampling （2）n-step prediction and n-step return （3）Forward view TD, backward TD
5.3	Sarsa：同轨策略 时序差分控制 Sarsa: on-policy TD control	1	（1）同轨策略的时序差分控制 Sarsa 算法 （2）n-步 Sarsa 算法 （1）Sarsa: On-policy Temporal-Difference control （2）n-step Sarsa
5.4	Q 学习：离轨策略 时序差分控制 Q-learning: off-policy TD control		（1）Q 学习：离轨策略时序差分控制 （1）Q-learning: off-policy TD control

续表

章节序号 Chapter Number	章节名称 Chapters	课时 Class Hour	知 识 点 Key Points
5.5	演员-评论员方法 Actor-Critic methods	2	（1）强化学习的三种形式：基于策略、基于价值函数、演员-评论员 （2）策略的目标函数、策略优化、策略梯度 （3）蒙特卡洛策略梯度 （4）演员-评论员算法 （1）Three types of RL: Policy-based, value-based RL, actor-critic （2）Policy objective function, policy optimization, policy gradient （3）Monte Carlo policy gradient （4）Actor-critic algorithms

第六章 深度强化学习方法（Methods for Deep reinforcement learning）

章节序号 Chapter Number	章节名称 Chapters	课时 Class Hour	知 识 点 Key Points
6.1	深度 Q 网络 Deep Q-Network	2	（1）Q 学习 （2）深度 Q 网络 （3）深度 Q 网络的变体 （4）分布式 DQN 及多步学习 （1）Q-learning （2）Deep Q-Network （3）Variants of DQN （4）Distributional DQN and multi-step learning
6.2	深度强化学习的策略梯度梯度方法 Policy gradient method for deep reinforcement learning	2	（1）随机/确定性策略梯度 （2）演员-评论员方法 （3）自然策略梯度与可信域优化 （1）Stochastic/deterministic policy gradient （2）Actor-Critic method （3）Natural policy gradient and trust region optimization

章节序号 Chapter Number	章节名称 Chapters	课时 Class Hour	知 识 点 Key Points
6.3	深度强化学习的泛化概念 The concept of generalization of reinforcement learning	2	(1) 特征选择 (2) 算法选择和函数逼近选择 (3) 目标函数调整与分级学习 (1) Feature selection (2) Choice of the learning algorithm and function approximation (3) Modifying the objective function and hierarchical learning

第七章 深度强化学习的前沿问题(Frontiers of Deep Reinforcement Learning)

章节序号 Chapter Number	章节名称 Chapters	课时 Class Hour	知 识 点 Key Points
7.1	模仿学习与示教学习 Imitation learning and learning by demonstration	2	(1) 模仿学习 (2) 示教学习 (1) Imitation learning (2) Learning by demonstration
7.2	注意与记忆 Attention and memory		(1) 记忆网络 (2) 微分神经计算机 (1) Memory network (2) Differentiable neural computer
7.3	自主学习 Learn to learn	2	(1) 自主学习的强化学习实现 (2) Learn to learn with RL
7.4	应用于真实世界问题的挑战 Challenges of applying RL to real-world problems		(1) 理解强化学习解决真实世界问题的主要难点问题 (1) Understanding the difficulties of applying RL to solve a real-world problem

大纲指导者：郑南宁教授（西安交通大学人工智能学院）

大纲制定者：薛建儒教授（西安交通大学人工智能学院）、张雪涛副教授（西安交通大学人工智能学院）

大纲审定：西安交通大学人工智能学院本科专业知识体系建设与课程设置第二版修订工作组

6.6 "生成式 AI 与大语言模型"课程大纲

课程名称：生成式 AI 与大语言模型

Course：Generative AI and Large Language Model

先修课程：工科数学分析、线性代数与解析几何、概率统计与随机过程、自然语言处理

Prerequisites：Mathematical Analysis for Engineering, Linear Algebra and Analytic Geometry, Probability Theory and Stochastic Process, Natural Language Processing

学分：2

Credits：2

6.6.1 课程目的和基本内容（Course Objectives and Basic Content）

本课程是人工智能学院本科专业必修课。

This course is a compulsory course for undergraduates in College of Artificial Intelligence.

人工智能是以人工的方法用机器（计算机）模拟、延伸和扩展人类智能的学科，具有多学科交叉、高度复杂、强渗透性的学科特点。本课程将探索和学习人工智能领域中生成式 AI 和大语言模型的基本理论，从介绍生成式 AI 和语言模型的基本概念和早期模型开始，然后深入研究深度学习语言模型如循环神经网络和长短期记忆网络。在这个基础上，课程将探讨表征学习、上下文学习、最前沿的大语言模型（如 BERT 和 GPT 等）以及它们如何推动了自然语言处理技术的发展。

本课程的目的是掌握生成式 AI 和大语言模型的核心概念、原理和应用领域。了解大语言模型的优势和局限性，通过实践能够使用大语言模型进行文本生成、对话系统设计等任务，并能够思考和讨论生成式 AI 对社会和伦理的影响。

Artificial intelligence is a discipline that simulates, extends, and expands human intelligence through artificial methods using machines (computers). It has multidisciplinary intersection, high complexity, and strong permeability as its discipline characteristics. This course will explore and learn the basic theories of generative AI and large language models in the field of artificial intelligence. Starting with the introduction of basic concepts and early models of generative AI and language models, it then delves into deep learning language models such as Recurrent Neural Networks and Long Short-Term Memory Networks. Based on this, we will discuss representation learning, context learning, and cutting-edge large language models (such as BERT and GPT, etc.), as well as how they have propelled the development of natural language processing technology.

The purpose of this course is to master the core concepts, principles, and application areas of generative AI and large language models. Understand the advantages and limitations of large language models, be able to use large language models for text generation, dialogue system design, and other practices through hands-on exercises, and be able to think about and discuss the impact of generative AI on society and ethics.

6.6.2 课程基本情况(Course Arrangements)

课程名称	生成式 AI 与大语言模型 Generative AI and Large Language Model							
开课时间	一年级		二年级		三年级		四年级	
^	秋	春	秋	春	秋	春	秋	春
课程定位	本科生"人工智能核心"课程群必修课							
学　　分	2 学分							
总学时	32 学时 (授课 32 学时、实验 0 学时)							
授课学时分配	课堂讲授(32 学时)							
先修课程	工科数学分析、线性代数与解析几何、概率统计与随机过程、自然语言处理							
后续课程								

人工智能核心

必修(学分):
- 人工智能概论(2)
- 机器学习(3)
- 自然语言处理(2)
- 计算机视觉与模式识别(3)
- 强化学习(2)
- 生成式 AI 与大语言模型(2)

选修(学分): /

续表

教学方式	课堂教学、作业、讨论、报告
考核方式	课程结束笔试成绩占70%，作业成绩和考勤等占30%
参考教材	斋藤康毅．深度学习进阶：自然语言处理［M］．陆宇杰，译．北京：人民邮电出版社，2020.
参考资料	1. Goodfellow I，Bengio Y，Courville A. 深度学习［M］．赵申剑，等译，北京：人民邮电出版社，2017. 2. Lane H，Howard C，Hapke H M．自然语言处理实战［M］．史亮，等译，北京：人民邮电出版社，2020. 3. Jurafsky D，Martin J H．自然语言处理综论［M］．2版．冯志伟，孙乐 译．北京：电子工业出版社，2018. 4. 宗成庆．统计自然语言处理［M］．2版．北京：清华大学出版社，2013. 5. 周志华．机器学习［M］．北京：清华大学出版社，2016.
其他信息	

6.6.3 教学目的和基本要求（Teaching Objectives and Basic Requirements）

（1）熟悉大语言模型的发展背景和主要发展阶段，从早期的基于规则和统计的方法到现代的深度学习模型的转变，理解其在人工智能领域中的重要性和影响。

（2）掌握大语言模型的基础理论，包括其结构、训练方法、微调策略等，以及如何进行模型训练和微调来达到特定的任务目标。

（3）理解生成式 AI 的基本原理和核心概念，理解如何从大量数据中学习并生成新的、与训练数据类似的输出，以及相应的复杂的生成模型。

（4）理解上下文学习在大语言模型中的作用和技术手段，以及上下文对于理解和生成文本的重要性。

（5）熟悉大语言模型在自然语言处理的各种应用，包括使用模型进行文本生成、机器翻译、文本摘要和情感分析等，并理解这些应用的实际工作原理。

（6）理解训练大语言模型的主要挑战，包括计算资源需求、对大量高质量标注数据的依赖以及如何在有限的资源和数据下进行有效的模型训练。

（7）掌握如何使用各种评估指标来评估大语言模型的性能，包括了解常见的错误类型以及如何通过改进模型设计和训练策略来克服这些错误。

（8）理解大语言模型可能带来的社会和道德影响，包括模型偏见的来源、对隐私的可能影响以及如何设计和使用模型以最大程度地减少这些负面影响。

6.6.4 教学内容及安排(Syllabus and Arrangements)

第一章 绪论(Introduction)

章节序号 Chapter Number	章节名称 Chapters	课时 Class Hour	知 识 点 Key Points
1	绪论 Introduction	2	(1) 生成式 AI 和大语言模型的概述 (2) 语言模型的历史发展 (3) 语言模型的应用 (4) 不同类型的语言模型的比较 (5) 大语言模型在改进 AI 性能中的角色 (1) Overview of Generative AI and Large Language Models (2) Development of language models (3) Application fieldsof language models (4) Comparison of different types of language models (5) Role of large language models in improving AI performance

第二章 生成式 AI(Generative AI)

章节序号 Chapter Number	章节名称 Chapters	课时 Class Hour	知 识 点 Key Points
2	生成式 AI Generative AI	4	(1) 生成式模型与判别模型的比较 (2) 概率生成模型：高斯混合模型、隐马尔可夫模型 (3) 深度生成模型：变分自编码器、生成对抗网络 (4) 深度生成模型的应用和案例分析 (5) 生成式模型的评估和挑战 (1) Comparison between generative models and discriminative models (2) Probabilistic generative models：Gaussian Mixture Model (GMM), Hidden Markov Model(HMM) (3) Deep generative models：Variational Autoencoder (VAE),Generative Adversarial Network (GAN) (4) Applications and case studies of deep generative models (5) Evaluation and challenge of generative models

第三章 表征学习的基本概念与方法（Basic Concepts and Methods of Representation Learning）

章节序号 Chapter Number	章节名称 Chapters	课时 Class Hour	知 识 点 Key Points
3	表征学习的基本概念与方法 Basic concepts and methods of representation learning	4	（1）特征表示和表征学习的定义 （2）词嵌入方法和应用 （3）句子和文本表示的技术和进展 （4）不同类型的词嵌入模型比较（Word2Vec、GloVe、FastText 等） （5）表征学习在其他 AI 领域的应用 (1) Definition of feature representation and representation learning (2) Methods and applications of word embedding (3) Techniques and advancements in sentence and text representation (4) Comparison of different types of word embedding models（Word2Vec, GloVe, FastText, etc.） (5) Applications of representation learning in other AI fields

第四章 上下文学习在大语言模型中的应用（Applications of Context Learning in Large Language Models）

章节序号 Chapter Number	章节名称 Chapters	课时 Class Hour	知 识 点 Key Points
4	上下文学习在大语言模型中的应用 Applications of context learning in large language models	4	（1）序列建模和上下文信息的重要性 （2）循环神经网络与长短期记忆 （3）注意力机制与上下文关注 （4）上下文学习在大型预训练模型中的角色 (1) Importance of sequence modeling and context information (2) Recurrent Neural Networks（RNN）and Long Short-Term Memory（LSTM） (3) Attention mechanism and context-awareness (4) Role of context learning in large pre-trained models

第五章 大语言模型(Large Language Models)

章节序号 Chapter Number	章节名称 Chapters	课时 Class Hour	知识点 Key Points
5	大语言模型 Large language models	8	(1) 变换器模型的基本理论 (2) BERT (3) GPT-1、GPT-2、GPT-3 和 GPT-4 (4) 大语言模型的最新进展和研究趋势 (1) Basictheory of transformer models (2) Bidirectional Encoder Representations from Transformers(BERT) (3) Generative Pretraining Transformer(GPT) 1,2,3 and 4 (4) Latest advancements and research trends in large language models

第六章 大语言模型的实战与应用(Practical Applications of Large Language Models)

章节序号 Chapter Number	章节名称 Chapters	课时 Class Hour	知识点 Key Points
6	大语言模型的实战与应用 Practical applications of large language models	8	(1) 大语言模型的训练 (2) 大语言模型的微调 (3) 大语言模型的性能评估和模型选择 (4) 大语言模型在自然语言处理中的应用 (5) 大语言模型的实际问题和挑战 (1) Training of large language models (2) Fine-tuning of large language models (3) Performance evaluation and model selection of large language models (4) Applications of large language models in natural language processing (5) Practical issues and challenges of large language models

第七章 生成式AI的社会和伦理影响（Social and Ethical Impacts of Generative AI）

章节序号 Chapter Number	章节名称 Chapters	课时 Class Hour	知 识 点 Key Points
7	生成式AI的社会和伦理影响 Social andethical impacts of generative AI	2	（1）生成式AI的社会影响 （2）生成式AI的伦理问题 （3）生成式AI的法律问题和监管方案 （4）公平性、透明度和隐私在生成式AI中的角色 （5）AI在未来的可能性和挑战 (1) Social impacts of generative AI (2) Ethical issues of generative AI (3) Legal issues and regulatory schemes of generative AI (4) Roles of fairness, transparency, and privacy in generative AI (5) Possibilities and challenges for AI in future

大纲指导者：郑南宁教授（西安交通大学人工智能学院）

大纲制定者：丁宁教授（西安交通大学人工智能学院）、辛景民教授（西安交通大学人工智能学院）

大纲审定：西安交通大学人工智能学院本科专业知识体系建设与课程设置第二版修订工作组

第 7 章

"认知与神经科学"课程群

7.1 "认知心理学基础"课程大纲

课程名称：认知心理学基础
Course：Cognitive Psychology
先修课程：无
Prerequisites：None
学分：2
Credits：2

7.1.1 课程目的和基本内容(Course Objectives and Basic Content)

本课程是人工智能学院本科专业必修课。

This course is a compulsory course for undergraduates in College of Artificial Intelligence.

认知心理学作为心理学的主流研究领域之一,已经开始与越来越多的学科产生交叉。其中,人工智能是一个重要的交叉领域。本课程围绕认知心理学的研究方法、心理学中基本的认知过程等展开讨论,主要介绍基本认知过程中涉及的实验范式和近期研究进展。

本课程的主要内容共 12 章。第一章是导言与认知心理学研究方法,介绍认知心理学的基本概念、基本问题和经典的研究方法。第二章到第八章从基本概念、经典理论和实证研究三个方面介绍基本认知加工过程,包括感知觉、模式识别、注意、意识和记忆。第九章到第十二章从基本概念、经典理论和实证研究三个方面介绍高级认知加工过程,包括知识表征、推理、决策、问题解决、创造性和智力。

本课程引导学生对经典的认知心理学实验进行深入理解,课程以课堂讲授模式为主,辅以文献阅读、课堂讨论等教学手段,着重训练学生的理论思维,旨在更好地帮助学生建立关于认知心理学的知识框架,并能够运用基本理论和方法分析解释现实生活中的各种心理现象,更好地为人工智能研究服务。

As the mainstream research fields in psychology, cognitive psychology has begun to contribute to more and more disciplines, especially in the field of artificial intelligence (AI). This course focuses on the research methods of cognitive psychology and the basic cognitive processes in psychology, and mainly introduces the experimental paradigms involved in the basic cognitive processes and the recent research in cognitive psychology.

This course consists of 12 chapters. Chapter 1 is the introduction and the research methods of cognitive psychology, including basic concepts, basic problems and classical research methods of cognitive psychology. Chapter 2 to Chapter 8 focus on the basic cognitive processing from the perspective of basic concepts, classical theories and empirical research, including sensation and perception, pattern recognition, attention, consciousness and memory. Chapter 9 to Chapter 12 focus on the advanced cognitive processing from the perspective of basic concepts, classical theories and empirical research, including knowledge representation, reasoning, decision-making, problem-solving, creativity, and human intelligence.

The course guides students to have a deep understanding of the classical cognitive psychology experiments. Teaching method is teacher presentation supplemented by literature reading and discussion in class. The course is aimed at training students' theoretical thinking and helping students to establish a general framework of cognitive psychology, so that students can use the basic theory and method of cognitive psychology to analyze psychological phenomenon in real life and ultimately serve AI research.

7.1.2 课程基本情况(Course Arrangements)

课程名称	认知心理学基础 Cognitive Psychology								
开课时间	一年级		二年级		三年级		四年级		
	秋	春	秋	春	秋	春	秋	春	
课程定位	本科生"认知与神经科学"课程群必修课								
学　分	2 学分								
总学时	36 学时 (授课 28 学时、实验 8 学时)								
授课学时分配	课堂授课 (28 学时)								
先修课程	无								
后续课程	计算神经工程								
教学方式	课堂教学、小组讨论、实验								
考核方式	课程结束笔试成绩占 70%,实验成绩占 25%,考勤占 5%								
参考教材	1. 王甦,汪安圣.认知心理学[M].北京:北京大学出版社,1992. 2. Solso R L. Cognitive Psychology[M].8 版.北京:机械工业出版社,2010.								
参考资料	Feldman R S.普通心理学[M].11 版.北京:人民邮电出版社,2015.								
其他信息									

认知与神经科学	
必修 (学分)	认知心理学基础(2) 计算神经工程(2)
选修 (学分)	/

7.1.3 教学目的和基本要求(Teaching Objectives and Basic Requirements)

(1) 理解认知心理学的基本问题和研究方法;
(2) 掌握感觉和知觉的概念及相关的经典实验研究及理论;
(3) 掌握模式识别的概念、理论模型和经典的模式识别效应;
(4) 了解注意的定义、功能和机制以及自动化加工;
(5) 了解意识的本质与功能、意识状态、无意识的基本概念;
(6) 重点掌握记忆的模型和短时记忆的特点;
(7) 重点掌握记忆的理论和长时记忆研究的主要发现;
(8) 了解遗忘与识记研究的主要发现;
(9) 重点掌握知识表征的两种主要形式;
(10) 重点掌握演绎推理和归纳推理的概念以及推理错误的原因;

(11) 掌握决策研究的经典发现；

(12) 了解问题解决和创造性的概念和理论。

7.1.4 教学内容及安排(Syllabus and Arrangements)

第一章 导言与认知心理学研究方法(Introduction and Research Methods)

章节序号 Chapter Number	章节名称 Chapters	课时 Class Hour	知　识　点 Key Points
1.1	导言 Introduction	1	(1) 认知心理学的基本概念与基本问题 (2) 认知心理学的历史与产生基础 (1) The fundamental concepts and problems in cognitive psychology (2) The history and foundation of cognitive psychology
1.2	认知心理学研究方法 Research Methods	1	(1) 认知心理学的基本范式 (2) 认知心理学的研究方法 (1) The basic paradigm of cognitive psychology (2) The research methods in cognitive psychology

第二章 感觉与知觉(Sensation and Perception)

章节序号 Chapter Number	章节名称 Chapters	课时 Class Hour	知　识　点 Key Points
2.1	感觉 Sensation	1	(1) 绝对阈值与差别阈值 (2) 视觉 (3) 听觉 (4) 其他感觉 (5) 感觉研究举例 (1) Absolute threshold and difference threshold (2) Vision (3) Auditory sensation (4) Other sensation (5) Examples of sensation research

续表

章节序号 Chapter Number	章节名称 Chapters	课时 Class Hour	知 识 点 Key Points
2.2	知觉 Perception	1	(1) 知觉加工 (2) 知觉恒常性 (3) 错觉 (4) 知觉研究举例 (1) Perceptual processing (2) Perceptual constancy (3) Perceptual illusion (4) Examples of perception research

第三章　模式识别(Pattern Recognition)

章节序号 Chapter Number	章节名称 Chapters	课时 Class Hour	知 识 点 Key Points
3.1	模式识别 Pattern recognition	2	(1) 模式识别的概念 (2) 模式识别的理论模型 (3) 经典的模式识别效应 (1) The definition of pattern recognition (2) Theoretical models for pattern recognition (3) Classic pattern recognition effects

第四章　注意(Attention)

章节序号 Chapter Number	章节名称 Chapters	课时 Class Hour	知 识 点 Key Points
4.1	注意 Attention	2	(1) 注意的研究历史 (2) 注意的定义与实质特征 (3) 注意的功能和模型 (4) 控制加工与自动化加工 (5) 注意和意识的关系 (1) Research history of attention (2) The definition and substantive characteristics of attention (3) The functions and models of attention (4) Controlled process and automatic process (5) The relationship between attention and consciousness

第五章 意识(Consciousness)

章节序号 Chapter Number	章节名称 Chapters	课时 Class Hour	知 识 点 Key Points
5.1	意识 Consciousness	2	(1) 意识与无意识 (2) 睡眠 (3) 梦 (4) 其他意识状态 (5) 意识的经典认知研究 (1) Consciousness and unconsciousness (2) Sleep (3) Dream (4) Other states of consciousness (5) The classical cognitive study of consciousness

第六章 记忆模型与短时记忆(Memory Model and Short-Term Memory)

章节序号 Chapter Number	章节名称 Chapters	课时 Class Hour	知 识 点 Key Points
6.1	记忆模型 Memory models	1	(1) 记忆的存储结构模型 (2) 记忆的加工水平理论 (1) Storage structure model of memory (2) Levels-of-processing theory of memory
6.2	短时记忆 Short-term memory	2	(1) 短时记忆 (2) 工作记忆模型 (1) Short-term memory (2) Working memory model

第七章　长时记忆系统(Long-Term Memory System)

章节序号 Chapter Number	章节名称 Chapters	课时 Class Hour	知　识　点 Key Points
7.1	长时记忆 Long-term memory	2	(1) 长时记忆的存储机制 (2) 外显记忆与内隐记忆 (3) 陈述性记忆 (4) 非陈述性记忆 (5) 长时记忆的编码与提取 (1) Storage mechanism of long-term memory (2) Explicit memory and implicit memory (3) Declarative memory (4) Non-declarative memory (5) Encoding and retrieval of long term memory

第八章　遗忘与识记(Forgetting and Remembering)

章节序号 Chapter Number	章节名称 Chapters	课时 Class Hour	知　识　点 Key Points
8.1	遗忘 Forgetting	1	(1) 遗忘的概念：艾宾豪斯遗忘曲线 (2) 遗忘的原因：遗忘理论 (1) The concept of forgetting: Ebbinghaus forgetting curve (2) The reason of forgetting: theories of forgetting
8.2	识记 Remembering	1	(1) 增强记忆的要素 (2) 记忆术 (1) Factors of memory enhancement (2) Mnemonic techniques

第九章　知识表征（Representation of Knowledge）

章节序号 Chapter Number	章节名称 Chapters	课时 Class Hour	知　识　点 Key Points
9.1	知识的语义表征 The verbal representation of knowledge	2	（1）概念表征 （2）命题表征 （3）连接主义的知识表征 （4）视觉的语义表征 (1) Conceptual representation (2) Propositional representation (3) The representation of knowledge in connectionism (4) Verbal representation of vision
9.2	知识的意象表征 The imagery representation of knowledge	2	（1）意象的定义 （2）意象的研究历史 （3）意象表征的三种理论 （4）意象表征的分类 （5）意象的功能 (1) Definition of imagery (2) History of imagery research (3) Three theories of imagery representation (4) Classification of imagery representation (5) Function of imagery

第十章　推理（Reasoning）

章节序号 Chapter Number	章节名称 Chapters	课时 Class Hour	知　识　点 Key Points
10.1	推理 Reasoning	2	（1）推理概述 （2）演绎推理 （3）归纳推理 （4）概率推理 (1) Summary of reasoning (2) Deductive reasoning (3) Inductive reasoning (4) Probabilistic reasoning

第十一章 决策(Decision Making)

章节序号 Chapter Number	章节名称 Chapters	课时 Class Hour	知识点 Key Points
11.1	决策 Decision making	2	(1) 贝叶斯推理与决策 (2) 决策框架 (3) 不确定条件下的人类决策 (4) 生态理性 (1) Bayesian reasoning and decision making (2) Decision making framework (3) Human decision making under uncertainty (4) Ecological rationality

第十二章 问题解决与创造性(Problem Solving and Creativity)

章节序号 Chapter Number	章节名称 Chapters	课时 Class Hour	知识点 Key Points
12.1	问题解决 Problem solving	1	(1) 问题解决的类型和特征 (2) 问题解决的理论 (3) 问题解决的过程 (4) 问题解决的策略 (5) 影响问题解决的其他心理因素 (1) Types and characteristics of problem solving (2) Theories of problem solving (3) Procedures problem solving (4) Strategies of problem solving (5) Other influence psychological factors of problem solving
12.2	创造性 Creativity	1	(1) 创造性的定义和测量 (2) 创造的过程 (3) 创造性的理论 (4) 创造性的影响因素 (1) Definition and measurement of creativity (2) Process of creation (3) Theories of creativity (4) Influence factors of creativity

续表

章节序号 Chapter Number	章节名称 Chapters	课时 Class Hour	知识点 Key Points
12.3	智力 Human intelligence	1	(1) 智力的定义 (2) 智力的主要理论 (3) 晶体智力和流体智力的区别 (4) 智力差异的形成原因 (5) 智力的测量方法 (1) The definition of intelligence (2) The theories of intelligence (3) The difference between crystallized intelligence and fluid intelligence (4) The determinations of variations in intellectual ability (5) Intelligence accessing tests

7.1.5 实验环节(Experiments)

序号 Num.	实验内容 Experiment Content	课时 Class Hour	知识点 Key Points
1	视知觉和注意实验 Visual perception and attention experiments	3	(1) 方向知觉(朝向判断实验) (2) 运动知觉(生物运动实验) (3) 颜色知觉(颜色知觉实验) (4) 面孔加工(人脸情绪识别实验) (5) 视觉空间注意(多物体追踪实验) (1) Direction perception (direction judgment experiment) (2) Motion perception (biological motion experiment) (3) Color perception (color perception experiment) (4) Face processing (face emotion recognition experiment) (5) Visual spatial attention (multiple object tracking experiment)

续表

序号 Num.	实验内容 Experiment Content	课时 Class Hour	知识点 Key Points
2	记忆实验 Memory experiments	3	(1) 视觉空间工作记忆(视觉空间记忆实验) (2) 工作记忆容量(变化盲实验) (3) 如何操纵工作记忆负荷(N-back 实验) (4) 长时记忆与遗忘(回忆和再认实验) (1) Visual spatial working memory (visual spatial memory experiment) (2) Working memory capacity (change blindness experiment) (3) How to manipulate the working memory load (N-back experiment) (4) Long-term memory and forgetting (recall and recognition experiments)
3	知识表征实验 Knowledge representation experiments	2	(1) 知识的视觉表征(心理旋转实验) (2) 知识的言语表征(人工字形学习实验) (1) Visual representation of knowledge (mental rotation experiment) (2) Speech representation of knowledge (artificial orthography learning experiment)

大纲指导者：郑南宁教授(西安交通大学人工智能学院)

大纲制定者：赵晶晶教授(陕西师范大学心理学院)、刘剑毅副教授(西安交通大学人工智能学院)

大纲审定：西安交通大学人工智能学院本科专业知识体系建设与课程设置第二版修订工作组

7.2 "计算神经工程"课程大纲

课程名称：计算神经工程

Course：Computational Neural Engineering

先修课程：认知心理学基础、人工智能概论、概率统计与随机过程

Prerequisites：Introduction to Cognitive Psychology，Introduction to Artificial Intelligence，Probability Theory and Stochastic Process

学分：2

Credits：2

7.2.1 课程目的和基本内容（Course Objectives and Basic Content）

本课程是人工智能学院本科专业必修课。

This course is a compulsory course for undergraduates in College of Artificial Intelligence.

计算神经工程是一门跨学科的新兴学科，综合了信息科学、物理学、数学、生物学、认知心理学等众多领域的成果。由于其发展的时间并不长，至今有关计算神经工程学尚没有一个具有权威性的定义。一般认为，计算神经工程把大脑看成一个信息处理的器件，以数学和计算为主要手段探索神经系统的功能，揭示神经系统编码和处理信息的机制，研究如何通过脑机接口、神经修复等途径对神经系统的功能缺失与异常等问题寻找有效的解决方法。

迄今为止，人们对大脑的探索还只停留在冰山一角，因此计算神经工程的研究面临的是一个充满未知的新领域，必须在基本原理和计算理论方面进行更深刻的探索。对人脑神经系统的结构、信息加工、记忆和学习机制进行计算建模和仿真，有助于揭示人脑的工作机理和提出智能科学的新思想、新方法。此外，脑机接口技术作为神经工程研究领域的一个重要方向，主要探索大脑意念与外部设备进行直接双向通信的创新技术，为基于意念的"脑控"和"控脑"应用提供支持。科幻电影中脑控的机甲战士和控脑的阿凡达成为脑机接口技术未来发展成果的形象体现。

信息科学与现代神经科学真正结合起来是一个很大的挑战。计算神经工程是二者之间的重要桥梁，在类脑计算、人工智能和脑机接口的发展中起关键作用。本课程围绕计算神经工程介绍其基本概念、方法和理论，主要包括神经科学基础、脑信号获取原理与技术、脑信号分析与建模、类脑智能以及脑机接口等。理解和掌握计算神经工程基本原理和理论，有助于学生运用物理、数学以及工程学的概念和分析工具研究大脑的功能，同时有利于培养出人工智能工程技术人才。

Computational Neural Engineering（CNE）is a new interdisciplinary discipline， which combines the achievements of information science, physics, mathematics, biology, cognitive psychology and many other fields. Because of its short development, there is no authoritative definition of Computational Neural Engineering up to now. It is generally believed that Computational Neural Engineering regards the brain as a device for information processing, and uses the mathematics and computation as the main means to explore the function of the nervous system, reveal the mechanism of information processing in the nervous system, and study how to find effective solutions to the problems of functional deficiency and abnormality of the nervous system through Brain Computer Interface(BCI), nerve repair and so on.

So far, the exploration of the brain is only at the tip of the iceberg, so the research of Computational Neural Engineering is facing a new field full of unknown, and it is necessary to make more profound exploration in basic principles and computational theory. The computational modeling and simulation of the structure, information processing, memory and learning mechanism of human brain nervous system, can help us to reveal the working mechanism of human brain and put forward new ideas and new methods of intelligent science. In addition, as an important direction in the field of neural engineering, Brain Computer Interface explores innovative technologies for direct two-way communication between brain and external devices. Brain-controlled machine-armored warriors and Avatar in science fiction movies represent the future development of Brain Computer Interface technology.

The combination of information science and modern neuroscience is a great challenge. Computational Neural Engineering is an important bridge between them and plays a key role in the development of brain-like computing, artificial intelligence and Brain Computer Interface. This course introduces the basic concepts, methods and theories of Computational Neural Engineering, including fundamentals of neuroscience, principles and technologies of brain signal acquisition, brain signals analysis and modeling, brain-like intelligence and Brain Computer Interface. Understanding and mastering the basic principles and theories of Computational Neural Engineering is helpful for students to use concepts and analysis tools of physics, mathematics and engineering to study the functions of the brain, and to cultivate artificial intelligence engineering and technology talents.

7.2.2 课程基本情况(Course Arrangements)

课程名称	计算神经工程 Computational Neural Engineering							
开课时间	一年级		二年级		三年级		四年级	
^	秋	春	秋	春	秋	春	秋	春
课程定位	本科生"认知与神经科学"课程群必修课							
学　　分	2 学分							
总 学 时	32 学时(授课 32 学时、实验 0 学时)							
授课学时分配	课堂讲授(32 学时)							
先修课程	认知心理学基础、人工智能概论、概率统计与随机过程							
后续课程								
教学方式	课堂教学、大作业、小组讨论							
考核方式	课程结束笔试成绩占 80%,大作业占 10%,考勤占 10%							
参考教材	1. Sanchez J C,Principe J C. Brain-Machine Interface Engineering[M]. Willison,VT: Morgan & Claypool Publishers,2007. 2. Dayan P,Abbott L F. Theoretical Neuroscience: Computational and Mathematical Modeling of Neural Systems[M]. Cambridge: The MIT Press,2001.							
参考资料	1. Rajesh P N. 脑机接口导论[M]. 陈民铀,译. 北京:机械工业出版社,2016. 2. Sanei S,Chambers J A. EEG Signal Processing[M]. NewYork: John Wiley & Sons, 2008. 3. Jonathan R W,Elizabeth W W. Brain-Computer Interfaces: Principles and Practice[M]. Oxford: Oxford University Press,2012.							
其他信息								

认知与神经科学
必修(学分): 认知心理学基础(2)、计算神经工程(2)
选修(学分): /

7.2.3 教学目的和基本要求(Teaching Objectives and Basic Requirements)

(1) 了解神经生物学和脑科学基本概念;
(2) 了解记录大脑信号和刺激大脑的基本方法;
(3) 掌握脑信号(Spikes,EEG,fMRI)处理与分析基本方法;
(4) 熟悉神经信息编解码基本概念和建模方法;
(5) 熟悉脉冲神经网络模型;
(6) 熟悉脑机接口的基本原理、主要类型和应用范例。

7.2.4 教学内容及安排(Syllabus and Arrangements)

第一章 概述(Introduction)

章节序号 Chapter Number	章节名称 Chapters	课时 Class Hour	知 识 点 Key Points
1.1	概述 Overview	2	(1) 神经工程概念 (2) 神经工程基本内容 (3) 神经工程历史与现状 (1) Concept of neural engineering (2) Basic contents of neural engineering (3) History and present situation of neural engineering

第二章 神经生物学与脑科学基础(Basis of Neurobiology and Brain Science)

章节序号 Chapter Number	章节名称 Chapters	课时 Class Hour	知 识 点 Key Points
2.1	神经元基本生物物理特性 Basic biophysical properties of neurons	2	(1) 神经元 (2) 突触 (3) 锋电位 (1) Neurons (2) Synapses (3) Spikes
2.2	神经连接的调节 Regulation of neural connections		(1) 突触可塑性 (1) Synaptic plasticity
2.3	大脑组织、解剖结构和功能 Brain tissue, anatomy, and function	1	(1) 中枢神经系统 (2) 外周神经系统 (3) 皮层 (1) Central nervous system (2) Peripheral nervous system (3) Cortex

第三章　记录大脑信号和刺激大脑（Recording the Brain Signal and Stimulating the Brain）

章节序号 Chapter Number	章节名称 Chapters	课时 Class Hour	知　识　点 Key Points
3.1	记录大脑信号 Recording brain signals	3	（1）锋电位序列 （2）脑电图 （3）脑磁图 （4）功能磁共振成像 （1）Spike train （2）EEG （3）MEG （4）fMRI
3.2	刺激大脑 Stimulating brain	1	（1）微电极 （2）经颅磁刺激 （3）神经芯片
3.3	同步记录和刺激 Synchronous recording and stimulation		（1）Microelectrode （2）Transcranial magnetic stimulation （3）Neurochip

第四章　锋电位信号处理与分析（Spikes Signal Processing and Analysis）

章节序号 Chapter Number	章节名称 Chapters	课时 Class Hour	知　识　点 Key Points
4.1	锋电位信号预处理 Spikes signal preprocessing	1	（1）阈值法 （2）窗口法 （1）Threshold method （2）Window method
4.2	锋电位信号分析 Spikes signal analysis	2	（1）特征识别 （2）信息解码 （1）Feature recognition （2）Information decoding

第五章 脑电信号处理与分析(EEG Signal Processing and Analysis)

章节序号 Chapter Number	章节名称 Chapters	课时 Class Hour	知 识 点 Key Points
5.1	脑电信号预处理 EEG signal preprocessing	2	(1) 伪迹去除 (2) 主成分分析 (3) 独立分量分析 (4) 自适应滤波 (1) Artifacts removal (2) Principal component analysis (3) Independent component analysis (4) Adaptive filtering
5.2	脑电信号分析 EEG signal analysis	4	(1) 时/频域分析 (2) 时间序列分析 (3) 复杂度分析 (4) 同步性分析 (5) 因果性分析 (6) 共空间模式分析 (7) 模式分类 (1) Time/frequency domain analysis (2) Time series analysis (3) Complexity analysis (4) Synchronization analysis (5) Causality analysis (6) Common spatial patterns (7) Pattern classification

第六章 功能磁共振信号处理与分析(fMRI Signal Processing and Analysis)

章节序号 Chapter Number	章节名称 Chapters	课时 Class Hour	知 识 点 Key Points
6.1	fMRI信号预处理 fMRI signal preprocessing	1	(1) 层时间校正 (2) 头动校正 (3) 配准 (1) Slice timing correction (2) Realignment (3) Co-register

续表

章节序号 Chapter Number	章节名称 Chapters	课时 Class Hour	知 识 点 Key Points
6.2	fMRI 信号分析 fMRI signal analysis	3	(1) 血液动力学响应 (2) 一般线性模型 (3) 统计检验 (4) 脑连接分析 (5) 脑疾病分析 (1) Hemodynamic response function (2) General linear model (3) Statistical test (4) Brain connectivity analysis (5) Brain disease analysis
6.3	fMRI 信号解码 fMRI signal decoding	2	(1) 多体素模式分析 (2) 单体素模型 (1) Multi-voxel pattern analysis (2) Voxel-wised model

第七章 神经形态计算(Neuromorphic Computing)

章节序号 Chapter Number	章节名称 Chapters	课时 Class Hour	知 识 点 Key Points
7.1	神经形态计算概念 Concept of neuromorphic computing	1	(1) 脑启发的计算模型 (2) 类脑芯片 (1) Brain-inspired computing model (2) Brain like chip
7.2	脉冲神经元网络 Spike neural network	2	(1) 脉冲神经元模型 (2) 突触连接学习规则 (3) 脉冲神经元网络结构 (4) 脉冲神经元网络应用 (1) Spike neural model (2) Synaptic connection learning rules (3) Spike neural network structure (4) Applications of spike neural network

第八章 脑机接口(Brain-Computer Interfaces)

章节序号 Chapter Number	章节名称 Chapters	课时 Class Hour	知识点 Key Points
8.1	脑机接口概念 Concept of brain computer interface	3	(1) 脑机接口原理 (2) 脑机接口范式 (3) 脑机接口类型 (1) Principle of brain computer interface (2) Paradigms of brain computer interface (3) Types of brain computer interface
8.2	脑机接口应用 Applications of brain computer interface	2	(1) 医学领域应用 (2) 非医学领域应用 (3) 安全性和隐私性 (1) Medical applications (2) Non-medical applications (3) Security and privacy
8.3	脑机接口伦理 Ethics of brain computer interface		

大纲指导者：郑南宁教授(西安交通大学人工智能学院)

大纲制定者：陈霸东教授(西安交通大学人工智能学院)

大纲审定：西安交通大学人工智能学院本科专业知识体系建设与课程设置第二版修订工作组

第 8 章

"先进机器人技术"课程群

8.1 "机器人学基础"课程大纲

课程名称：机器人学基础

Course：Introduction to Robotics

先修课程：线性代数与解析几何、现代控制工程、现代物理与人工智能

Prerequisities：Linear Algebra and Analytic Geometry, Modern Control Engineering, Physics for Artificial Intelligence

学分：2

Credits：2

8.1.1 课程目的和基本内容（Course Objectives and Basic Content）

本课程是人工智能学院本科专业必修课。

This course is a compulsory course for undergraduates in College of Artificial Intelligence.

机器人学是一门高度交叉的前沿学科，其内涵和外延都在不断地发展变化。本课程立足已被广泛应用的典型工业机器人，系统地介绍机器人建模与控制的相关基础知识。理解和掌握本课程的内容，不仅可以使学生全面系统地了解已被广泛应用的各类机器人系统，也能够为日后进一步学习研究智能机器人与无人系统打下基础。

通过本课程的学习，可以使学生系统地了解机器人学的基础知识，特别是机器人建模、控制等方面的相关知识，较为深入地学习串联关节型机器人的运动学和动力学分析建模，为进一步的学习和应用打下基础。本课程以课堂教学为主，配合教学实验等环节，系统地培养学生进行机器人分析、建模、控制的能力，以及机器人系统设计和

分析仿真的基本技能。

本课程以机器人的运动学和动力学建模为主线，系统地介绍了刚体的运动描述、空间坐标变换、正向/逆向运动学、动力学分析等基础知识，以及典型机器人系统的设计与分析。第一章概括介绍了机器人学的发展历程；第二章重点介绍了刚体的运动描述；第三章到第六章以关节型串联机器人（机械臂）为背景，详细介绍了机器人运动学/动力学分析的基础知识；第七章介绍了机械臂的运动轨迹生成；第八章介绍了机械臂和移动平台的设计；第九章对工业机器人的控制进行了简单介绍。

Robotics is a highly interdisciplinary and cutting-edge discipline, its intension and extension are evolving constantly. Based on the industrial manipulator that have been widely used, this course systematically introduces the basic knowledge of robotic modeling and control. By understanding and mastering the content of this course, students will comprehensively and systematically understand the robotic systems that have been widely used, and build the foundation for study of intelligent robots and unmanned systems in the future.

This course helps students systematically understand the basic theory of robotics, especially the knowledge of robotic modeling and control, deeply learn the kinematics and dynamics of articulated robots, which build a knowledge framework for further study and application. This course develops the students' ability to perform robotic analysis, modeling and control, trains the students' skill in the design and application of robotic system by classroom teaching in addition with experiments.

This course focuses on the kinematics and dynamics modeling of robots. It systematically introduces the basic knowledge of rigid body motion description, space coordinate transformation, forward/inverse kinematics and dynamics, and the design and analysis of typical robot systems. Chapter 1 gives an overview of the evolution of robotics. Chapter 2 focuses on the motion description of rigid bodies. Chapter 3 to Chapter 6 provide a detailed introduction to the kinematics and dynamics analysis of the manipulator. Chapter 7 introduces the trajectory generation of manipulators; Chapter 8 introduces the mechanism design of manipulators and mobile platform; Chapter 9 gives a brief introduction to the control of industrial robots.

8.1.2 课程基本情况（Course Arrangements）

课程名称	机器人学基础 Introduction to Robotics							
开课时间	一年级		二年级		三年级		四年级	
	秋	春	秋	春	秋	春	秋	春
课程定位	本科生"先进机器人技术"课程群必修课							
学　　分	2学分							
总 学 时	36学时（授课学时32学分、实验4学时）							
授课学时分配	课堂讲授（32学时）							
先修课程	线性代数与解析几何、现代控制工程、现代物理与人工智能							
后续课程	认知机器人							
教学方式	课堂教学、大作业、实验							
考核方式	闭卷考试成绩占60%、作业成绩占10%、实验成绩占30%							
参考教材	1. Craig J J. 机器人学导论[M]. 4版. 北京：机械工业出版社，2018. 2. 蔡自兴. 机器人学[M]. 3版. 北京：清华大学出版社，2015.							
参考资料								
其他信息								

先进机器人技术	
必修 （学分）	机器人学基础(2)
	多智能体与人机混合智能(2)
选修 （学分）	认知机器人(2)
	先进自动驾驶技术与系统(2)

8.1.3 教学目的和基本要求（Teaching Objectives and Basic Requirements）

（1）了解机器人学的发展、现状和趋势；
（2）掌握机器人位姿和运动的描述；
（3）掌握机器人的运动学正解和逆解方法；
（4）掌握机器人的速度与静力分析方法；
（5）掌握机器人的动力学分析与建模方法；
（6）掌握基本的运动轨迹生成方法；
（7）了解机器人系统设计与控制方法。

8.1.4 教学内容及安排(Syllabus and Arrangements)

第一章 绪论（Introduction）

章节序号 Chapter Number	章节名称 Chapters	课时 Class Hour	知 识 点 Key Points
1.1	绪论 Introduction	2	（1）机器人学的发展历史、现状、趋势 （2）机器人的相关定义、特点、分类 （3）工业机器人与工业自动化 （4）机器人与人工智能 (1) History, status, and trend (2) Definition, characteristics and classification (3) Industrial robotics and industrial automation (4) Robotics and artificial intelligence

第二章 刚体的空间描述与变换（Spatial Description and Transformation of Rigid Bodies）

章节序号 Chapter Number	章节名称 Chapters	课时 Class Hour	知 识 点 Key Points
2.1	齐次变换矩阵 Homogeneous transform matrix	2	（1）刚体自由度 （2）旋转矩阵和齐次变换矩阵 （3）变换方程 (1) DoF of rigid body (2) Rotation matrix and homogeneous transform matrix (3) Transform equations
2.2	其他表达 Other representations	2	（1）朝向的固定角和欧拉角表达 （2）矩阵指数表达 （3）欧拉参数表达 (1) Fixed angle and Euler angle (2) Exponential coordinate (3) Euler parameter

第三章　机械臂的正向运动学(Forward Kinematics of Manipulators)

章节序号 Chapter Number	章节名称 Chapters	课时 Class Hour	知　识　点 Key Points
3.1	D-H 参数描述 Denavit-Hartenberg parameters	2	(1) 机构的自由度 (2) 构形空间 (3) 连杆与关节 (4) D-H 参数描述 (1) DoF of mechanism (2) Configuration space (3) Links and joints (4) Denavit-Hartenberg parameter
3.2	运动学正解 Forward kinematics	2	(1) 连杆变换 (2) D-H 参数描述的变种 (3) 典型的关节型机械臂运动学正解 (4) 包含闭链机构的机械臂运动学正解 (1) Link transformations (2) Variants of the D-H parameter (3) Forward kinematics of articulated robots (4) Forward kinematics of robots with closed chain

第四章　机械臂的逆向运动学(Inverse Kinematics of Manipulators)

章节序号 Chapter Number	章节名称 Chapters	课时 Class Hour	知　识　点 Key Points
4.1	逆向运动学的闭式解 Closed-form solution of inverse kinematics	2	(1) 解的存在性、多解性问题 (2) 闭式解的代数解法 (3) 闭式解的几何解法 (1) Existence of solutions, multiple solutions (2) Algebraic solution (3) Geometric solution

续表

章节序号 Chapter Number	章节名称 Chapters	课时 Class Hour	知 识 点 Key Points
4.2	典型工业机械臂的运动学逆解 Inverse kinematic solution of the typical manipulators	2	(1) 三个关节轴交于一点的六轴机械臂 (2) 具有三个平行关节轴的六轴机械臂 (3) 少于六轴的机械臂 (1) 6-axes manipulator with three axes intersect (2) 6-axes manipulator with three axes parallel (3) n-axes manipulator($n<6$)

第五章　速度与静力(Velocities and Static Forces)

章节序号 Chapter Number	章节名称 Chapters	课时 Class Hour	知 识 点 Key Points
5.1	刚体的运动描述 Motion description of rigid bodies	2	(1) 线速度与角速度 (2) 连杆之间的速度传播 (1) Linear and angular velocities (2) Velocity propagation from link to link
5.2	雅可比 Jacobians	2	(1) 雅可比矩阵及其计算 (2) 奇异、冗余 (1) Jacobians matrix (2) Singularity, redundancy
5.3	逆向运动学的数值解 Numerical inverse kinematics	1	(1) 雅可比伪逆法 (2) 雅可比转置法 (1) Jacobian pseudo-inverse (2) Jacobian transpose
5.4	静力学分析 Static forces	1	(1) 连杆之间的静力传播 (2) 力域的雅可比 (3) 速度-静力的对偶关系 (1) Static forces propagation from link to link (2) Jacobians in the force domain (3) Kineto-statics duality

第六章　机械臂动力学（Manipulator Dynamics）

章节序号 Chapter Number	章节名称 Chapters	课时 Class Hour	知　识　点 Key Points
6.1	惯性张量 Inertia tensor	2	（1）刚体的线加速度与角加速度 （2）刚体的质量分布与惯性张量 (1) Linear acceleration and angular acceleration (2) Mass distribution and inertia tensor
6.2	迭代牛顿-欧拉动力学方程 Iterative Newton-Euler dynamic formulation	1	（1）牛顿-欧拉方程 （2）迭代牛顿-欧拉动力学方程 (1) Newton-Euler formulation (2) Iterative Newton-Euler dynamic formulation
6.3	机械臂动力学的拉格朗日方程 Lagrangian formulation of manipulator dynamics	2	（1）拉格朗日方程 （2）机械臂动力学的拉格朗日方程 (1) Lagrangian formulation (2) Lagrangian formulation of manipulator dynamics
6.4	机械臂动力学模型的特性 Properties of dynamic model	1	（1）重排动力学方程 （2）机械臂动力学模型的特性 (1) Rewriting dynamics formulation (2) Properties of dynamic model

第七章　轨迹生成（Trajectory Generation）

章节序号 Chapter Number	章节名称 Chapters	课时 Class Hour	知　识　点 Key Points
7.1	关节空间的轨迹生成 Joint-space schemes	1	（1）路径规划与轨迹生成的问题描述 （2）基于关节空间的轨迹生成 (1) Path planning and trajectory generation (2) Trajectory generation in joint-space

续表

章节序号 Chapter Number	章节名称 Chapters	课时 Class Hour	知 识 点 Key Points
7.2	笛卡儿空间的轨迹生成 Cartesian-space schemes	1	(1) 基于笛卡儿空间的轨迹生成； (2) 笛卡儿空间和关节空间生成轨迹的优缺点； (1) Trajectory generation in Cartesian-space (2) Pros and cons in Cartesian and joint space

第八章 机构设计(Mechanism Design)

章节序号 Chapter Number	章节名称 Chapters	课时 Class Hour	知 识 点 Key Points
8.1	机械臂的机构设计 Mechanism design of manipulators	1	(1) 主要技术指标 (2) 臂与腕的机构设计 (3) 典型的工业机器人系统 (4) 机械臂的冗余结构和闭链结构 (1) Main technical specifications (2) Mechanism design of arms and wrists (3) Typical industrial robot systems (4) Redundancies and Closed-chain structures
8.2	移动机器人简介 Introduction to mobile robots	1	(1) 运动问题 (2) 稳定性、机动性、可控性 (3) 轮式移动机器人 (1) Locomotion (2) Stability, maneuverability, controllability (3) Wheeled mobile robots

第九章 机器人的控制(Robot Control)

章节序号 Chapter Number	章节名称 Chapters	课时 Class Hour	知 识 点 Key Points
9.1	机器人的运动控制 Motion control of robot	1	(1) 单关节机器人的运动控制 (2) 多关节机器人的运动控制 (1) Motion control of single joint robot (2) Motion control of multi-joint robot

续表

章节序号 Chapter Number	章节名称 Chapters	课时 Class Hour	知识点 Key Points
9.2	机器人的力控制 Force control of robot	1	(1) 被动柔顺与主动柔顺 (2) 刚性控制、阻抗控制 (3) 力与位置混合控制 (1) Passive compliance and active compliance (2) Stiffness control, impedance control (3) Hybrid position/force control

8.1.5 实验环节(Experiments)

序号 Num.	实验内容 Experiment Content	课时 Class Hour	知识点 Key Points
1	开链工业机器人运动学实验 Experiment for kinematics of open chain robots	2	(1) 使用 D-H 参数描述机械臂 (2) 机械臂的正向与逆向运动学 (3) 机械臂的工作空间分析 (1) Defining a manipulator by using Denavit and Hartenberg parameters (2) Forward and inverse kinematics of manipulator (3) Analyzing the workspace of a manipulator
2	闭链机器人运动学实验 Experiment for closed chain robots	2	(1) 使用 URDF 描述一款 Delta 机器人 (2) Delta 机器人的正向与逆向运动学 (3) 绘制 Delta 机器人的工作空间 (1) Defining a Delta robot by using unified Robot description Format (2) Forward and inverse kinematics of Delta robot (3) Plotting the workspace of Delta robot

大纲指导者:郑南宁教授(西安交通大学人工智能学院)

大纲制定者:徐海林高级工程师(西安交通大学人工智能学院)

大纲审定：西安交通大学人工智能学院本科专业知识体系建设与课程设置第二版修订工作组

8.2 "多智能体与人机混合智能"课程大纲

课程名称：多智能体与人机混合智能
Course：Multi-agent and Human-machine Hybrid Intelligence
先修课程：人工智能概论、强化学习、认知心理学基础、计算神经工程
Prerequisites：Introduction to Artificial Intelligence, Reinforcement Learning, Introduction to Cognitive Psychology, Computational Neural Engineering
学分：2
Credits：2

8.2.1 课程目的和基本内容（Course Objectives and Basic Content）

本课程是人工智能学院本科专业必修课，课程由两个主题组成，包括主题1：多智能体，主题2：人机混合智能。

This course is a compulsory course for undergraduates in College of Artificial Intelligence. It consists of two topics, topic 1 is Multi-agent, topic 2 is Human-machine Hybrid Intelligence.

主题1 旨在理解多智能体系统的基础理论和研究成果，回顾多智能体的出现及发展。20世纪70年代末期，智能体概念初现并迅速发展，90年代多智能系统体涌现出自主性、社会能力、反应性等特点，因此成为机器人领域、控制领域等众多领域的研究热点。多智能体系统通过信息融合与协作可以完成超出它们各自能力范围的任务，使得系统整体能力大于个体能力之和，其根本任务是其对多个个体的信息进行融合以及协同控制。对此，本课程从多智能体如何共同学习和工作两个角度分别讲述多智能体强化学习、多智能体系统协同控制的代表性方法，并通过介绍蚁群、粒子群等典型集群智能算法，展示由个体聚集成群体而涌现出更高水平智能的现象。

Topic 1 aims to understand the basic theories and research results of multi-agent systems, review the emergence and development of multi-agents. In the late 1970s, the concept of Agent was first developed and developed rapidly. In the 1990s, many intelligent systems emerged with autonomy, social ability, and reactivity. Therefore, they became the research hotspots in many fields such as robotics and control. Multi-agent systems can accomplish tasks beyond their respective capabilities through information fusion and collaboration, making the overall system capacity greater than the sum of individual capabilities. The fundamental task is to integrate and coordinate the information of multiple individuals. In this regard, this course introduces the representative methods of multi-intelligence reinforcement learning and cooperative control of multi-intelligence systems from the perspectives of how multi-intelligence learn and work together, and demonstrates the phenomenon of higher level of intelligence emerging from the aggregation of individuals into a group by introducing typical cluster intelligence algorithms such as ant colony and particle swarm.

主题 2 旨在理解将人的作用引入到智能系统的人机混合智能，它把人对模糊、不确定问题分析与响应的高级认知机制与机器智能系统紧密耦合，使得两者相互适应、协同工作，形成双向的信息交流与控制。把人的感知、认知能力和计算机强大的运算和存储能力相结合，可以形成"1+1＞2"的增强智能形态，从而实现大规模的非完整、非结构化知识信息的处理，同时避免由于当前人工智能技术的局限性带来的决策风险和系统失控等问题。启发学生在修完人工智能核心专业课程后，理解认知计算的基本框架，进一步思考如何将机器学习、知识库和人类决策更好地结合起来，使计算机在人类参与度降低的情况下仍能以较高的准确度和置信度完成大部分工作。

Topic 2 aims to understand the human-machine hybrid intelligence that introduces the human role into the intelligent system. It closely couples the advanced cognitive mechanism of fuzzy and uncertain problem analysis and response with the machine intelligence system, so that the two adapt and work together to form two-way information exchange and control. Combining human perception, cognitive ability and powerful computing and storage capabilities of the computer can form an enhanced intelligent form of "1+1＞2", thereby realizing large-scale processing of non-holistic and unstructured knowledge information. Avoid problems such as decision-making risks and system out-of-control due to the limitations of current artificial intelligence technologies. Inspire students to understand the basic framework of cognitive computing after completing the core courses of artificial intelligence, and further think about how to better combine machine learning, knowledge base and human decision making, and the computers can accomplish most of the work with high accuracy and confidence, even the human participation is reduced.

8.2.2 课程基本情况(Course Arrangements)

课程名称	多智能体与人机混合智能 Multi-agent and Human-machine Hybrid Intelligence								
开课时间	一年级		二年级		三年级		四年级		先进机器人技术
	秋	春	秋	春	秋	春	秋	春	
课程定位	本科生"先进机器人技术"课程群必修课								必修 (学分)	机器人学基础(2)
学 分	2学分									多智能体与人机混合智能(2)
总学时	32学时(授课32学时、实验0学时)								选修 (学分)	认知机器人(2)
授课学时分配	课堂讲授(30学时),大作业讨论(2学时)									先进自动驾驶技术与系统(2)
先修课程	人工智能概论、强化学习、认知心理学基础、计算神经工程								
后续课程									
教学方式	课堂教学、大作业与实验、小组讨论、综述报告								
考核方式	课程结束笔试成绩占50%,大作业成绩占40%,考勤占10%								
参考教材	Zheng N N. Hybrid-augmented intelligence: collaboration and cognition[J]. Frontiers of IT & EE, 2017, 18(2): 153-179.								
参考资料	1. 视听觉信息的认知计算项目组. 视听觉信息的认知计算[M]. 杭州:浙江大学出版社,2020. 2. Mitchell M. AI 3.0[M]. 王飞跃,李玉珂,王晓,等译. 成都:四川科学技术出版社,2021.								
其他信息									

8.2.3 教学目的和基本要求(Teaching Objectives and Basic Requirements)

主题1 多智能体

(1) 掌握多智能体系统的概念与特性,了解多智能体系统的发展和现状;
(2) 掌握多智能体强化学习的典型方法;
(3) 理解多智能体协同控制的基本思想;
(4) 掌握集群智能的典型优化方法;
(5) 熟悉多智能体的典型应用。

主题2 人机混合智能

(1) 理解人机混合智能的基本思想;

(2) 掌握受脑认知启发的混合增强智能基本理论与方法；
(3) 掌握人机协作的混合增强智能基本理论与方法；
(4) 熟悉人机混合智能的典型应用。

8.2.4 教学内容及安排(Syllabus and Arrangements)

主题1 多智能体(Multi-agent)

章节序号 Chapter Number	章节名称 Chapters	课时 Class Hour	知 识 点 Key Points
1	多智能体系统 Introduction to multi-agent system	2	(1) 多智能体系统基本概念 (2) 多智能体系统基本框架 (3) 多智能体系统发展历史与现状 　　多智能体系统是指由多个智能体组成的系统，每个智能体都具有一定的自主性和智能性，可以通过相互交互和合作来完成特定的任务。多智能体系统的优点在于可以充分利用各个智能体的特长，提高系统的效率和性能，并且具有较高的灵活性和鲁棒性。但是，由于多智能体系统中涉及到多个智能体之间的相互作用和协调，因此其设计和优化也面临着较大的挑战。未来，随着人工智能技术的不断发展和完善，多智能体系统将会得到更广泛的应用和发展 (1) Basic concepts of multi-agent systems (2) Basic framework of multi-agent system (3) History and current status of multi-agent systems development 　　A multi-agent system is a system composed of multiple intelligence, each of which has a certain degree of autonomy and intelligence and can accomplish specific tasks through mutual interaction and cooperation. The advantage of multi-agent system is that it can make full use of the strengths of each agent, improve the efficiency and performance of the system, and have high flexibility and robustness. However, since multi-agent systems involve interactions and coordination among multiple agents, their design and optimization also face greater challenges. In the future, with the continuous development and improvement of artificial intelligence technology, multi-agent systems will be more widely used and developed

第8章 "先进机器人技术"课程群

续表

章节序号 Chapter Number	章节名称 Chapters	课时 Class Hour	知 识 点 Key Points
2	多智能体强化学习 Multi-agent reinforcement learning	6	(1) 多智能体强化学习理论基础 (2) 完全合作任务多智能体强化学习方法 (3) 完全竞争任务多智能体强化学习方法 (4) 混合竞争与合作任务多智能体强化学习方法 (5) 多智能体强化学习的典型应用与发展趋势 　　多智能体强化学习研究多个智能体在一个共享环境中协同学习的问题。与传统的单智能体强化学习相比，多智能体强化学习需要考虑智能体之间的相互作用和影响，因此更具挑战性。在多智能体强化学习中，智能体需要通过与环境的交互来学习最优策略，同时还需要考虑其他智能体的行为对自身策略的影响。为了解决这些问题，研究者们提出了许多方法，如基于博弈论的方法、基于集成学习的方法、基于分布式学习的方法等。多智能体强化学习是未来人工智能发展的重要方向之一，其研究将会对人类社会产生深远的影响 (1) Theoretical foundations of multi-agent reinforcement learning (2) Multi-agent reinforcement learning in fully cooperative setting (3) Multi-agent reinforcement learning in fully competitive setting (4) Multi-agent reinforcement learning in mixture cooperative and competitive setting (5) Typical applications and future directions of marl 　　Multi-agent reinforcement learning studies the problem of multiple agents learning collaboratively in a shared environment. Compared with traditional single-agent reinforcement learning, multi-agent reinforcement learning is more challenging because it needs to consider the interactions and influences among the agents. In multi-agent reinforcement learning, agents need to learn optimal strategies through interaction with the environment, and also need to consider the effects of other agents' behaviors on their own strategies. In order to solve these problems, researchers have proposed many methods, such as game theory-based methods, integrated learning-based methods, and distributed learning-based methods. Multi-intelligent body reinforcement learning is one of the important directions for the development of artificial intelligence in the future, and its research will have a far-reaching impact on human society

续表

章节序号 Chapter Number	章节名称 Chapters	课时 Class Hour	知 识 点 Key Points
3	多智能体系统的协同控制 Cooperative control of multi-agent systems	4	（1）多智能体系统协同控制基础理论 （2）分布式控制与估计 （3）多智能体系统一致性控制 （4）多智能体系统协同编队控制 　　多智能体协同控制是指多个智能体在完成任务时相互协作，通过信息交换和合作来实现任务目标。在多智能体系统中，每个智能体都具有一定的自主决策能力和控制能力，可以根据任务需求和环境变化进行相应的调整。多智能体协同控制可以应用于各种领域，例如无人机编队控制、智能交通系统、工业自动化等。在实际应用中，多智能体协同控制可以提高系统的效率和鲁棒性，减少人工干预，降低操作成本。然而，多智能体协同控制也存在一些挑战，例如信息共享、决策一致性、安全性等问题。因此，如何设计合理的协同策略和优化算法是多智能体协同控制研究的重要方向 （1）Basic theory of cooperative control of multi-agent systems （2）Distributed control and estimation （3）Multi-agent system consistency control （4）Multi-agent formation control 　　Multi-agent collaborative control refers to the fact that multiple agents collaborate with each other in accomplishing the task, and realize the task goal through information exchange and cooperation. In a multi-agent system, each intelligent body has a certain degree of autonomous decision-making ability and control ability, which can be adjusted according to the task requirements and environmental changes. Multi-agent cooperative control can be applied to various fields, such as UAV formation control, intelligent transportation system, industrial automation and so on. In practical applications, multi-agent cooperative control can improve the efficiency and robustness of the system, reduce manual intervention, and lower operating costs. However, there are some challenges in multi-agent cooperative control, such as information sharing, decision consistency, security and other issues. Therefore, how to design reasonable cooperative strategies and optimization algorithms is an important direction in the research of multi-agent cooperative control

续表

章节序号 Chapter Number	章节名称 Chapters	课时 Class Hour	知　识　点 Key Points
4	集群智能 Swarm intelligence	4	(1) 集群智能的基本概念 (2) 集群智能的优化方法：蚁群优化、粒子群优化、蜂群优化 (3) 集群智能的典型应用 　　集群智能利用计算机算法来模拟群体智能的表现和决策过程。它的基本思想是将大量的智能个体组合成一个有机的整体，从而达到超越单个个体的效果。集群智能可以应用于各种领域，如数据挖掘、图像识别、智能交通等。其中，最为著名的应用是蚁群算法和粒子群算法。蚁群算法模拟了蚂蚁在寻找食物时的行为，通过模拟信息素的传递和蚂蚁的行动来寻找最优解。粒子群算法则模拟了鸟群在寻找食物时的行为，通过模拟粒子的速度和位置变化来寻找最优解。集群智能技术具有高效性、自适应性和鲁棒性等优点，可以有效地解决复杂问题，为人类社会带来更多的福利 (1) Basic concepts of swarm intelligence (2) Optimization in swarm intelligence: Particle Swarm Optimization (PSO), Ant Colony Optimization (ACO), Bees Colony Algorithm (BCA) (3) Typical applications of swarm intelligence 　　Swarm intelligence uses computer algorithms to simulate the performance and decision-making process of group intelligence. Its basic idea is to combine a large number of intelligent individuals into an organic whole, so as to achieve the effect of surpassing a single individual. Swarm intelligence can be applied to various fields, such as data mining, image recognition, intelligent transportation and so on. Among them, the most famous applications are ant colony algorithm and particle swarm algorithm. The ant colony algorithm simulates the behavior of ants when searching for food, and searches for the optimal solution by simulating the transmission of pheromones and the actions of ants. The particle swarm algorithm, on the other hand, simulates the behavior of a flock of birds while searching for food, and finds the optimal solution by simulating the change in speed and position of the particles. Swarm intelligence technology has the advantages of high efficiency, adaptivity and robustness, which can effectively solve complex problems and bring more benefits to human society

主题 2　人机混合智能(Human-machine Hybrid Intelligence)

章节序号 Chapter Number	章节名称 Chapters	课时 Class Hour	知　识　点 Key Points
1	人机混合智能概述 Introduction of human-machine hybrid intelligence	2	(1) 人工智能与人类智能 (2) 经典机器学习方法的局限性 (3) 人机混合智能的基本框架 　　人工智能与人类智能具有不同的优势,两者具有高度的互补性,将人的作用或人的认知模型引入到人工智能系统中,是人工智能或机器智能的可行的、重要的成长模式 (1) Artificial intelligence and human intelligence (2) Limitations of existing machine learning methods (3) Basic framework of hybrid augmented intelligence 　　Artificial intelligence and human intelligence have different advantages, the two have a high degree of complementarity, the introduction of human role or human cognitive model into artificial intelligence system, is a feasible and important growth model of artificial intelligence or machine intelligence

第 8 章 "先进机器人技术"课程群

续表

章节序号 Chapter Number	章节名称 Chapters	课时 Class Hour	知 识 点 Key Points
2	受脑认知启发的 混合增强智能 Brain cognition inspired hybrid-augmented intelligence	6	(1) 生物智能与脑认知过程的信息处理 (2) 长期记忆与注意力模型 (3) 学习与知识演化 (4) 直觉推理 (5) 因果推理与溯因推理 (6) 从机器学习到学习机器 　　人脑所具有的自然生物智能形式，为提高机器对复杂动态环境或情景的适应性，以及非完整、非结构化信息处理、自主学习能力和构建基于脑认知启发的混合增强智能提供了重要启示。受脑认知启发的混合增强智能是指通过模仿人脑功能提升计算机的感知、推理和决策能力的智能软件或硬件，以更准确地建立像人脑一样感知、推理和响应激励的智能计算模型，尤其是建立因果模型、直觉推理和联想记忆的新计算框架 (1) Information processing of biological intelligence and brain cognitive processes (2) Long-term memory and attention models (3) Learning and knowledge evolution (4) Intuitive reasoning (5) Causal reasoning and abductive reasoning (6) From Machine Learning to Learning Machine 　　The natural form of biological intelligence of the human brain provides important enlightenment for improving the adaptability of machines to complex dynamic environments or situations, as well as incomplete and unstructured information processing, self-learning ability and construction of hybrid augmented intelligence based on brain cognitive inspiration. Brain cognition inspired hybrid-augmented intelligence refers to intelligent software or hardware that enhances a computer's ability to perceive, reason, and make decisions by mimicking the functions of the human brain, in order to more accurately model intelligent computation that perceives, reasons, and responds to incentives as the human brain does, and in particular, to build new computational frameworks for causal modeling, intuitive reasoning, and associative memory

续表

章节序号 Chapter Number	章节名称 Chapters	课时 Class Hour	知 识 点 Key Points
3	人机协同的混合增强智能 Human-machine collaboration hybrid-augmented intelligence	4	(1) 人机交互智能的人为因素 (2) 人机交互认知与学习 (3) 人机交互推理与智能决策 　　人机协同的混合增强智能是指需要人参与交互的一类智能系统,人始终是这类智能系统的一部分。把人的感知、认知能力和计算机强大的运算和存储能力相结合,从而实现大规模的非完整、非结构化知识信息处理,同时避免由于当前人工智能技术的局限性带来的决策风险和系统的失控等问题 (1) The human factor of human-machine interaction (2) Human-machine interaction cognition and learning (3) Human-machine interactive reasoning and decision-making 　　Human-machine collaboration hybrid-augmented intelligence refers to a class of intelligent systems that require human participation in interaction, and humans are always part of such intelligent systems. Combining human perception and cognitive abilities with the powerful computing and storage capabilities of computers, thus realizing large-scale information processing of incomplete and unstructured knowledge, while avoiding decision-making risks and system loss of control brought about by the limitations of current artificial intelligence technologies

续表

章节序号 Chapter Number	章节名称 Chapters	课时 Class Hour	知 识 点 Key Points
4	人机混合智能的典型应用 Typical application of human-machine hybrid-augmented intelligence	2	（1）受脑认知和神经科学启发的混合增强智能算法：AlphaGo、人工智能驱动的科学研究、面向任务的多智能体协作 （2）人在回路的混合增强智能系统：面向智能教育的人机交互知识学习、产业复杂性与风险管理、人-机协同的工业智能决策支持系统、人-机交互的智能医疗诊断系统、人在回路的智慧农业系统、人在回路的工业智能检测与装配 　　人工智能是一种引领许多领域产生颠覆性变革的使能技术，合理并有效地利用人工智能技术，意味着价值创造和竞争优势。人机协同的混合增强智能是新一代人工智能的典型特征。智能机器已经成为人类的伴随者，人与智能机器的交互、混合是未来社会的发展形态 (1) Hybrid-augmented intelligence algorithms inspired by brain cognition and neuroscience: AlphaGo, AI for Science, Task-oriented multi-agent collaboration (2) Human-in-the-loop hybrid augmented intelligence system: ChatGPT and reinforcement learning from human feedback, human-machine interaction knowledge learning for intelligent education, industry complexity and risk management, human-machine collaborative industrial intelligent decision support system, human-machine interaction for intelligent medical diagnosis system, human-in-the-loop smart agricultural systems, human-in-the-loop industrial intelligent inspection and assembly 　　Artificial Intelligence is an enabling technology leading to disruptive changes in many fields, and rational and effective utilization of AI technology means value creation and competitive advantage. Hybrid augmented intelligence with human-machine collaboration is a typical feature of the new generation of AI. Intelligent machines have become the accompaniment of human beings, and the interaction and mixing of human beings and intelligent machines is the development form of the future society

大纲指导者：郑南宁教授（西安交通大学人工智能学院）

大纲制定者：郑南宁教授（西安交通大学人工智能学院）、张雪涛副教授（西安交通大学人工智能学院）、马永强助理教授（西安交通大学人工智能学院）

大纲审定：西安交通大学人工智能学院本科专业知识体系建设与课程设置第二版修订工作组

8.3 "认知机器人"课程大纲

课程名称：认知机器人

Course：Cognitive Robotics

先修课程：人工智能概论、机器学习、自然语言处理、计算机视觉与模式识别、强化学习、认知心理学基础、计算神经工程、机器人学基础

Prerequisites: Introduction to Artificial Intelligence, Machine Learning, Natural Language Processing, Computer Vision and Pattern Recognition, Reinforcement Learning, Introduction to Cognitive Psychology, Computational Neural Engineering, Introduction to Robotics

学分：2

Credits：2

8.3.1 课程目的和基本内容（Course Objectives and Basic Content）

本课程是人工智能学院本科专业选修课。

This course is an elective course for undergraduates in College of Artificial Intelligence.

人工智能与机器人技术在过去十年取得一系列巨大进展并在各行各业广泛应用，然而构建具有类人能力的机器人仍是一项极具挑战的任务。认知机器人研究为该任务提供了一种全新的解决方式，从人类、动物等自然生物系统中汲取灵感，构建具有类人行为、认知与社交能力的智能系统，以及智能系统在物理、社会环境中的学习与生

长发育的范式,以适应复杂的物理环境,完成复杂的操作任务。认知机器人的术语早在20世纪90年代就提出,但目前还未形成完善的理论体系,仍有大量开放性问题需要广大研究者深入探索。本课程旨在介绍认知机器人的基本概念、模型、方法、系统与应用等基本内容,内容涵盖以下三个方面。首先介绍认知机器人的定义、发展历史、基本类型及其系统组成;其次介绍认知机器人的典型认知架构与计算过程;最后重点介绍认知机器人的基本模型与方法,包括认知视觉、空间导航、人机协同、语言交互、知识表示与推理、概念抽象等。本课程采用课堂讲授、文献阅读、课堂展示、综述报告等教学手段,促使学生了解认知机器人的基本概念与知识,初步掌握认知机器人系统开发的基本能力。

Artificial intelligence and robotics have made a series of great progress and widely implemented in the past decade. Nevertheless, building robots with human-like capabilities is still a challenge. Cognitive robotics offer a novel and insightful way to address this challenge. Cognitive robotics take inspiration from natural cognitive systems like humans and animals to build intelligent systems with human-like behaviors, cognition, and social abilities, as well as the learning and growth paradigm in complex environments. The term and field of "cognitive robotics" have their origins in the 1990s, but it has not yet formed a perfect theoretical architecture, and there are still a large number of open questions that need to be deeply explored by the researchers. This course aims to introduce the basic concepts, models, methods, systems and applications of cognitive robots, covering the following three parts. First, the definition, history, common types and system composition of cognitive robots are introduced. Second, basic cognitive architectures and computing processes of cognitive robots are introduced. Third, a series of basic models and methods of cognitive robots are introduced, including cognitive vision, navigation, human-robot collaboration, language communication, knowledge representation and reasoning, abstract concept, and so on. This course adopts classroom teaching, supplemented by paper reading, classroom presentation and discussion, technical report and other methods, in order to training students' abilities to well understand the basic concepts, knowledge of cognitive robots, and develop a basic system of cognitive robots.

8.3.2 课程基本情况(Course Arrangements)

课程名称	认知机器人 Cognitive Robotics								
开课时间	一年级		二年级		三年级		四年级		先进机器人技术
	秋	春	秋	春	秋	春	秋	春	
课程定位	本科生"先进机器人技术"课程群选修课								必修(学分): 机器人学基础(2)、多智能体与人机混合智能(2)
学分	2学分								
总学时	32学时(授课32学时、实验0学时)								选修(学分): 认知机器人(2)、先进自动驾驶技术与系统(2)
授课学时分配	课堂讲授(32学时)								
先修课程	人工智能概论、机器学习、自然语言处理、计算机视觉与模式识别、强化学习、认知心理学基础、计算神经工程、机器人学基础								
后续课程									
教学方式	课堂讲授、文献阅读、综述报告、课堂展示								
考核方式	平时成绩占30%,综述报告占40%,课堂展示占20%,考勤占10%								
参考教材	Cangelosi A, Asada M. Cognitive Robotics[M]. Cambridge: MIT Press, 2022.								
参考资料	IEEE Robotics & Automation Society[EB/OL]. https://www.ieee-ras.org/cognitive-robotics/resources.								
其他信息									

8.3.3 教学目的和基本要求(Teaching Objectives and Basic Requirements)

(1) 了解认知机器人的定义与范畴;
(2) 了解认知机器人的基本类型与系统组成;
(3) 了解典型认知架构的组成与计算过程;
(4) 熟悉认知视觉的基本原理与模型;
(5) 熟悉认知机器人导航的典型方法;
(6) 熟悉认知机器人操作与人机协同基本方法;
(7) 熟悉认知机器人的典型语言模型及特点;
(8) 熟悉知识表示与推理基本方法;
(9) 了解认知机器人的抽象概念与推理模型。

8.3.4 教学内容及安排(Syllabus and Arrangements)

章节序号 Chapter Number	章节名称 Chapters	课时 Class Hour	知 识 点 Key Points
1	认知机器人学概述 Cognitive robotics overview	4	(1) 认知机器人的基本定义与范畴 (2) 常见的认知机器人类型 (3) 认知机器人的系统组成 　　了解认知机器人的基本概念与类型,理解认知机器人的系统组成以及与传统机器人的区别 (1) Basic definition of cognitive robots (2) Common types of cognitive robots (3) System composition of cognitive robots 　　Understand the basic concepts and types of cognitive robots, the system composition of cognitive robots and their differences from traditional robots
2	认知架构 Cognitive architectures	4	(1) 认知科学基础 (2) 认知架构分类 (3) 认知架构特点 (4) 典型认知架构及应用 　　设计和实现适当的认知架构是建立认知机器人的核心步骤,涉及感知、注意、行为、学习、推理、记忆等人类认知过程的机制,指定了认知机器人的组件结构及组件动态关联的方式 (1) The foundations of cognitive science (2) The types of cognitive architecture (3) Desirable characteristics of a cognitive architecture (4) Typical cognitive architectures and their applications 　　The design and implementation of an appropriate cognitive architecture is an essential step in building a cognitive robot, which involves the mechanisms in human cognitive processes such as perception, attention, behavior, learning, reasoning, and memory, and specifies the component structure and dynamic association of cognitive robots

续表

章节序号 Chapter Number	章节名称 Chapters	课时 Class Hour	知　识　点 Key Points
3	认知视觉 Cognitive vision	4	（1）认知视觉的基本概念与范畴 （2）认知视觉的基本原理与模型 　　理解传统计算机视觉的局限与认知视觉的研究目标，系统地建立认知视觉的一系列基本模型，实现类人的认知机器人视觉系统 （1）Basic concepts and categories of cognitive vision （2）Basic principles and models of cognitive vision 　　Understand the limitations of traditional computer vision and the objectives of cognitive vision, systematically establish a series of basic models of cognitive vision, and realize a human-like vision system for cognitive robots
4	认知机器人导航 Cognitive robot navigation	4	（1）认知机器人导航的心理学与神经科学基础 （2）机器人空间认知的计算理论 　　认知机器人的导航能力受到动物空间导航研究的极大影响，借鉴这一领域的心理学和神经科学研究成果，可以为机器人建立更稳定、更通用的智能导航系统，提高机器人的自主性和适应性 （1）Psychological and neuroscientific foundations of cognitive robot navigation （2）Computational theories on robot spatial cognition 　　The navigation ability of cognitive robots is greatly affected by animal spatial navigation research, and drawing on the results of psychological and neuroscience research in this field, a more stable and general intelligent navigation system can be established for robots and improve the autonomy and adaptability of robots

续表

章节序号 Chapter Number	章节名称 Chapters	课时 Class Hour	知 识 点 Key Points
5	认知机器人操作与人机协同 Cognitive robot manipulation and human-robot collaboration	4	（1）认知机器人操作基本方法 （2）认知决策与控制 （3）人机交互的基本方法 （4）机器人的社会认知能力 　　操作是机器人与物理环境交互的基本能力，建立认知机器人的操作、决策与控制等基本方法，实现以人为中心的人机协同作业，并赋予机器人与物理环境中其他个体互动的能力，使机器人能够像人一样完成更加复杂的操作任务 （1）Basic manipulation methods of cognitive robots （2）Cognitive control and decision-making （3）Basic methods of human-robot interaction （4）Social cognition of robots 　　Operation is the basic ability of robots to interact with the physical environment. The basic methods of manipulation, decision-making, control, human-robot collaboration, and social cognition are established for cognitive robots to finish complex tasks like humans
6	语言与交流 Language and communication	4	（1）人类语言发展与学习 （2）发育式机器人语言模型 （3）基于NLP的机器人语言模型 （4）基于机器学习的机器人语言模型 　　交流是一个将语言和注视、手势等非语言行为结合的多模态信息交互理解过程。对于认知机器人，语言和非语言交流技能是其与人互动所必需的基本心理认知能力，而且人和机器人都必须拥有类似语言的通信系统 （1）Language development and learning in Humans （2）Developmental robot language models （3）NLP-based robot language models （4）Machine-learning robot language models 　　Communication is a multimodal information interaction understanding process that combines language with non-verbal behaviors such as gaze and gestures. For cognitive robots, verbal and non verbal communication skills are the basic psychological cognitive abilities necessary for their interaction with people, and both humans and robots must have language-like communication systems

续表

章节序号 Chapter Number	章节名称 Chapters	课时 Class Hour	知识点 Key Points
7	知识表示与推理 Knowledge representation and reasoning	4	（1）符号化知识表示与问答 （2）本体论与知识库 （3）认知机器人的知识表示与推理系统 （4）神经符号学习与推理 　　知识表示和推理是认知机器人在开放环境中完成模糊指定任务的关键信息处理能力，补充机器学习决策和控制机制，将机器学习方法的黑盒推理替换为合理的可解释的推理链 （1）Symbolic knowledge representation and question answering （2）Ontologies and encyclopedic knowledge bases （3）Knowledge representation and reasoning systems for cognitive robots （4）Neurosymbolic learning and reasoning 　　Knowledge representation and reasoning are key information processing capabilities for cognitive robots to complete specified tasks in an open environment, complement machine learning decision-making and control mechanisms, and replace the black-box reasoning of machine learning methods with reasonable and interpretable reasoning chains

续表

章节序号 Chapter Number	章节名称 Chapters	课时 Class Hour	知 识 点 Key Points
8	抽象概念 Abstract concepts	4	（1）抽象概念发展的教育学、神经科学和心理学视角 （2）抽象单词的认知机器人模型 （3）数值概念的认知机器人模型：发展与表示 （4）情绪的认知机器人模型 　　人类智能的特征之一是对抽象概念进行思考和推理的能力。抽象概念构成了人类语言的重要组成部分，抽象思维和推理发展的具体理论构成了设计能够进行抽象和符号处理的智能体的理论基础。如何使认知机器人具备类似能力是这个领域当前面临的挑战之一 （1）Education, neuroscience, and psychology views on the development of abstract concepts （2）Cognitive robotics models of abstract words （3）Cognitive robotics models of numerical concepts: development and representation （4）Cognitive robotics models of emotions 　　One of the characteristics of human intelligence is the ability to think and reason about abstract concepts. Abstract concepts form an important part of human language, and concrete theories of the development of abstract thinking and reasoning form the theoretical basis for designing agents capable of abstract and symbolic processing. How to make cognitive robots have similar capabilities is one of the current challenges in this field

　　大纲指导者：郑南宁教授（西安交通大学人工智能学院）

　　大纲制定者：兰旭光教授（西安交通大学人工智能学院）、张雪涛副教授（西安交通大学人工智能学院）、杨勐副教授（西安交通大学人工智能学院）

　　大纲审定：西安交通大学人工智能学院本科专业知识体系建设与课程设置第二版修订工作组

8.4 "先进自动驾驶技术与系统"课程大纲

课程名称：先进自动驾驶技术与系统
Course：Advanced Autonomous Driving Technology and System
先修课程：工科数学分析、线性代数与解析几何、概率统计与随机过程、计算机视觉与模式识别
Prerequisites：Mathematical Analysis for Engineering，Linear Algebra and Analytic Geometry，Probability Theory and Stochastic Processes，Computer Vision and Pattern Recognition
学分：2
Credits：2

8.4.1 课程目的和基本内容（Course Objectives and Basic Content）

本课程是人工智能学院本科专业选修课。

This course is an elective course for undergraduates in College of Artificial Intelligence.

自动驾驶系统是一种通过计算机与人工智能技术来实现车辆自主行驶的智能系统，涉及计算机视觉、车辆工程、自动控制等多个领域。本课程将全面系统地介绍自动驾驶的理论基础、关键技术、算法实现、仿真测试与验证平台等内容。课程主要分为两部分，第一部分介绍自动驾驶的传感器系统、环境感知与理解、道路使用者行为识别与预测等，旨在让学生熟悉自动驾驶系统使用的传感器系统以及对交通场景的感知和理解；第二部分讨论自动驾驶的地图构建与定位、决策规划与控制、仿真与测试和学习型自动驾驶系统，旨在让学生学习如何通过定位、规划、控制和测试等技术实现车辆的自主驾驶。

本课程通过对自动驾驶基本理论、设计方法和应用技术的学习，帮助学生深入了解自动驾驶领域经典和新兴算法、核心关键技术以及系统开发平台。课程采用集中授课与小组学习相结合的模式，并辅之以小组讨论、日常作业等教学手段，旨在增强学生解决自动驾驶技术所面临的挑战和问题的能力，培养跨学科的综合素养。课程将通过论文阅读报告和综述报告等环节进一步加强学生独立分析问题、解决问题的能力，培养综合设计及创新能力，培养注重实事求是、严肃认真的科学作风和良好的实验习惯，为今后的工作打下良好的基础。

随着人工智能技术的飞速发展,自动驾驶在交通领域成为一种革命性的创新技术,其基本原理和方法应用在汽车领域,不仅使得车辆能够实现智能化、自主化地行驶,而且提高了交通安全性、便捷性和乘车舒适性。因此理解和掌握自动驾驶的基本概念、原理和方法对于培养未来自动驾驶领域的专业人才至关重要。这不仅能激发学生去探索新的理论和技术,而且通过运用自动驾驶技术,学生能够进入一个充满挑战和潜力的研究领域。

Automatic driving system is an intelligent system that realizes the autonomous driving of vehicles through computer and artificial intelligence technology, involving computer vision, vehicle engineering, automatic control and other fields. This course will comprehensively and systematically introduce the theoretical basis, key technologies, algorithm implementation, simulation test and verification platform of autonomous driving, etc. It is mainly divided into two parts. The first part introduces the sensor system of autonomous driving, environmental perception and understanding, road user behavior recognition and prediction, etc., aiming to familiarize students with the sensor system used by the autonomous driving system and the perception and understanding of traffic scenes. The second part discusses map construction and positioning, decision-making planning and control, simulation and testing, and learning automatic driving system for autonomous driving.

This course helps students gain an in-depth understanding of classic and emerging algorithms, core key technologies and system development platforms in the field of autonomous driving through the study of basic theories, design methods and application technologies of autonomous driving. The course adopts a combination of intensive teaching and group learning, supplemented by group discussions, daily homework and other teaching methods, aiming to enhance students' ability to solve the challenges and problems faced by autonomous driving technology and cultivate interdisciplinary comprehensive quality. The course will further strengthen students' ability to independently analyze and solve problems through links such as thesis reading reports and review reports, cultivate comprehensive design and innovation capabilities, cultivate a pragmatic, serious scientific style and good experimental habits, and lay a solid foundation for future work.

With the rapid development of artificial intelligence technology, autonomous driving has become a revolutionary innovative technology in the transportation field. Its basic principles and methods are applied in the automotive field, which not only enables vehicles to achieve intelligent and autonomous driving, but also improves the traffic safety, convenience and ride comfort. Therefore, understanding and mastering the basic concepts, principles and methods of autonomous driving is very important for cultivating professionals in the field of autonomous driving in the future. Not only does this inspire students to explore new theories and technologies, but by using autonomous driving, students are able to enter a field of research that is full of challenges and potential.

8.4.2 课程基本情况(Course Arrangements)

课程名称	先进自动驾驶技术与系统 Advanced Autonomous Driving Technology and System								
开课时间	一年级		二年级		三年级		四年级		先进机器人技术
	秋	春	秋	春	秋	春	秋	春	
课程定位	本科生"先进机器人技术"课程群选修课								必修 (学分)	机器人学基础(2)
学　　分	2学分									多智能体与人机混合智能(2)
总 学 时	32学时 (授课32学时、实验0学时)								选修 (学分)	认知机器人(2)
授课学时分配	课堂讲授(32学时)									先进自动驾驶技术与系统(2)
先修课程	工科数学分析、线性代数与解析几何、概率统计与随机过程、计算机视觉与模式识别								
后续课程									
教学方式	课堂教学、课后作业								
考核方式	考核成绩(大作业)占70%,平时作业成绩占20%,考勤占10%								
参考教材	郑南宁,陈仕韬,杜少毅,等. 自动驾驶高级教程[M]. 北京:清华大学出版社,2023.								
参考资料	1. 李桂成. 计算方法[M]. 3版. 北京:电子工业出版社,2018. 2. 孙文瑜,徐成贤,朱德通. 最优化方法[M]. 北京:高等教育出版社,2010. 3. 张学工,汪小我. 模式识别[M]. 北京:电子工业出版社,2019. 4. Szeliski R. Computer Vision[M]. 2nd. Berlin:Springer Nature,2022. 5. Maurer M. Autonomous Driving Technical,Legal and Social Aspects[M]. Brelin:Springer Nature,2016. 6. Liu S. Creating Autonomous Vehicle System[M]. 2nd. Berlin:Springer Nature,2022.								
其他信息									

8.4.3 教学目的和基本要求(Teaching Objectives and Basic Requirements)

(1) 了解自动驾驶的发展历史、面临的挑战、基本架构和技术分级,熟悉自动驾驶需要解决的基本组成和问题;

(2) 了解自动驾驶传感器系统的构成,熟悉各类传感器的基本功能原理以及优缺点,掌握多传感器时间同步与联合标定方法;

(3) 熟悉交通场景感知与理解的架构,掌握车道线检测与跟踪、交通场景目标检测、交通标识识别、交通场景语义分割、视觉位置识别等算法;

(4) 掌握弱势道路使用者检测和行为识别、意图预测和运动预测的方法;

(5) 掌握基于统计建模和深度学习的车辆行为预测的主要方法,了解周围车辆驾驶风格识别方法;

(6) 掌握自动驾驶中高精度地图的几种常用坐标系统和标准,熟悉高精度语义地图信息生成和编辑;

(7) 掌握里程计定位、基于组合惯性导航系统的融合定位和基于 SLAM 的空间定位方法,熟悉无依托定位和有依托定位的方法;

(8) 掌握自动驾驶中五种行为决策方法的基本原理,熟悉每种决策方法所用的模型架构;

(9) 熟悉基于优化、搜索以及采样的自动驾驶运动规划方法,掌握 A * 路径规划方法和多模型的自动驾驶轨迹规划方法;

(10) 熟悉乘用汽车的典型运动学和动力学模型,理解自动驾驶横向和纵向运动控制算法的基本原理,掌握路径跟踪控制和自适应巡航控制等算法;

(11) 了解车辆动力学模型构建与表征、传感器仿真与建模、情景生成等关键技术方法;熟悉自动驾驶注入式仿真和平行仿真的基本原理及系统实现;

(12) 了解数据驱动的学习型自动驾驶的基本原理,熟悉不同类型的端到端自动驾驶方法;

8.4.4 教学内容及安排(Syllabus and Arrangements)

第一章 绪论(Introduction)

章节序号 Chapter Number	章节名称 Chapters	课时 Class Hour	知 识 点 Key Points
1.1	自动驾驶简介 Introduction to autonomous driving	2	(1) 为什么要发展自动驾驶 (2) 发展历程的五个阶段 (3) 自动驾驶技术的分级 (1) Why develop autonomous driving (2) Five stages of the development progress (3) Levels of autonomous driving technology

续表

章节序号 Chapter Number	章节名称 Chapters	课时 Class Hour	知 识 点 Key Points
1.2	自动驾驶的基本组成和科学问题 Basic components and scientific issues of autonomous driving		（1）自动驾驶的基本组成 （2）基本科学问题与主要难题 （1）Basic components of autonomous driving （2）Fundamental scientific issues and primary challenges

第二章 自动驾驶的传感器系统（Sensor Systems for Autonomous Driving）

章节序号 Chapter Number	章节名称 Chapters	课时 Class Hour	知 识 点 Key Points
2.1	传感器系统简介 Introution to sensor system		（1）环境感知传感器 （2）车辆状态感知传感器 （1）Environmental perception sensors （2）Vehicle state perception sensors
2.2	相机的原理与应用 Camera principles and applications	2	（1）相机的基本原理 （2）相机的内参与外参 （3）相机的标定 （4）全景相机的原理与应用 （5）事件相机的原理与应用 （1）Basic principles of camera （2）Intrinsic and extrinsic parameters of camera （3）Camera calibration （4）The principle and application of panoramic camera （5）The principle and application of event camera

续表

章节序号 Chapter Number	章节名称 Chapters	课时 Class Hour	知 识 点 Key Points
2.3	激光雷达数据采集与处理 LiDAR data acquisition and processing	1	（1）激光雷达工作原理 （2）激光雷达与相机的联合标定 （3）多激光雷达联合标定 (1) Working principle of LiDAR (2) Joint calibration of LiDAR and camera (3) Multi-LiDAR joint calibration
2.4	毫米波雷达数据采集与处理 Millimeter wave radar data acquisition and processing		（1）毫米波雷达的工作原理 （2）毫米波雷达与相机的联合标定 (1) The principle of millimeter-wave radar (2) Joint calibration of millimeter wave radar and camera

第三章 交通场景感知与理解（Traffic Scene Perception and Understanding）

章节序号 Chapter Number	章节名称 Chapters	课时 Class Hour	知 识 点 Key Points
3.1	概述 Overview	2	（1）交通场景感知与理解简介 (1) Introduction to traffic scene perception and understanding
3.2	车道线检测 Lane detection		（1）传统车道线检测方法 （2）基于深度学习的车道线检测方法 (1) Traditional lane line detection method (2) Lane line detection method based on deep learning

续表

章节序号 Chapter Number	章节名称 Chapters	课时 Class Hour	知 识 点 Key Points
3.3	交通场景检测与语义分割 Traffic scene detection and semantic segmentation	2	（1）基于图像的2D/3D目标检测与场景分割 （2）基于点云数据的3D目标检测与场景分割 （3）多传感器信息融合的3D目标检测与可行驶区域分割 (1) Image-based 2D/3D object detection and scene segmentation (2) 3D object detection and scene segmentation based on point cloud data (3) 3D object detection and drivable area segmentation based on multi-sensor information fusion
3.4	交通场景的层次化表征与视觉位置识别 Hierarchical representation of traffic scenes and visual place recognition		（1）交通场景视觉信息的层级结构表征 （2）基于词袋模型的视觉位置识别 （3）应用卷积神经网络的视觉位置识别 （4）基于transformer的视觉位置识别 (1) Hierarchical structure representation of traffic scene visual information (2) Visual localization based on bag-of-words model (3) Visual place recognition based on CNN (4) Visual position recognition based on transformer

第四章 弱势道路使用者的行为识别与预测（Behavior Recognition and Prediction of Vulnerable Road Users）

章节序号 Chapter Number	章节名称 Chapters	课时 Class Hour	知 识 点 Key Points
4.1	弱势道路使用者的检测与行为识别和运行预测 Vulnerable road user detection and behavior recognition	2	（1）弱势道路使用者的检测 （2）弱势道路使用者的姿态识别 （3）弱势道路使用者行为识别 （4）弱势道路使用者运动预测 (1) Detection of vulnerable road users (2) Gesture recognition for vulnerable road users (3) Behavior identification of vulnerable road users (4) Movement prediction of vulnerable road users

续表

章节序号 Chapter Number	章节名称 Chapters	课时 Class Hour	知识点 Key Points
4.2	弱势道路使用者意图预测 Prediction of vulnerable road user intentions		（1）基于运动特征的弱势道路使用者意图预测 （2）基于姿态估计的弱势道路使用者意图预测 （3）基于环境或他人交互的弱势道路使用者意图预测 （4）基于特征融合的弱势道路使用者意图预测 (1) Intention prediction of vulnerable road users based on motion characteristics (2) Intention prediction of vulnerable road users using pose estimation (3) Intention prediction of vulnerable road users based on environment or interaction with others (4) Intention prediction of vulnerable road users based on feature fusion

第五章 自动驾驶周围车辆行为预测与驾驶风格识别（Vehicle Behavior Prediction and Driving Style Recognition around Autonomous Driving）

章节序号 Chapter Number	章节名称 Chapters	课时 Class Hour	知识点 Key Points
5.1	周围车辆的行为和轨迹预测 Behavior and trajectory prediction of surrounding vehicles	2	（1）基于统计模型的行为预测 （2）基于深度学习的轨迹预测 (1) Behavior prediction based on satistical model (2) Trajectory prediction based on deep learning
5.2	周围车辆驾驶风格识别 Driving style recognition of surrounding vehicles		（1）应用混合高斯的驾驶风格识别 （2）基于短期观测或时序数据的实时驾驶风格分类方法 (1) Driving style recognition using mixture Gaussian (2) Real-time driving style classification method based on short-term observation or time series data

第六章　自动驾驶的地图（Autonomous Driving Map）

章节序号 Chapter Number	章节名称 Chapters	课时 Class Hour	知　识　点 Key Points
6.1	概述 Overview	2	（1）自动驾驶的地图简介 （2）地图的坐标系统 (1) Introduction to maps for autonomous driving (2) Map coordinate system
6.2	高精度语义地图 的信息生成与编辑 Information generation and editing of HD semantic maps		（1）高精度语义地图的信息生成 （2）基于人机交互的高精度语义地图编辑 (1) Informaton generation of HD semantic map (2) HD semantic map editing based on human-machine interaction
6.3	其他高精度地图标准 Other standards of HD maps		（1）OpenDrive 地图标准 （2）Lanelet 地图标准 （3）NDS 地图标准和多层级地图 (1) OpenDrive map standard (2) Lanelet map standard (3) NDS map standard and multi-level maps

第七章　自动驾驶的定位系统（Positioning System for Autonomous Driving）

章节序号 Chapter Number	章节名称 Chapters	课时 Class Hour	知　识　点 Key Points
7.1	概述 Overview		（1）自动驾驶的定位系统简介 (1) Introduction to positioning system for autonomous driving

续表

章节序号 Chapter Number	章节名称 Chapters	课时 Class Hour	知 识 点 Key Points
7.2	组合导航系统 Integrated navigation system	2	(1) 系统组成 (2) 基于扩展卡尔曼滤波的组合导航算法 (1) System configuration (2) Integrated navigation algorithm based on Kalman filter
7.3	同步定位与建图 Simultaneous Localization And Mapping(SLAM)		(1) 自动驾驶定位的里程计方法 (2) SLAM 系统框架 (3) 视觉或激光 SLAM (1) Odometry method for autonomous driving localization (2) SLAM system framework (3) Visual or laser SLAM
7.4	基于车载传感器的无依托定位 Support-free positioning based on on-board sensors	1	(1) 面向自动驾驶定位的高精度地图构建 (2) 地图压缩与编码 (3) 应用先验高精度地图的点云匹配定位 (4) 软硬件协同设计方法 (1) HD map construction for autonomous driving positioning (2) Map compression and encoding (3) Point cloud matching and positioning using prior HD maps (4) Software-hardware co-design method
7.5	协同其他交通元素的有依托定位 Relying on positioning in coordination with other traffic elements		(1) 基于交通标识的协同定位 (2) 基于交通参与者行为分析的协同定位 (1) Co-location based on traffic signs (2) Co-location based on traffic participant behavior analysis

第八章　自动驾驶的行为决策（Behavioral Decision Making in Autonomous Driving）

章节序号 Chapter Number	章节名称 Chapters	课时 Class Hour	知　识　点 Key Points
8.1	传统的行为决策 Traditional behavior decisions	3	（1）基于有限状态机的行为决策 （2）基于部分可观测马尔可夫决策过程的行为决策 （3）应用博弈论的交互行为决策 （4）应用蒙特卡洛方法的行为决策 （1）Behavior decision based on finite state machine （2）Behavior decision based on partially observable Markov decision process （3）Interactive behavior decision using game theory （4）Behavior decision using Monte Carlo methods
8.2	基于学习的行为决策 Learning-based behavior decisions		（1）基于模仿学习的行为决策 （2）基于强化学习的行为决策 （3）逆强化学习的行为决策 （1）Behavior decision based on imitation learning （2）Behavior decision based on reinforcement learning （3）Behavior decision based on inverse reinforcement learning

第九章　自动驾驶的运动规划（Motion Planning for Autonomous Driving）

章节序号 Chapter Number	章节名称 Chapters	课时 Class Hour	知　识　点 Key Points
9.1	自动驾驶运动规划的基本方法 Basic methods of motion planning for autonomous driving	2	（1）自动驾驶的运动规划简介 （2）基于优化的规划 （3）基于搜索的规划 （4）基于采样的规划 （1）Introduction to motion planning for autonomous driving （2）Optimization-based planning （3）Search-based planning （4）Sampling-based planning

续表

章节序号 Chapter Number	章节名称 Chapters	课时 Class Hour	知 识 点 Key Points
9.2	多层混合A*路径规划 Multilayer hybrid A* path planning		(1) 算法框架 (2) 区域划分和局部终点生成模块 (3) 路径搜索模块 (1) Algorithm framework (2) Region division and local endpoint generation module (3) Route search module
9.3	多模型的自动 驾驶轨迹规划 Multi-model trajectory planning for autonomous driving	1	(1) 算法框架 (2) 路径规划模块 (3) 速度生成模块 (1) Algorithm framework (2) Path planning module (3) Speed generation module

第十章 自动驾驶的控制(Control of Autonomous Driving)

章节序号 Chapter Number	章节名称 Chapters	课时 Class Hour	知 识 点 Key Points
10.1	自动驾驶的控制简介 Introduction to the control of autonomous driving		(1) 自动驾驶的控制简介 (2) 自动驾驶的车辆模型 (3) 自动驾驶的运动控制 (1) Introduction to the control of autonomous driving (2) Vehicle model for autonomous driving (3) Motion control in autonomous driving

续表

章节序号 Chapter Number	章节名称 Chapters	课时 Class Hour	知　识　点 Key Points
10.2	自动驾驶的路径跟踪与预测控制 Path tracking and predictive control for autonomous driving	3	（1）自动驾驶的路径跟踪与车道保持 （2）自动驾驶的预测控制算法 （3）自动驾驶的自适应巡航 （1）Path tracking and lane keeping for autonomous driving （2）Predictive control algorithm for autonomous driving （3）Adaptive cruise control for autonomous driving
10.3	基于学习的自动驾驶控制方法 Learning-based autonomous driving control methods		（1）基于模仿学习的自动驾驶控制 （2）基于强化学习的自动驾驶控制 （1）Automatic driving control based on imitation learning （2）Automatic driving control based on reinforcement learning

第十一章　自动驾驶的仿真、测试与验证（Simulation, Testing and Verification of Autonomous Driving）

章节序号 Chapter Number	章节名称 Chapters	课时 Class Hour	知　识　点 Key Points
11.1	自动驾驶仿真测试简介 Introduction to autonomous driving simulation and testing	1	（1）自动驾驶测试的机遇与挑战 （2）自动驾驶仿真系统功能 （3）虚实融合的自动驾驶仿真测试 （1）Opportunities and challenges of autonomous driving testing （2）Functions of autonomous driving simulation systems （3）Virtual-real integrated simulation for autonomous driving

章节序号 Chapter Number	章节名称 Chapters	课时 Class Hour	知 识 点 Key Points
11.2	车辆动力学与车载传感器仿真建模 Simulation modeling of vehicle dynamics and vehicle sensors		（1）车辆动力学模型构建与表征 （2）车载传感器仿真与建模 （1）Vehicle dynamics model construction and representation （2）Simulation and modeling of on-board sensors
11.3	自动驾驶的仿真与测试 Simulation system for autonomous driving	2	（1）自动驾驶仿真与车辆在环测试 （2）自动驾驶的注入式仿真系统 （3）真实数据驱动的测试场景生成 （4）自动驾驶的平行仿真 （1）Autonomous driving simulation and vehicle-in-the-loop testing （2）Injection-based simulation system for autonomous driving （3）Real data-driven test scenario generation （4）Parallel simulation for autonomous driving
11.4	自动驾驶常用数据集与开源工具 Commonly used datasets and open source tools for autonomous driving		（1）常用数据集 （2）开源工具 （1）Commonly used dataset （2）Open source tools

第十二章　基于数据驱动的学习型自动驾驶系统（Data-driven Learning Approach for Autonomous Driving System）

章节序号 Chapter Number	章节名称 Chapters	课时 Class Hour	知　识　点 Key Points
12.1	数据驱动的学习型自动驾驶系统基本原理 Fundamentals of data-driven learning autonomous driving systems	2	（1）基于数据驱动的学习型自动驾驶系统简介 （2）基于浅层卷积神经网络的自动驾驶 （3）基于高精度地图的学习型自动驾驶 （4）基于多模态输入的学习型自动驾驶 (1) Introduction to data-driven learning autonomous driving system (2) Autonomous driving based on shallow convolutional neural networks (3) Learning-based autonomous driving based on high-precision maps (4) Learning-based autonomous driving based on multimodal inputs
12.2	端到端的学习型系统 End-to-end learning autonomous driving system		（1）端到端自动驾驶的车辆控制模式 （2）端到端自动驾驶的学习方法 （3）端到端自动驾驶的注意力机制 （4）端到端自动驾驶的仿真与评估 (1) Vehicle control modes for end-to-end autonomous driving (2) Learning approach for end-to-end autonomous driving (3) An attention mechanism for end-to-end autonomous driving (4) Simulation and evaluation of end-to-end autonomous driving

大纲指导者：郑南宁教授（西安交通大学人工智能学院）

大纲制定者：郑南宁教授（西安交通大学人工智能学院）、杜少毅教授（西安交通大学人工智能学院）、陈仕韬助理教授（西安交通大学人工智能学院）、薛建儒教授（西安交通大学人工智能学院）、杨静副教授（西安交通大学自动化科学与工程学院）

大纲审定：西安交通大学人工智能学院本科专业知识体系建设与课程设置第二版修订工作组

第 9 章

"人工智能与社会"课程群

9.1 "人工智能的科学理解"课程大纲

课程名称：人工智能的科学理解——控制论与人工智能，智能系统的信念、知识和模型

Course：A Scientific Understanding of Artificial Intelligence—Cybernetics and AI, Beliefs, Knowledge and Models of AI Systems

先修课程：概率统计与随机过程、现代控制工程、人工智能概论、机器学习、理论计算机科学的重要思想

Prerequisites：Probability Theory and Stochastic Process, Modern Control Engineering, Introduction to Artificial Intelligence, Machine Learning, Great Ideas in Theoretical Computer Science

学分：1

Credits：1

9.1.1 课程目的和基本内容（Course Objectives and Basic Content）

本课程是人工智能学院本科专业选修课，课程由两个主题（Topic）组成，包括主题1：控制论与人工智能和主题2：智能系统的信念、知识和模型。

This course is an elective course for undergraduates in College of Artificial Intelligence. It consists of two topics: Cybernetics and AI; Beliefs, Knowledge and Models of AI Systems.

主题1 旨在更好地理解人工智能与计算机、控制论之间的联系，阐述反馈（feedback）、控制（control）以及行为模拟在人工智能系统中的重要作用，回顾和重新认识维纳的控制论对人工智能发展的贡献，了解麦卡洛克和匹茨提出的第一个人工神经细胞

模型("MP 模型",1943),该模型给出了基于仿生学结构模拟的方法探讨实现人工智能的途径,介绍细胞自动机的自我复制机制;并从 1948 年维纳的 *Cybernetics* 和艾什比 1954 年的 *Design of A Brain* 出发,讨论如何从行为模拟出发研究人工智能。控制论作为人工智能早期理论基础之一,对人工智能的发展产生巨大的推动作用。1950 年,英国科学家图灵发表论文 *Computing Machinery and Intelligence*,这一人工智能领域的开山之作论述了图灵测试,开启了用计算机模拟人的智能的研究时代。现代人工智能是与计算机、控制论一起成长的,但需要强调的是:为创造可以思维的机器而开展的科学探索是由控制论引发的,或者更明确地说,是由思维机械化的观念引发的,这种观念来源于对大脑思维的分析而得到的灵感和启发,并已成为当今切实的科学框架。

Topic 1 aims to provide a better understanding of the connection between artificial intelligence, computers and cybernetics, to illustrate the important role of feedback, control and behavioral simulation in AI systems, and to review and re-recognize the contribution of Wiener's *Cybernetics* (1948) to the development of AI, learn about the first artificial nerve cell model (MP model, 1943) proposed by McCulloch and Pitts, and discuss how to study artificial intelligence based on behavioral simulation from Wiener's *Cybernetics* and Ashby's *Design of A Brain* (1954). Cybernetics, as one of the early theoretical foundations of artificial intelligence, has played a great role in promoting the development of AI. In 1950, Alan Turing published his paper *Computing Machinery and Intelligence*, a pioneering work in the field of AI, which discussed Turing testing and opened the era of computer simulation of human intelligence. Modern artificial intelligence has grown with computers and cybernetics, but it needs to be emphasized that scientific exploration for creating thinking machines was initiated by cybernetics or, more specifically, by the idea of mechanization of thinking, which was inspired by the analysis of brain thinking and has become a practical scientific framework by now.

主题 2 旨在如何从信念、知识和模型的角度去理解认知、思维和人工智能三者的关系,阐述人类如何认识事物这一基本问题,探讨信念、知识和模型的定义,信念评价的科学方法,知识与大模型的内在关联,以及如何构建可信的通用人工智能系统,启发学生在修完人工智能核心专业课程后,进一步思考如何使"一个物理组织或系统(人工智能或机器人)具有信念"。人工智能系统的信念可以在一种"贝叶斯信念网络"中进行计算,也可以通过系统数据库中命题的添加、修改或删除来进行更新,修改可以由程序员或人工智能系统的"自修改"完成,如用机器学习的方法实现;而与真实物理世界交互的自治机器人(autonomous robots),如自动驾驶汽车(autonomous vehicles),可以根据各种传感器提供的环境数据不断更新它们的世界模型,形成信念。

Topic 2 aims to explain the relationship among cognition,thinking and AI from the perspective of belief,expound the basic problem of how humans know objects, explore the definitions of belief,knowledge and model,the scientific method of belief evaluation,the inherent correlation between knowledge and large models, as well as how to construct a basic framework of trusted Artificial General Intelligence(AGI), so as to inspire students to think further about how to enable "a physical organization or system(AI system or robots) to have belief ",after learning the core courses of AI. The beliefs of AI systems can be calculated in a "Bayesian belief network" or updated by adding, modifying or deleting propositions in the system database. The modifications can be accomplished by programmers or "self-modifying" of AI systems,such as machine learning. However,autonomous robots that interact with the real physical world,autonomous vehicles,for example,can constantly update their world models and form beliefs based on environmental data provided by various sensors.

9.1.2 课程基本情况(Course Arrangements)

课程名称	人工智能的科学理解——控制论与人工智能、智能系统的信念、知识和模型 A Scientific Understanding of Artificial Intelligence—Cybernetics and AI, Beliefs, Knowledge and Models of AI Systems							
开课时间	一年级		二年级		三年级		四年级	
^	秋	春	秋	春	秋	春	秋	春
课程定位	本科生"人工智能与社会"课程群选修课							
学分	1 学分							
总学时	16 学时 (授课 16 学时、实验 0 学时)							
授课学时分配	课堂讲授(12 学时), 文献阅读与小组讨论(4 学时)							

	人工智能与社会
必修(学分)	/
选修(学分)	人工智能的科学理解(1)
^	人工智能的哲学基础与伦理(1)
^	人工智能的社会风险与法律(1)

续表

先修课程	概率统计与随机过程、现代控制工程、人工智能概论、机器学习、理论计算机科学的重要思想
后续课程	
教学方式	课堂讲授、文献阅读与小组讨论、大作业
考核方式	课程结束笔试成绩占60%，大作业占40%
参考教材	1. 维纳.控制论(或关于在动物和机器中控制和通信的科学)[M].郝季仁,译.2版.北京：科学出版社,2009. 2. Nilsson N J.理解信念[M].王飞跃,等译.北京：机械工业出版社,2016.
参考资料	1. Zheng N N,et al. Hybrid-Augmented Intelligence：Collaboration and Cognition[J]. Frontiers of Information Technology & Electronic Engineering,2017,18(2)：153-179. 2. 彭永东.控制论的发生与传播研究[M].太原：山西教育出版社,2011.
其他信息	

9.1.3 教学目的和基本要求(Teaching Objectives and Basic Requirements)

主题1 控制论与人工智能

（1）理解控制论的基本概念与控制的属性；了解控制论的基本方法与研究对象，理解信息与通信中的时间序列的基本分析方法；

（2）通过控制论中的生理和心理因素，讨论机器与生命体（计算机与神经系统）的类比；

（3）理解反馈、控制与行为的机器模拟之间的关系；

（4）理解机器智能行为产生与环境交互的关系，讨论具身智能的环境交互与行为进化；

（5）通过生命体自组织行为与分布式认知，理解智能自主系统的工作方式；

（6）了解控制论对人工智能等学科发展的贡献。

主题2 智能系统的信念、知识和模型

（1）掌握信念的基本概念，了解理论、陈述性知识和程序性知识、模型等概念；

（2）认识信念的作用，了解信念通过何种机理实现其作用；

（3）理解信念的评价过程，熟悉批判性思维的三要素、信念网络；

（4）理解信念强度的概率表示方法，了解给信念指定概率的两种途径，了解贝叶斯信念网络；

（5）理解大模型的知识表征与推理过程；

（6）熟悉通用人工智能的基本概念、实现途径与引发的伦理问题。

9.1.4 教学内容及安排(Syllabus and Arrangements)

主题 1 控制论与人工智能(Cybernetics and AI)

章节序号 Chapter Number	章节名称 Chapters	课时 Class Hour	知 识 点 Key Points
1	控制论的基本概念、研究对象与控制的属性 Basic concepts and objects of cybernetics, and attributes of control	1.5	(1) 控制概念的属性 (2) 控制与行为的因果关联 (3) 控制论的统计理论基础 (4) 控制系统的负反馈与熵减少过程 　　理解控制论的出现所带来的方法论,对于现代学科的产生和发展的重要性;控制概念的最基本属性是"目的性","控制"与"行为"密切关联、互为因果。控制系统是一个与周围环境密切联系的系统,与自发地趋于热平衡的系统和过程不同,控制系统通过自身的"反馈"可以减少系统的"无组织程度",即控制系统中经常发生熵减少的过程 (1) Attributes of the concept of control (2) Causal link between control and behavior (3) Statistical theoretical basis of cybernetics (4) Negative feedback and entropy reduction of control systems 　　Understanding the methodologies brought about by cybernetics is of great importance to the emergence and development of modern disciplines. The most basic attribute of the concept of control is "purposiveness", and "control" is closely related to "behavior" and mutually causal. Control system is a kind of system which is closely related to the surrounding environment. Different from the system and process that tend toward heat balance spontaneously, the control system can reduce the "unorganized degree" of the system through its own "feedback", that is, the process of entropy reduction often occurs in the control system

续表

章节序号 Chapter Number	章节名称 Chapters	课时 Class Hour	知 识 点 Key Points

2	机器与生命体（计算机与神经系统）的类比
	The analogy between machine and living body (computer and nervous system)
		1.5	(1) 控制论中的生理和心理因素
			(2) 计算机与神经系统的类比
			(3) 推理是计算吗
			(4) 人与机器协同的混合智能
			  理解控制论中的生理和心理因素行为；理解控制论的另一重要概念，即生命与非生命体没有本质的不同，它们都遵循着统一的物理化学规律；功能模拟与人工智能、仿生的关系；符号语言和推理如何使人类智能扩展到非生命计算系统；如何实现人机协作的混合增强智能
			
			(1) Physiological and psychological factors in cybernetics
			(2) Analogy between computer and nervous system
			(3) Is reasoning a type of calculation
			(4) Hybrid intelligence based on human-machine collaboration
			  Understanding the behavior of physiological and psychological factors in cybernetics; understanding another important concept of cybernetics, that is, there is no essential difference between life and non-life, they all follow a unified physical and chemical law; the relationship between functional simulation, artificial intelligence and bionics; how symbolic language and reasoning extend human intelligence to non-living computing systems; and how to realize hybrid augmented intelligence in the framework of man-machine collaboration

3	反馈、控制与行为的机器模拟
	Feedback, control and machine simulation of behavior
		1.5	(1) 控制论中的反馈与控制
			(2) 控制论的创见-行为功能模拟
			(3) 控制、通信、反馈和信息的关系
			  从控制和信息的角度理解目的性是智力和生命的一个本质特征；分析反馈系统的稳定性和计算机的记忆、运算和控制装置的特点，以及控制、通信、反馈和信息的关系
			
			(1) Feedback and control in cybernetics
			(2) The creative idea of cybernetics-behavioral function simulation
			(3) The relationship between control, communication, feedback and information
			  Understanding purposiveness from the perspective of control and information is an essential feature of intelligence and life; analyzing the stability of feedback system and the characteristics of computer memory, operation and control devices, as well as the relationship between control, communication, feedback and information

续表

章节序号 Chapter Number	章节名称 Chapters	课时 Class Hour	知 识 点 Key Points
4	机器-环境交互与具身智能 Machine-environment interaction, and embodied intelligence	1.5	(1) 机器-环境交互的基本概念与方法 (2) 机器如何通过环境交互产生智能行为 (3) 具身智能的基本框架 (4) 具身智能体如何进行环境交互与行为进化 　　了解机器-环境交互的基本概念与方法,讨论机器如何在与环境交互过程中产生智能行为;从智能体与环境的交互和智能体的行为进化等角度理解具身智能的基本框架,分析具身智能与行为主义的内在关联 (1) Basic concepts and methods of machine-environment interaction (2) How machines generate intelligence through environmental interaction (3) Basic framework of embodied intelligence (4) How embodied agents interact with the environment and evolve their behavior 　　Understanding the basic concepts and methods of machine-environment interaction, and discussing how machines generate intelligence through environmental interaction; understanding the basic framework of embodied intelligence from the interaction between agent and environment, as well as the behavioral evolution of agent, and analyzing the inherent relationship between embodied intelligence and behaviorism
5	智能体的自组织与分布式认知 Self-organization and distributed cognition of agent	0.5	(1) 分布式认知的概念 (2) 自组织与社群智能 (3) 多智能体分布式强化学习 　　理解分布式认知的基本概念;生命的出现与自指相关,从生命体自组织行为分析社群智能与智能自主系统;理解多智能体在合作环境与非合作环境下的强化学习 (1) The concept of distributed cognition (2) Self-organization and societal intelligence (3) Distributed reinforcement learning of multi-agent 　　Understanding the concept of distributed cognition; The emergence of life is related with self-reference, therefore analyzing societal intelligence and intelligent autonomous system from the self organizing behavior of life; Understanding multi-agent reinforcement learning in both cooperative and non-cooperative environments

主题2　智能系统的信念、知识和模型（Beliefs, knowledge and models of AI System）

章节序号 Chapter Number	章节名称 Chapters	课时 Class Hour	知　识　点 Key Points
1	信念、知识和模型 Belief, knowledge and model	1	（1）信念的概念及其性质 （2）陈述性知识和程序性知识的概念及性质 （3）模型的概念 （4）信念获得的途径及心理结构 　　理解信念、知识与模型的关系，讨论信念获得的两个途径，即感觉、对已相信结果的解释和衍生结果；讨论信念的心理结构，由感觉获得信念的过程给出信念的两种心理结构，即结果和解释 (1) Concept and nature of belief (2) Concept and nature of declarative knowledge and procedural knowledge (3) Concept of model (4) Ways to acquire belief and the psychological structure of belief 　　Understanding the relationship between belief, knowledge and model, discussing two ways to acquire beliefs, namely, feeling, interpretation of believed results and derivative results. Discussing the psychological structure of belief-the process of acquiring belief from sensation gives two psychological structures of belief, i. e. result and explanation
2	信念的作用：预测与选择 The role of beliefs: prediction and choice of action	0.5	（1）信念的作用 （2）信念如何帮助预测和选择行动 （3）信念如何帮助解释观察 （4）信念的深度分级 　　理解信念的作用：预测和选择行动，讨论信念的深度分级 (1) The role of beliefs (2) How beliefs help predict and choose action (3) How beliefs help explain observations (4) Deep hierarchy of beliefs 　　Understanding the role of beliefs: prediction and choice of action, discussing the deep grading of beliefs

续表

章节序号 Chapter Number	章节名称 Chapters	课时 Class Hour	知　识　点 Key Points
3	信念的评价与 信念网络 Evaluating beliefs and network of beliefs	1	（1）理解信念的评价过程 （2）讨论批判性思维的三要素：测试结果、创造解释和解释消除 （3）结合实例理解信念网络及其中各层之间的相互关系 　　理解评价信念、批判性思维、测试结果、创造解释、解释消除、信念网络的基本概念 (1) Understand the evaluation process of beliefs (2) Discuss the three elements of critical thinking: test results, creative interpretation and explanatory elimination (3) Understand the relationship between belief networks and their layers with examples 　　Understanding how to evaluate beliefs, test consequences, construct critical thinking, and basic concepts of creating interpretation, eliminating interpretation and belief network
4	信念强度的 概率表示与推理 The probability representation of belief intensity and reasoning	1	（1）给出信念强度的概率表示方法 （2）了解信念指定概率的两大途径 （3）讨论贝叶斯信念网络以及因果推理和证据推理 　　人工智能系统的信念可以在一种"贝叶斯信念网络"中进行计算，也可以通过人工智能系统的"自修改"完成 (1) Give the probability representation of belief intensity (2) Two ways to understand the probability of belief assignment (3) Discuss Bayesian belief networks, causal and evidential reasoning 　　The beliefs of AI systems can be calculated in a Bayesian belief network, and the updating of beliefs can also be accomplished by "self-modifying" of AI systems

续表

章节序号 Chapter Number	章节名称 Chapters	课时 Class Hour	知识点 Key Points
5	人工智能的 大模型基础 The large model foundation of AI	1.5	(1) 大模型的基本概念 (2) 大模型中的知识表征与推理 (3) 大模型的知识涌现机理 　　了解大模型的基本概念；讨论知识与大模型的内在关联；从上下文学习、人在回路反馈增强等角度理解大模型的技术，理解知识表征与推理的过程；讨论大模型的知识涌现机理 (1) Basic concepts of large models (2) Knowledge representation and reasoning of large models (3) The mechanism of knowledge emergence in large models 　　Understanding the basic concepts of large models; discussing the inherent correlation between knowledge and large models; analyzing the techniques of large models from the in-context learning and human-in-loop feedback enhancement, and understanding the process of knowledge representation and reasoning; discussing the mechanism of knowledge emergence in large models
6	可信通用 人工智能的基本框架 Basic framework of trusted AGI	0.5	(1) 通用人工智能的概念与特点 (2) 可信通用人工智能的基本框架 (3) 通用人工智能的伦理问题 　　了解通用人工智能的的基本概念与特点；讨论如何从具有智能、大模型等角度构建可信通用人工智能的基本框架；熟悉通用人工智能引发的伦理问题 (1) The concept and characteristics of AGI (2) Basic framework of trusted AGI (3) Ethical issues of AGI 　　Mastering the concepts and characteristics of AGI; Understanding the general framework for building a trusted AGI system; Understanding the ethical issues caused by AGI

大纲指导者：郑南宁教授（西安交通大学人工智能学院）

大纲制定者：郑南宁教授（西安交通大学人工智能学院）、周三平副教授（西安交通大学人工智能学院）

大纲审定：西安交通大学人工智能学院本科专业知识体系建设与课程设置第二版修订工作组

9.2 "人工智能的哲学基础与伦理"课程大纲

课程名称：人工智能的哲学基础与伦理
Course：Philosophical Foundation and Ethics of Artificial Intelligence
先修课程：无
Prerequisites：None
学分：1
Credits：1

9.2.1 课程目的和基本内容（Course Objectives and Basic Content）

本课程是人工智能学院本科专业选修课。
This course is an elective course for undergraduates in College of Artificial Intelligence.

课程旨在培养学生对人工智能研究中所涉及的核心概念、主要方法背后的哲学基础形成深刻的理解，建立起学生对本学科研究的宏观构架；同时引导学生积极思考人工智能技术应用领域的扩展所引发的伦理问题，培养负责任的科学家和工程师。

本课程将分别介绍人工智能所涉及的基本哲学问题和在现实应用中引发的道德难题。具体而言，在课程的前半部分，将讨论当前人工智能哲学的研究传统与发展趋势，并着重讨论当前思想界对强人工智能提出的理论反驳和辨护，以及人与机器的意识问题，启发学生追问和反思人工智能中的核心概念，鼓励学生对现有人工智能哲学的基本认识提出挑战。在课程的后半部分,将向学生展示在人工智能的最新应用中已经或可能显现的道德难题,介绍在学术界、产业界和政策规划中已经确立的伦理规范并反思其背后的价值基础,进而引导学生思考人工智能与人类道德的一般关系,其中包括:人工智能与社会正义的关系;算法与机器学习的伦理问题;人类道德规范是否应嵌入人工智能系统;人工智能系统如何做出道德决策、如何承担道德责任;以及强人工智能是否具有道德能动性等问题。

The course aims to train students to have a deep understanding of the core concepts involved in the research of artificial intelligence and the philosophical basis behind the main approaches, and to establish a macro framework for the research of this discipline. At the same time, it guides students to actively think about the ethical issues caused by the expansion of the application of artificial intelligence technology, and trains responsible scientists and engineers.

This course will introduce the basic philosophical issues involved in artificial intelligence and the moral dilemmas arising from its practical application. Specifically, the first half of the course discusses the current research of philosophy of artificial intelligence, and focuses strong artificial intelligence and the consciousness of human and machine, which aims at inspiring students to understand the key concepts of artificial intelligence, encouraging them to challenge the basic understanding of the existing philosophy of artificial intelligence. The second half of the course shows the latest applications of artificial intelligence and the moral dilemmas of artificial intelligence; introduces the established ethical norms in academia, industry and policy planning and reflects on the value basis behind them; then lead students to think about artificial intelligence and general relations of human morality, which include: the relationship between artificial intelligence and social justice; ethical issues of algorithms and machine learning; whether human ethics should be embedded in the system of artificial intelligence; how does the artificial intelligence system make the moral decision, how to assign the moral responsibility; and whether strong artificial intelligence has moral agency.

9.2.2 课程基本情况(Course Arrangements)

课程名称	人工智能的哲学基础与伦理 Philosophical Foundation and Ethics of Artificial Intelligence							
开课时间	一年级		二年级		三年级		四年级	人工智能与社会
^	秋	春	秋	春	秋	春	秋 春	必修(学分) /
课程定位	本科生"人工智能与社会"课程群选修课							选修(学分) 人工智能的科学理解(1) 人工智能的哲学基础与伦理(1) 人工智能的社会风险与法律(1)
学 分	1学分							
总 学 时	16学时							
授课学时分配	课堂讲授(16学时)							
先修课程	无							
后续课程								
教学方式	课堂教学、小组讨论、课堂报告							
考核方式	论文考核成绩占70%,平时成绩占10%,课堂报告成绩占20%							
参考教材	无							
参考资料	1. Copeland J. Artificial Intelligence: A Philosophical Introduction[M]. Wiley-Blackwell. 1993. 2. Carter M. Minds and Computers: The Philosophy of Artificial Intelligence[M]. Edinburgh: Edinburgh University Press, 2007. 3. José Luis B. Cognitive science: An Introduction to the Science of Mind[M]. Cambridge: Cambridge University Press, 2020. 4. Wallach W, Allen C, Machines M. Teaching Robots Right from Wrong[M]. Oxford: Oxford University Press, 2009.							
其他信息								

9.2.3 教学目的和基本要求(Teaching Objectives and Basic Requirements)

(1) 掌握人工智能哲学研究的代表人物及其核心思想;
(2) 了解当前认知科学哲学对人工智能的基本观点;
(3) 了解强人工智能的界定和相关哲学论证;
(4) 掌握当前人工智能发展所面临的核心道德困境及各国相关伦理规范;

(5) 熟悉算法歧视的概念、来源和规避方式；
(6) 熟悉机器学习的基本伦理问题；
(7) 了解自动驾驶、智能医疗、自主武器系统等领域的伦理问题；
(8) 理解人工智能系统的道德决策、道德责任及道德能动性问题。

9.2.4 教学内容及安排（Syllabus and Arrangements）

导论（Introduction）

章节序号 Chapter Number	章节名称 Chapters	课时 Class Hour	知 识 点 Key Points
0.1	导论 Introduction	2	(1) 当前人工智能哲学的研究议题 (2) 人工智能关涉哪些伦理问题 (1) Current research topics in the philosophy of AI (2) Ethical issues related with AI

第一章 人工智能的一般哲学问题（General Philosophical Issues of AI）

章节序号 Chapter Number	章节名称 Chapters	课时 Class Hour	知 识 点 Key Points
1.1	图灵测试 Turing Test	1	(1) 图灵测试的基本陈述 (2) 对图灵测试的三个反驳：人类中心主义、残忍的专家反驳、香农-麦卡锡反驳 (1) Basic statements of Turing test (2) Three objections to the test: anthropocentrism, fiendish expert objections, The Shannon-McCarthy objection
1.2	中文屋论证 The Chinese Room Argument	1	(1) 4种版本的中文屋论证：香草版、户外版、模拟器版、健身房版 (1) Four principal versions of the Chinese room argument: the vanilla version, the outdoor version, the simulator version, and the gymnasium version

续表

章节序号 Chapter Number	章节名称 Chapters	课时 Class Hour	知 识 点 Key Points
1.3	超计算 Hypercomputation	1	(1) 谕示机 (2) 邱奇-图灵谬误 (3) 人工智能与等价谬误 (1) Oracle machines (2) The Church-Turing fallacy (3) AI and the equivalence fallacy
1.4	彭罗斯对人工智能的"哥德尔反驳" Penrose's 'Gödel Objection' to AI	0.5	(1) 哥德尔不完备性定理 (2) 彭罗斯对哥德尔定理的改造 (1) Gödel's incompleteness theorem (2) Penrose's transformation of Gödel's theorem
1.5	德雷福斯对早期人工智能的批判 Dreyfus's critique of early AI	0.5	(1) 德雷福斯对符号主义人工智能的批判及其现象学构造 (1) Dreyfus's critique of symbolic AI and his phenomenological construction

第二章　强人工智能的哲学问题（Philosophy of Strong AI）

章节序号 Chapter Number	章节名称 Chapters	课时 Class Hour	知 识 点 Key Points
2.1	强人工智能的界定 The definition of strong AI	1	(1) 弱人工智能与强人工智能 (2) 强人工智能的辩护与反驳 (1) Strong AI and weak AI (2) The defense and rebuttal of strong AI
2.2	强人工智能是否具有类人的"思维能力" Whether strong AI has human-like thinking ability	0.5	(1) 人类意识的认知科学与哲学研究前沿 (2) 人类意识与机器意识的对照研究 (1) The frontiers of cognitive science and philosophy of human consciousness (2) Human consciousness V. S. Machine consciousness

章节序号 Chapter Number	章节名称 Chapters	课时 Class Hour	知　识　点 Key Points
2.3	强人工智能是否具有类人的"情感" Whether strong AI has human-like emotion	0.5	(1) 明斯基的"情感机器" (1) Marvin Minsky and "The Emotion Machine"

第三章　人工智能的伦理规范问题（The Ethical Norms of AI）

章节序号 Chapter Number	章节名称 Chapters	课时 Class Hour	知　识　点 Key Points
3.1	人工智能的道德约束机制 The moral restraint mechanism of AI	1	(1) 阿西莫夫的机器人原则 (2) IEEE 的人工智能伦理报告 (3) 德国《自动驾驶伦理报告》 (1) Asimov's principle of robotics (2) The ethical report of AI by IEEE (3) Germany's *Ethics Commission Automated and Connected Driving*
3.2	人类道德规范与人工智能系统 Human ethics and AI systems	1	(1) 人工智能是否需要人类道德规范 (2) 人工智能应载入"谁"的道德规范 (3) 如何确保人工智能道德的动态进步 (1) Whether AI need human ethics (2) Whose ethic code should be embedded in AI (3) How to ensure the dynamic progress of AI ethics
3.3	人工智能算法的伦理问题 The ethics of AI algorithms	1	(1) 算法是否有道德价值负载 (2) 算法歧视的概念、现实问题和规避方式 (3) 如何在智能系统设计中消除算法歧视 (1) Whether the algorithm is value-laden (2) The concept，specific problems and avoiding ways of algorithmic discrimination (3) How to eliminate algorithmic discrimination in intelligent system design

续表

章节序号 Chapter Number	章节名称 Chapters	课时 Class Hour	知　识　点 Key Points
3.4	机器学习的伦理问题 The ethics of machine learning	1	(1) 机器学习与排名算法及预测算法 (2) 机器学习的三个阶段中的伦理问题：数据收集、模型建构、模型使用 (1) Machine learning and ranking algorithms, predictive algorithms (2) Ethical issues in the three phases of machine learning: data collection, model construction, model use
3.5	人工智能应用领域中的伦理问题 Ethical issues in the applications of AI	1	(1) 自动驾驶汽车的伦理问题 (2) 智能医疗的伦理问题 (3) 自主武器系统的伦理问题 (1) The ethics of self-driving cars (2) The ethics of smart healthcare (3) The ethics of autonomous weapon system

第四章　人工智能系统的道德困境（The Moral Dilemma of AI Systems）

章节序号 Chapter Number	章节名称 Chapters	课时 Class Hour	知　识　点 Key Points
4.1	人工智能系统如何做出道德决策 How do AI systems make moral decisions	1	(1) 什么是道德决策 (2) 人类的道德决策与机器的道德决策 (1) What is moral decision (2) Human moral decision and machine moral decision
4.2	人工智能系统如何承担道德责任 How do AI systems assign moral responsibility	1	(1) 什么是道德责任 (2) 制造商、使用者与机器自身的责任分配 (1) What is moral responsibility (2) The division of responsibility between the manufacturer, the user and the machine itself

续表

章节序号 Chapter Number	章节名称 Chapters	课时 Class Hour	知　识　点 Key Points
4.3	人工智能系统是否能够成为道德能动者 Whether AI system can become moral agent	1	（1）什么是道德能动性 （2）人工智能作为道德能动者的反驳与辩护 （3）人工智能与道德进步 （1）What is moral agency （2）The objections and justifications on AI as moral agent （3）AI and moral progress

大纲指导者：郑南宁教授（西安交通大学人工智能学院）

大纲制定者：白惠仁研究员（浙江大学哲学学院）

大纲审定：西安交通大学人工智能学院本科专业知识体系建设与课程设置第二版修订工作组

9.3 "人工智能的社会风险与法律"课程大纲

课程名称：人工智能的社会风险与法律

Course：Social Risk and Law of Artificial Intelligence

先修课程：无

Prerequisites：None

学分：1

Credits：1

9.3.1 课程目的和基本内容（Course Objectives and Basic Content）

本课程是人工智能学院本科专业选修课。

This course is an elective course for undergraduates in College of Artificial Intelligence.

人工智能技术为社会经济发展提供了重要的机遇，同时对未来社会形态的塑造提出了巨大的挑战。因此，本课程的目的：帮助人工智能专业学生深刻理解人工智能时代的新特征，把握未来社会发展的新趋势，熟悉人工智能的社会风险和法律规制，培养适应于新时代的思维能力和广阔的全局视野。

本课程将从人工智能与人类社会的基本关系作为切入点，聚焦于：人工智能在未来社会秩序的塑造中所扮演的角色，包括经济、法律、社会治理等方面的影响；以及社会对人工智能的反向规制作用，包括智能系统的相关政策、人机共生关系及社会风险控制等问题。本课程将以一种面向未来的视角看待人工智能与社会之间的双向互动以及由此可能产生的各种理论和实践难题，从而启发学生将技术发展置于更宏观的人类社会视野中，使学生充分理解智能革命所带来的经济问题、社会问题及法律问题等，思考人工智能融入社会生活所造就的思维变革，把握人类社会对人工智能发展的边界限制。更进一步，通过本课程的学习，希望可以启发学生对未来社会形态的想象，进而反向作用于学生在专业学习和研究时的方向选择，使得学生以一种长远的整体视野和深切的人文关怀看待所从事的专业工作。

Artificial intelligence provides important opportunities for social and economic development and poses great challenges to the shaping of future social forms. Therefore, the purpose of this course is to help students deeply understand the new characteristics of the era of artificial intelligence, grasp the new trend of future social development, master the basic social risk and legal regulations of artificial intelligence, and cultivate the thinking ability and broad global vision to adapt to the new era.

This course will start from the basic relationship between artificial intelligence and human society, focusing on the role of artificial intelligence in shaping the future social order, including the impact of economic, legal and social governance; then turn to social regulation of artificial intelligence, including intelligent system related policies, human-machine symbiosis and social risk control, etc. This course will provide a future-oriented perspective on artificial intelligence and its related theoretical and practical problems, so as to make the students fully understand intelligent revolution brought about by the economic, social problems and legal problems, then grasp the development of human society on the artificial intelligence boundary constraints. Furthermore, through the study of this course, it is hoped that it can inspire students' imagination of future social forms, so that students can view their professional work with a long-term overall vision and deep humanistic care.

9.3.2 课程基本情况(Course Arrangements)

课程名称	人工智能的社会风险与法律 Social Risk and Law of Artificial Intelligence								
开课时间	一年级		二年级		三年级		四年级		人工智能与社会
	秋	春	秋	春	秋	春	秋	春	
					✓				必修 (学分) /
课程定位	本科生"人工智能与社会"课程群选修课								选修 (学分)	人工智能的科学理解(1)
学　分	1学分								人工智能的哲学基础与伦理(1)
总学时	16学时(授课16学时、实验0学时)								人工智能的社会风险与法律(1)
授课学时分配	课堂讲授(16学时)								
先修课程	无								
后续课程									
教学方式	课堂教学、小组讨论、课堂报告								
考核方式	论文考核成绩占70%,平时成绩占20%,课堂报告成绩占10%								
参考教材	无								
参考资料	1. 腾讯研究院. 人工智能[M]. 北京:中国人民大学出版社,2017. 2. Caloi R,Froomkin M A,Kerr I. 人工智能与法律的对话[M]. 陈吉栋,等译. 上海:上海人民出版社,2018. 3. Cath C,Wachter S,Mittelstadt B,et al. Artificial intelligence and the "good society": the US, EU, and UK approach[J]. Science and Engineering Ethics,2018,24(2):505-528. 4. Allen G,Chan T. Artificial intelligence and national security[R]. Cambridge,MA:Belfer Center for Science and International Affairs,2017. 5. Agrawal A,Gans J,Goldfarb A. Economic Policy for Artificial Intelligence[J]. Innovation Policy and the Economy,2019,19(1):139-159.								
其他信息									

9.3.3 教学目的和基本要求(Teaching Objectives and Basic Requirements)

(1) 了解当前人文社会科学对人工智能的研究状况;
(2) 熟悉不同国家、文化对人工智能发展的影响;
(3) 掌握人工智能产业发展现状与当前主要应用领域;
(4) 掌握世界主要国家人工智能技术和产业发展的政策现状;

(5) 了解人工智能可能造成的技术性失业问题；
(6) 了解人工智能技术创新对经济发展的作用；
(7) 熟悉人工智能可能带来的责任归属、隐私权等法律问题；
(8) 了解现实的和未来可能的人机共生关系及社会治理方式。

9.3.4 教学内容及安排（Syllabus and Arrangements）

导论（Introduction）

章节序号 Chapter Number	章节名称 Chapters	课时 Class Hour	知 识 点 Key Points
0.1	导论 Introduction	2	(1) 历史上的颠覆性新技术与社会变革的关系 (2) 人工智能概念形成的社会思想渊源 (1) The relationship between breakthrough new technology and social change in human history (2) The social ideological origin of the formation of the concept of AI

第一章 人工智能的产业发展与政策规制（Industrial Development and Policy Regulation of AI）

章节序号 Chapter Number	章节名称 Chapters	课时 Class Hour	知 识 点 Key Points
1.1	人工智能产业发展现状 Current AI Industry	1	(1) 国内外人工智能研究领军高校、研究机构和企业现状介绍 (1) Introduction of the leading universities, research institutions and enterprises of AI research
1.2	当前人工智能的核心应用领域 The present applications of AI	1	(1) 自动驾驶汽车、智能机器人、智能医疗、自主武器系统 (1) Automated vehicle, intelligent robot, intelligent medicare, autonomous weapon system
1.3	人工智能发展的政策规制 Policy of AI	2	(1) 世界主要国家人工智能技术和产业发展的政策现状 (1) Policy of AI technology and industrial development in major countries

第二章 人工智能的社会风险(The Social Risk of AI)

章节序号 Chapter Number	章节名称 Chapters	课时 Class Hour	知 识 点 Key Points
2.1	人工智能与非传统安全 AI and the non-traditional security	1	(1) 什么是非传统安全 (2) 人工智能如何影响非传统安全 (1) What is so called non-traditional security (2) How does AI affect non-traditional security
2.2	人工智能的风险评估 The risk assessment of AI	1	(1) 技术的社会评估 (2) 如何将社会风险理论运用于人工智能 (1) Social assessment of technology (2) How to apply social risk theory to AI
2.3	智能时代的贫富分化 The gap between rich and poor in intelligent era	1	(1) 人工智能可能引发的社会财富分配不公正 (1) Unjust distribution of social wealth caused by AI

第三章 人工智能的法律挑战(The Legal Challenges of AI)

章节序号 Chapter Number	章节名称 Chapters	课时 Class Hour	知 识 点 Key Points
3.1	人工智能立法现状 The legislation of AI	1	(1) 各国对人工智能技术立法的现状 (1) The status quo of AI technology legislation
3.2	谁为机器人的行为负责 Who is responsible for the behavior of the robot	1	(1) 如何界定法律责任 (2) 机器人的法律责任问题 (1) How to define legal liability (2) The legal responsibility of the robot
3.3	人工智能与隐私保护 AI and privacy protection	1	(1) 隐私与隐私权的概念 (2) 人工智能的具体应用可能产生的隐私权问题 (1) The concept of privacy and the right to privacy (2) The possible privacy problems caused by AI's applications

续表

章节序号 Chapter Number	章节名称 Chapters	课时 Class Hour	知识点 Key Points
3.4	人工智能技术发展的法律边界 The legal boundaries of AI technology	1	(1) 如何限制无人机、智能自主武器系统等的滥用 (1) How to limit the abuse of autonomous systems and so on

第四章 人工智能与未来社会（AI and the Future Society）

章节序号 Chapter Number	章节名称 Chapters	课时 Class Hour	知 识 点 Key Points
4.1	人工智能与未来就业 AI and future employment	1	(1) 人工智能取代了哪些职业 (2) 人工智能可能创造的新就业机会 (3) 人工智能时代的人类行业设置与教育制度 (1) What professions has AI replaced (2) What kind of new jobs AI could create (3) Human industry setting and education system in the era of AI
4.2	人工智能与新的经济形态 AI and the new economy	1	(1) 历史上重大技术创新与经济发展间的关系 (2) 人工智能对共享经济的影响 (1) The relationship between major technological innovation and economic development in history (2) The impact of AI on the sharing economy
4.3	未来社会可能的人机共生方式 The possible human-machine symbiosis in the future society	1	(1) 文学、影视科幻作品中的人工智能与人类关系 (2) 人工智能与人的协同演化的可能性 (1) The relationship between AI and human beings in science fiction literature and movies (2) The possibility of co-evolution of AI and human

大纲指导者：郑南宁教授（西安交通大学人工智能学院）

大纲制定者：白惠仁研究员（浙江大学哲学学院）

大纲审定：西安交通大学人工智能学院本科专业知识体系建设与课程设置第二版修订工作组

第 10 章

"人工智能工具与平台"课程群

10.1 "深度学习工具与平台"课程大纲

课程名称：深度学习工具与平台
Course：Tools and Platform for Deep Learning
先修课程：计算机程序设计、数据结构与算法
Prerequisites：Computer Programming, Data Structure and Algorithm
学分：2
Credits：2

10.1.1 课程目的和基本内容（Course Objectives and Basic Content）

本课程是人工智能学院本科专业必修课。

This course is a compulsory course for undergraduates in College of Artificial Intelligence.

工具与平台技术每一次的进步都极大地推动着人工智能研究的发展，本课程旨在介绍人工智能系统平台设计及应用知识。课程以深度学习框架为主线，对深度学习模型全生命周期实践中的训练部署与分布式训练系统、推断部署、作业调试与分析、大规模 GPU 集群调度、AI 平台集群部署运维的基本理论与实践方法展开讨论。第一章介绍深度学习计算框架的原理与应用，重点是掌握当前主流 PyTorch 框架的编程及使用方法。第二章介绍矩阵运算与体系结构，了解深度学习平台的硬件实现。第三章介绍分布式训练系统，为大数据大模型的学习与训练打下基础。第四章介绍自动机器学习系统，重点掌握深度学习中超参数调优的系统手段。第五章介绍集群资源调度与管

理,了解其中的主要算法原理及主流平台使用方法。第六章介绍人工智能安全与隐私,了解技术手段应对人工智能伦理问题的一些代表性算法。

通过对人工智能工具与平台的基础框架和应用进行学习,帮助学生系统地掌握常用的工具平台原理及用法。课堂通过小组学习模式,训练学生使用深度学习框架进行程序设计的方法,通过实际使用工具,加深对人工智能的理解,通过实践进一步提高学生独立分析问题、解决问题的能力。

Every progress in tools and platform technology has greatly promoted the development of artificial intelligence research. This course aims to introduce the design and application knowledge of artificial intelligence system platform. Taking the deep learning framework as the main line, the course discusses the basic theories and practical methods of training deployment and distributed training system, inference deployment, job debugging and analysis, large-scale GPU cluster scheduling, AI platform cluster deployment and operation and maintenance in the full life cycle practice of the deep learning model. Chapter 1 introduces the principle and application of the deep learning computing framework, focusing on mastering the programming and usage of the current mainstream PyTorch framework. Chapter 2 introduces the matrix operation and architecture, and understands the hardware implementation of the deep learning platform. Chapter 3 introduces the distributed training system, laying the foundation for the learning and training of big data and big models. Chapter 4 introduces the automatic machine learning system, focusing on the systematic means of hyper-parameter tuning in deep learning. Chapter 5 introduces the cluster resource scheduling and management, and understands the main algorithm principles and the use of mainstream platforms. Chapter 6 introduces AI security and privacy, and understands some representative algorithms of technical means to address AI ethical issues.

Through learning the basic framework and application of AI tools and platforms, students can systematically master the principles and usage of commonly used tool platforms. In this course, students are trained to use the deep learning framework to program design through the group study manner. Through the actual use of tools, students' understanding of artificial intelligence can be deepened, and their ability to analyze and solve problems independently can be further improved through practice.

10.1.2 课程基本情况(Course Arrangements)

课程名称	深度学习工具与平台 Tools and Platform for Deep Learning								
开课时间	一年级		二年级		三年级		四年级		人工智能工具与平台
	秋	春	秋	春	秋	春	秋	春	
课程定位	本科生"人工智能工具与平台"课程群必修课								必修 (学分)	深度学习工具与平台(2)
学 分	2学分								选修 (学分)	3D深度感知(1)
总学时	44学时(授课20学时、实验24学时)									人工智能芯片设计导论(1)
授课学时 分配	课堂讲授(20学时)								
先修课程	计算机程序设计、数据结构与算法								
后续课程	计算机视觉与模式识别、自然语言处理、强化学习								
教学方式	课堂教学、大作业与实验、小组讨论								
考核方式	课程结束笔试成绩占60%,实验成绩占35%,考勤占5%								
参考教材									
参考资料	1. Chollet F. Python深度学习[M].张亮,译.北京:人民邮电出版社,2018. 2. Rogers S,Girolami M.机器学习基础教程[M].郭茂祖,等译.北京:机械工业出版社,2014. 3. Stroustrup B. C++程序设计语言[M].裘宗燕,译.北京:机械工业出版社,2018.								
其他信息									

10.1.3 教学目的和基本要求(Teaching Objectives and Basic Requirements)

(1) 了解深度学习框架的演进以及第一代、第二代框架的特点,掌握PyTorch框架编程的基本方法;

(2) 了解深度学习通用网络模型结构与矩阵运算的映射关系、深度学习常用硬件处理平台等知识点,掌握CPU、GPU、ASIC对矩阵运算的加速优化方法;

（3）了解模型并行、数据并行的基本算法原理，掌握使用 Horovod 开源平台进行深度学习模型数据并行训练的方法；

（4）了解超参数优化的常用算法原理，掌握使用 NNI 开源平台进行深度学习模型超参数调优的方法；

（5）了解镜像与容器技术原理，了解常用集群调度算法原理，了解使用 OpenPAI 开源平台在集群中进行深度学习作业训练和推理的方法；

（6）了解对抗攻击损失函数设计、白盒/黑盒攻击概念，掌握 FGSM 算法原理及典型对抗样本攻击与对抗防御的基本方法。

10.1.4 教学内容及安排(Syllabus and Arrangements)

第一章 深度学习计算框架的原理与应用(Principle and Application of Computational Frameworks for Deep Learning)

章节序号 Chapter Number	章节名称 Chapters	课时 Class Hour	知识点 Key Points
1.1	深度学习框架演进 Evolution of deep learning framework	1	（1）深度学习计算框架的目的 （2）第一代框架：Caffe 的特点 （3）第二代框架：自动求导、计算流图、编译图优化、动态数据流图、TensorFlow 及 PyTorch 的特点 （4）下一代深度学习框架 (1) Aims of computational framework for deep learning (2) 1st generation framework: characteristic of Caffe (3) 2nd generation framework: auto differentiation, computational flowchart, optimization of compilation graph, dynamic data flowchart, characteristics of TensorFlow and PyTorch (4) Next generation deep learning framework

续表

章节序号 Chapter Number	章节名称 Chapters	课时 Class Hour	知 识 点 Key Points
1.2	PyTorch 应用基础 Basics application of PyTorch	1.5	（1）PyTorch 环境的安装与配置：GPU 驱动、CUDA、cuDNN、选择 PyTorch 版本、配置环境变量 （2）定义一个神经网络：张量操作、卷积、池化、非线性激活、全连接等算子的应用 （3）神经网络的训练：数据加载与预处理、定义损失函数、反向传播、batch 及 epoch 的概念、选择优化器与超参 （4）神经网络的推理 （5）实例：CIFAR-10 目标识别 (1) Installation and configuration of PyTorch Environment：GPU drivers, CUDA, cuDNN, PyTorch version, environment variables (2) Define a neural network：tensor operation, convolution, pooling, non-linear activation, full connection operators etc. (3) Train a neural network：data loader and pre-processing, define the loss function, backward propagation, concepts of batch and epoch, select optimizer and hyper-parameters (4) Inference of neural network (5) Example：object classification using CIFAR-10
1.3	PyTorch 应用进阶 Advanced applications of PyTorch	1.5	（1）数据集的定制：torchvision 中的常用数据集、自定义数据集及 transforms （2）实现更多的深度神经网络：Inception、RNN、ResNet、DenseNet、GAN 等网络结构的 PyTorch 实现 （3）实例：SSD 目标检测 (1) Customize the dataset：datasets in torchvision, self-define the dataset and transforms (2) More DNNs：implement the network architectures of Inception、RNN、ResNet、DenseNet, and GAN etc. in PyTorch (3) Example：object detection using SSD algorithm

续表

章节序号 Chapter Number	章节名称 Chapters	课时 Class Hour	知识点 Key Points
1.4	可视化工具 Visualization tool	1	（1）TensorBoard 开源可视化平台 （2）TensorBoard 编程 （3）本地及远程可视化的实现 （1）TensorBoard open source visualization platform （2）Programming using TensorBoard （3）Implement visualization locally or remotely

第二章 矩阵运算与体系结构（Computer Architecture for Matrix Computation）

章节序号 Chapter Number	章节名称 Chapters	课时 Class Hour	知识点 Key Points
2.1	深度学习常见模型结构 Common model structures of deep learning	1	（1）全连接层映射到矩阵运算 （2）卷积层映射到矩阵运算 （3）循环网络层映射到矩阵运算 （4）注意力层映射到矩阵运算 （1）FC layer mapping to matrix operation （2）CNN mapping to matrix operation （3）RNN mapping to matrix operation （4）Attention mapping to matrix operation
2.2	CPU 体系结构与矩阵运算 CPU architecture and matrix operations	1	（1）CPU 体系结构、内存结构 （2）CPU 实现高效计算矩阵乘 （1）CPU architecture, memory architecture （2）Efficient calculation of matrix multiplication using CPU
2.3	GPU 体系结构与矩阵运算 GPU architecture and matrix operations	0.5	（1）GPU 体系结构、内存结构 （2）GPU 执行模型 （3）GPU 实现高效计算矩阵乘 （1）GPU architecture, memory architecture （2）GPU execution model （3）Efficient calculation of matrix multiplication using GPU

续表

章节序号 Chapter Number	章节名称 Chapters	课时 Class Hour	知 识 点 Key Points
2.4	ASIC体系结构与矩阵运算 ASIC architecture and matrix operations	0.5	（1）计算特点与设计思路 （2）节省内存消耗的核心：脉动阵列 (1) Calculation features and design principles (2) Key technique of saving memory consumption: systolic array

第三章 分布式训练系统（Distributed Training System）

章节序号 Chapter Number	章节名称 Chapters	课时 Class Hour	知 识 点 Key Points
3.1	分布式计算原理 Principle of distributed computation	0.5	（1）分布式计算的概念 （2）阿姆达尔定律 （3）古斯塔夫森定律 (1) Concept of distributed computation (2) The Amdahl's law (3) The Gustafson's law
3.2	通信后端原理 Principle of communication backends	1.5	（1）点对点通信：阻塞式通信、非阻塞式通信 （2）集合通信原语：broadcast、gather、scatter、reduce、allreduce、allgather （3）AllReduce算法实现 （4）各类通信后端：MPI、NCCL、Gloo (1) Peer to peer communication: blocked communication, non-blocked communication (2) Primitives of collective communication: broadcast, gather, scatter, reduce, allreduce, allgather (3) Implementation of AllReduce algorithm (4) Various communication backends: MPI、NCCL、Gloo

续表

章节序号 Chapter Number	章节名称 Chapters	课时 Class Hour	知 识 点 Key Points
3.3	模型并行 Model parallelism	1	（1）流水线的实现 （2）Gpipe 算法 （3）PipeDream 算法 （1）Implementation of pipelinealgotihm （2）The Gpipe algorithm （3）The PipeDream algorithm
3.4	数据并行 Data parallelism	1.5	（1）分布式 SGD 算法 （2）利用梯度稀疏压缩分布式训练中的通信带宽 （3）基于 RPC 的多机分布式训练 （1）Distributed SGD algorithm （2）Compress the communication bandwidth of distributed training using gradient sparsity （3）Multi-machine distributed training based on RPC
3.5	分布式训练平台 Distributed training platform	0.5	（1）Horovod 开源分布式深度学习平台 （2）Horovod 的应用 （1）The Horovod open source distributed deep learning platform （2）Using the Horovod platform

第四章　自动机器学习系统（AutoML System）

章节序号 Chapter Number	章节名称 Chapters	课时 Class Hour	知 识 点 Key Points
4.1	自动机器学习的背景 Background of AutoML	0.5	（1）参数与超参数的概念 （2）经典机器学习中的交叉验证方法 （1）Concepts of parameter and hyper-parameter （2）Cross-validation in classical machine learning

续表

章节序号 Chapter Number	章节名称 Chapters	课时 Class Hour	知　识　点 Key Points
4.2	自动机器学习的原理 Principle of AutoML	1.5	(1) 机器学习的机器学习 (2) 搜索空间,探索与利用 (3) 贝叶斯优化算法 (4) 基于强化学习的超参优化 (1) The machine learning of machine learning (2) Searching space, exploration and exploitation (3) Bayesian optimization algorithm (4) Hyper-parameter optimization using reinforcement learning
4.3	自动机器学习系统 AutoML system	1	(1) NNI 开源 AutoML 平台 (2) 基于 NNI 的超参优化 (1) NNI open source AutoML platform (2) Hyper-parameter optimization using NNI

第五章　集群资源调度与管理(Resource Scheduling and Management in Cluster)

章节序号 Chapter Number	章节名称 Chapters	课时 Class Hour	知　识　点 Key Points
5.1	镜像与容器技术 The image and container	0.5	(1) Docker 镜像的分层原理 (2) 容器与镜像的关系 (3) Dockerfile 配置文件的构成 (4) 实例：一个推理容器的部署 (1) Principle of layers-wise Docker image (2) Relationship of image and container (3) Constitution of Dockerfile (4) Example: deployment of an inference container

续表

章节序号 Chapter Number	章节名称 Chapters	课时 Class Hour	知识点 Key Points
5.2	多租调度系统的原理 Principle of multi-tenant scheduling system	1	(1) 公平性原理：单资源调度的最大最小公平算法、异构资源调度的 DRF 算法 (2) 拓扑亲和性原理：HiveD 算法 (3) 预测与早期反馈原理：Gandiva 调度机制 (1) Principle of fairness: the min-max fairness algorithm foruni-resource scheduling, the DRF algorithm for heterogeneous resource scheduling (2) Principle of topology affinity: theHiveD algorithm (3) Principle of prediction and early feedback: the scheduling mechanisms in Gandiva
5.3	异构集群管理系统 Heterogeneous cluster management system	0.5	(1) OpenPAI 开源人工智能异构资源调度管理平台 (2) OpenPAI 平台的部署与运维 (1) OpenPAI open source artificial intelligence heterogeneous resource scheduling and management system (2) Deployment and manipulation of theOpenPAI platform

第六章　人工智能安全与隐私(Security and Privacy of AI)

章节序号 Chapter Number	章节名称 Chapters	课时 Class Hour	知识点 Key Points
6.1	人工智能安全概述 AI security overview	0.5	(1) 人工智能应用中的安全与隐私问题 (2) 深度学习模型生命周期 (3) 人工智能安全的基本性质 (1) Security and privacy issues in AI applications (2) Deep learning model lifecycle (3) The fundamental properties of artificial intelligence security

章节序号 Chapter Number	章节名称 Chapters	课时 Class Hour	知　识　点 Key Points
6.2	对抗攻击 Adversarial attack	1	（1）基本内涵 （2）损失函数设计 （3）攻击原理与示例 （4）脆弱性分析 （5）FGSM 攻击方法介绍 （6）白盒攻击与黑盒攻击 （7）其他领域攻击示例 （1）Basic concept （2）Design of loss function （3）Principles and examples of attacks （4）Vulnerability analysis （5）FGSM introduction （6）White box attack and black box attack （7）Attack examples of other domains
6.3	对抗防御 Adversarial defense	0.5	（1）防御方式简介：被动、主动 （2）被动防御方法介绍：特征缩减、随机化 （3）主动防御方法介绍：对抗训练 （1）Defense methods：passive and proactive （2）Passive defense：features queeze，randomization （3）Proactive defense：adversarial training

10.1.5　实验环节（Experiments）

序号 Num.	实验内容 Experiment Content	课时 Class Hour	知　识　点 Key Points
1	深度学习框架及工具入门 Getting started with deep learning frameworks and tools	3	（1）深度学习框架及工作流程 （2）TensorBoard 的功能和使用方法 （3）Profiler 的功能和使用方法 （4）不同硬件和批大小条件下张量运算的开销 （1）Deep learning framework and workflow （2）Function and usage of TensorBoard （3）Function and usage of Profiler （4）Cost of tensor operations for different hardware and batch sizes

续表

序号 Num.	实验内容 Experiment Content	课时 Class Hour	知 识 点 Key Points
2	定制新的张量运算 Customization of new tensor operations	3	（1）在 Pytorch 平台中基于函数和模块实现张量运算 （2）在 Pytorch 平台中通过 C++Extension 编写 Python 函数模块，实现张量运算 （3）基于不同方法实现新的张量运算的性能差异 (1) Implement tensor operations based on function and module in Pytorch (2) Implement tensor operations by writing Python Function Modules through C++ Extension in Pytorch (3) Performance difference of implementing new tensor operations based on different methods
3	分布式训练方法 Distributed training method	3	（1）OpenMPI 的功能和使用方法 （2）Horovod 的功能和使用方法 （3）不同通信后端的基本原理和适用范围 （4）调用不同的通信后端实现数据并行的并行/分布式训练方法 (1) Function and usage of OpenMPI (2) Function and usage of Horovod (3) Basic principles and scope of application of different communication backends (4) Parallel/distributed training methods for data parallelism by calling different communication backends
4	自动机器学习系统 Automatic machine learning system	3	（1）自动机器学习中的基本概念 （2）NNI 库的功能和使用方法 （3）基于 NNI 的模型自动参数调优 (1) Fundamental concepts in automated machine learning (2) Function and usage of NNI library (3) Automatic parameter tuning of models based on NNI

续表

序号 Num.	实验内容 Experiment Content	课时 Class Hour	知 识 点 Key Points
5	对抗攻击方法 Adversarial attack method	3	(1) FGSM(Fast Gradient Sign Method)攻击方法的原理和使用方法 (2) 无目标及有目标对抗攻击实现 (3) 不同约束下攻击效果的差异对比 (1) Principle and usage of FGSM attack method (2) Untargeted and targeted adversarial attack implementation (3) Comparison of differences in attack effects under different constraints
6	基于Container的云上训练和推理 Configure containers for training and inference	3	(1) Container 机制 (2) Docker Engine 的功能和使用方法 (3) 基于 Container 的深度学习训练和推理 (1) Container mechanism (2) Function and usage of Docker engine (3) Deep learning training and inference using containers
7	基于OpenPAI的调度管理系统 Scheduling management system based on OpenPAI	6	(1) OpenPAI 平台部署和集群搭建 (2) 集群的使用和管理方法 (3) 不同调度策略对比和分析 (1) OpenPAI platform deployment and cluster construction (2) Cluster usage and management (3) Comparison and analysis of different scheduling strategies

大纲指导者：郑南宁教授(西安交通大学人工智能学院)

大纲制定者：刘剑毅副教授(西安交通大学人工智能学院)、张旭翀副教授(西安交通大学人工智能学院)、张玥工程师(西安交通大学人工智能学院)、高彦杰研究员(微软亚洲研究院)

大纲审定：西安交通大学人工智能学院本科专业知识体系建设与课程设置第二版修订工作组

10.2 "3D 深度感知"课程大纲

课程名称：3D 深度感知
Course：3D Depth Sensing
先修课程：计算机视觉与模式识别
Prerequisites：Computer Vision and Pattern Recognition
学分：1
Credits：1

10.2.1 课程目的和基本内容（Course Objectives and Basic Content）

本课程是人工智能学院本科专业选修课。
This course is an elective course for undergraduates in College of Artificial Intelligence.

课程以 3D 深度感知技术及设备（结构光 3D 深度相机）为主线，对高精度、高分辨率深度获取的基本理论与结构光编解码方法展开讨论，同时介绍结构光 3D 深度相机原理、组成及应用。第一章到第四章分别讨论结构光深度感知技术的基础理论、单目结构光深度感知原理方法、激光散斑结构光编码方法、深度解码方法，这部分内容的重点是 3D 深度获取的激光散斑结构光编解码方法。第五章是即插即用的嵌入式 3D 深度相机应用，包括 3D 点云生成。

本课程通过对基本理论、设计方法和应用技术的学习，帮助学生建立关于 3D 深度感知基本原理和应用设计方面的知识框架。课程采用小组学习模式，并辅之以研究性实验、课堂测验、小组讨论及综述报告等教学手段，训练学生用基本理论和方法分析解决实际问题的能力，掌握 3D 深度感知系统设计所必须的基本知识和技能。课程通过散斑编码图案的采集处理及 3D 深度感知系统设计实验使学生巩固和加深深度感知的理论知识，通过实践进一步加强学生独立分析问题、解决问题的能力，培养综合设计及创新能力，培养实事求是、严肃认真的科学作风和良好的实验习惯，为今后的工作打下良好的基础。

当前无人机、3D 打印、机器人、虚拟现实、智能手机、智能监控等领域的深入研究与发展,需要解决避障、3D 成像、自然交互、精确识别等难题,3D 深度感知技术作为关键共性技术有助于解决这些难题,将极大地释放和激发人们对相关研究领域的科学想象力和创造力。因此,理解和掌握好 3D 深度感知的基本概念、基本原理和方法,在遇到实际问题时,能激发学生去寻找新的理论和技术,也能使学生利用一种熟悉的工具进入一个生疏的研究领域。

The course focuses on 3D depth sensing technology and equipment(structured-light depth camera), and discusses the basic theories of high precision and high resolution depth acquisition and the structured-light codec methods. Moreover, it introduces the principle, structure and applications about the structured-light depth camera. Chapter 1 to Chapter 4 discuss the basic theories of structured-light depth sensing technology, monocular structured-light depth sensing principle, the laser speckle structured-light coding method, and the depth decoding method. The focus of this part is on the 3D depth acquisition according to laser speckle structured-light codec method. Chapter 5 is the applications about the plug-and-play embedded 3D depth camera, including 3D point cloud generation.

This course helps students build a knowledge framework for the basic principles of 3D depth sensing and application design through the study of basic theories, design methods, and applied techniques. The course adopts group study method, supplemented by experiments, in-class tests, discussions and reports, in order to train students the ability to solve practical problems with basic theories and methods and master the basic knowledge and skills for 3D depth sensing application design. The course includes several experiments on speckle coding patterns collection and pre-processing and 3D depth sensing system design in order to consolidate the students' theoretical knowledge of depth sensing technology, further strengthen their ability to analyze and solve problems independently, and develop their comprehensive abilities on system design and innovations as well as good habits for future work.

With the in-depth research and development in the fields of unmanned aerial vehicle, 3D printing, robot, virtual reality, smart phone, smart surveillance and so on, the problems of obstacle avoidance, 3D imaging, natural interaction and precision recognition is needed to be solved. As the key co-use technology, 3D depth sensing is very important for solving these problems. Therefore, understanding the basic concepts, principles and methods of depth sensing technology can stimulate students to find new theories and techniques when they encounter practical problems. Moreover, it also helps students to enter a strange research area with familiar tools.

10.2.2 课程基本情况(Course Arrangements)

课程名称	3D深度感知 3D Depth Sensing								
开课时间	一年级		二年级		三年级		四年级		
	秋	春	秋	春	秋	春	秋	春	
课程定位	本科生"人工智能工具与平台"课程群选修课								
学　分	1学分								
总学时	20学时(授课16学时、实验4学时)								
授课学时分配	课堂讲授(16学时)								
先修课程	计算机视觉与模式识别								
后续课程									
教学方式	课堂教学、小组讨论、大作业与实验								
考核方式	大作业占40%,实验编程成绩占50%,考勤占10%								
参考教材	Zanuttigh P,et al. Time-of-Flight and Structured Light Depth Cameras[M]. Berlin: Springer,2016.								
参考资料	1. 葛晨阳,姚慧敏,周艳辉.结构光深度相机系统实验指导书. 2. 万哲先.代数与编码[M].3版.北京:高等教育出版社出版,2007.								
其他信息									

人工智能工具与平台	
必修 (学分)	深度学习工具与平台(2)
选修 (学分)	3D深度感知(1)
	人工智能芯片设计导论(1)

10.2.3 教学目的和基本要求(Teaching Objectives and Basic Requirements)

(1) 掌握结构光深度感知技术的基础理论;
(2) 掌握单目结构光深度感知系统测距原理;
(3) 掌握激光散斑结构光编码方法;
(4) 掌握深度解码方法;
(5) 了解嵌入式3D深度相机原理,熟悉单幅点云生成;
(6) 利用单目结构光深度相机采集散斑数据,并计算生成深度和点云信息。

10.2.4 教学内容及安排(Syllabus and Arrangements)

第一章 绪论(Introduction)

章节序号 Chapter Number	章节名称 Chapters	课时 Class Hour	知 识 点 Key Points
1.1	绪论 Introduction	1	(1) 主被动式感知的区别 (2) 结构光深度感知技术的基础理论 (1) The differences between active and passive depth sensing (2) Basic principle of structured-light depth sensing technology

第二章 单目结构光深度感知原理方法(Monocular Structured-light Depth Sensing Principle)

章节序号 Chapter Number	章节名称 Chapters	课时 Class Hour	知 识 点 Key Points
2.1	单目结构光深度感知系统 Monocular structured-light depth sensing system	1	(1) 深度感知测量原理 (1) Depth sensing measurement
2.2	散斑图像预处理算法 Speckle pattern preprocessing algorithm		(1) 一致性增强预处理 (1) Consistency enhancement preprocessing
2.3	视差估计算法 Disparity estimation algorithm	1	(1) 块匹配、子像素插值 (1) Block matching, sub-pixel interpolation
2.4	单目深度计算方法 Monocular depth calculation method	1	(1) 视差、基线、距离 (1) Disparity, baseline, distance

第三章 激光散斑结构光编码方法（Laser Speckle Structured-light Coding Method）

章节序号 Chapter Number	章节名称 Chapters	课时 Class Hour	知 识 点 Key Points
3.1	随机序列 Random sequence	1	（1）随机序列、随机阵列、窗口唯一性 （1）Random sequence, random array, window uniqueness
3.2	散斑编码图案的生成 A generation of speckle coding pattern	1	（1）编码方法、编码特性 （1）Coding method, coding characteristics
3.3	激光散斑光学投射系统 Laser speckle optical projection system	1	（1）激光散斑光学投射系统 （1）Laser speckle optical projection system

第四章 深度解码方法（Depth Decoding Method）

章节序号 Chapter Number	章节名称 Chapters	课时 Class Hour	知 识 点 Key Points
4.1	深度解码系统架构 System architecture of depth decoding	4	（1）系统架构 （1）System architecture
4.2	深度解码功能模块 Function module of depth decoding		（1）功能模块 （1）Function module

第五章 嵌入式3D深度相机（Embedded 3D Depth Camera）

章节序号 Chapter Number	章节名称 Chapters	课时 Class Hour	知 识 点 Key Points
5.1	嵌入式3D深度相机 Embedded 3D depth camera	2	（1）嵌入式3D深度相机 （1）Embedded 3D depth camera

续表

章节序号 Chapter Number	章节名称 Chapters	课时 Class Hour	知　识　点 Key Points
5.2	单幅点云获取 The point cloud acquisition for single frame	2	（1）点云获取 （1）Point cloud acquisition

第六章　搭建单目结构光深度相机系统（Construction of Monocular Structured-light Depth Camera System）

章节序号 Chapter Number	章节名称 Chapters	课时 Class Hour	知　识　点 Key Points
6.1	采集散斑图像 Speckle image collection	1	（1）采集散斑图像 （1）Speckle image collection
6.2	C代码编程实现深度解码算法 Achieving deep decoding algorithm through C code programming		（1）C代码编程实现深度解码算法 （1）Achieving deep decoding algorithm through C code programming

10.2.5　实验环节（Experiments）

序号 Num.	实验内容 Experiment Content	课时 Class Hour	知　识　点 Key Points
1	单目深度感知测量 Monocular depth sensing measurement	4	（1）单目结构光深度感知系统 （2）散斑图像的预处理 （3）视差估计 （4）深度计算方法 （1）Monocular structured-light depth sensing system （2）Speckle pattern preprocessing （3）Disparity estimation （4）Depth calculation method

大纲指导者：郑南宁教授（西安交通大学人工智能学院）
大纲制定者：葛晨阳副教授（西安交通大学人工智能学院）、姚慧敏工程师（西安交通大学人工智能学院）
大纲审定：西安交通大学人工智能学院本科专业知识体系建设与课程设置第二版修订工作组

10.3 "人工智能芯片设计导论"课程大纲

课程名称：人工智能芯片设计导论
Course：Introduction to Processor Design for AI Applications
先修课程：数据结构与算法、电子技术与系统、计算机体系结构
Prerequisites：Data Structure and Algorithm, Electronic Technology and System, Computer Architecture
学分：1
Credits：1

10.3.1 课程目的和基本内容（Course Objectives and Basic Content）

本课程是人工智能学院本科专业选修课。
This course is an elective course for undergraduates in College of Artificial Intelligence.

课程讲解以人工智能应用为代表的领域专用计算架构设计的基本原则以及深度学习加速模块的设计方法，旨在锻炼学生综合应用数字系统结构与设计、计算机体系结构等课程的知识。课程教学部分主要包括以下环节：门级电路设计语言基本语法、FPGA设计方法以及深度神经网络加速器的基本知识，包括：重定时、电路展开/折叠、调度和资源分配、脉冲阵列等。

通过本课程，加深学生对基于指令和基于数据流驱动的计算机系统的全面理解，深入理解数字系统，特别是领域专用计算架构的运行原理和实现机制，培养学生综合应用所学知识解决实际问题的能力。本课程的教学重点在于对先修课程知识的融会贯通和综合应用。

The course explains the basic principles of domain-specific computing architecture(e. g. artificial intelligence application) and the design of deep learning acceleration module. The aim of this course is to train students to comprehensively apply the knowledge of digital system structure and design, computer architecture. The course mainly includes such teaching topics: basic syntax of gate-level circuit design language, FPGA design method, basic knowledge of deep neural network accelerator including retiming, folding, unfolding, scheduling and resource allocation, systolic array, etc.

Through this course, students will fully understand the instruction-based and data-driven computer systems, and deeply understand the digital system, especially the operating principle and implementation mechanism of domain-specific computer architecture. The focus of this course is on the integration of pre-requisite knowledge and development of students' practical ability.

10.3.2 课程基本情况(Course Arrangements)

课程名称	人工智能芯片设计导论 Introduction to Processor Design for AI Applications							
开课时间	一年级		二年级		三年级		四年级	
	秋	春	秋	春	秋	春	秋	春
课程定位	本科生"人工智能工具与平台"课程群选修课							
学 分	1学分							
总学时	16学时(授课16学时、实验0学时)							
授课学时分配	课堂讲授(16学时)							
先修课程	数据结构与算法、电子技术与系统、计算机体系结构							
后续课程								
教学方式	课堂教学、大作业与实验、小组讨论、综述报告							
考核方式	平时成绩占10%,课程作业成绩占70%,分组答辩占20%							
参考教材	1. Patterson D A, Hennessy J L. 计算机组成与设计-硬件/软件接口(RISC-V版本)[M]. 北京:机械工业出版社,2015. 2. Parhi K K. VLSI数字信号处理系统:设计与实现[M]. 陈弘毅,译. 北京:机械工业出版社,2004.							
参考资料								
其他信息								

人工智能工具与平台

必修(学分)	深度学习工具与平台(2)
选修(学分)	3D深度感知(1)
	人工智能芯片设计导论(1)

10.3.3 教学目的和基本要求（Teaching Objectives and Basic Requirements）

（1）了解数字系统结构与设计中的相关概念和原理；
（2）熟悉基于数据流驱动的深度神经网络加速器设计；
（3）掌握时序电路和组合逻辑电路的设计原则；
（4）掌握 Verilog、Python、HSL 等基本语言和 EDA 工具；
（5）掌握 FPGA 设计的方法学和基本工具；
（6）掌握深度神经网络加速器设计基本方法。

10.3.4 教学内容及安排（Syllabus and Arrangements）

绪论 课程要求与简介（Introduction）

章节序号 Chapter number	章节名称 Chapters	课时 Class Hour	知识点 Key points
0.1	课程简介 Introduction	2	（1）数字系统结构与设计中的相关概念和原理 （1）Concepts and principles in digital system structure and design
0.2	数字系统知识 Digital system knowledge	2	（1）数字逻辑 （2）计算机体系结构 （1）Digital logic （2）Computer architecture

第一章 Verilog 和 HLS 语法（Verilog and HLS Syntax）

章节序号 Chapter number	章节名称 Chapters	课时 Class Hour	知识点 Key points
1.1	Verilog 和 HLS 简介 Introduction of Verilog and HLS	4	（1）Verilog 和 HLS 语法 （2）组合逻辑、时序逻辑 （3）Testbench 的编写 （1）Verilog and HLS syntax （2）Combination logic, sequential logic （3）How to write Testbench

第二章　FPGA电路结构、开发板介绍和开发环境使用讲解（Introduction to FPGA Design and Development Environment）

章节序号 Chapter number	章节名称 Chapters	课时 Class Hour	知　识　点 Key points
2.1	FPGA设计和开发环境介绍 Introduction to FPGA design and development environment	2	（1）FPGA电路结构 （2）开发板和开发环境使用讲解 （1）FPGA circuit structure （2）Development PCB board and development environment introduction

第三章　数据流驱动的计算架构（Data-stream Computer Architecture）

章节序号 Chapter number	章节名称 Chapters	课时 Class Hour	知　识　点 Key points
3.1	数据流驱动计算架构的关键技术 Key technologies of data-stream computer architecture	2	（1）重定时技术及流水线 （2）电路折叠和展开 （3）调度和资源分配 （1）Retiming and Pipeline （2）Circuit Folding/Unfolding （3）Scheduling and Resource Allocation
3.2	脉动阵列电路设计 Systolic array design	2	（1）脉动阵列设计的一般方法 （2）脉动阵列典型设计举例：FIR滤波器、矩阵乘电路、Top-K排序、模式匹配等 （1）Principle of systolic array design （2）Typical design examples of systolic arrays：FIR filter，matrix multiplication circuit，top-K sorting，pattern matching，etc.

第四章 AI 芯片设计回顾和 DNN 加速器设计（AI Processor and DNN Accelerator Review）

章节序号 Chapter number	章节名称 Chapters	课时 Class Hour	知　识　点 Key points
4.1	AI 芯片和 DNN 加速器 AI chip and DNN accelerator	2	（1）AI 芯片设计回顾和 DNN 加速器讲解 （1）Review of AI processor and DNN accelerator

大纲制定者：任鹏举教授（西安交通大学人工智能学院）、赵文哲助理教授（西安交通大学人工智能学院）、夏天副教授（西安交通大学人工智能学院）

大纲审定：西安交通大学人工智能学院本科专业知识体系建设与课程设置第二版修订工作组

第 11 章

"专业综合性实验"课程群

11.1 "脑信号处理"实验课大纲

课程名称：脑信号处理
Course：Brain Signal Processing
先修课程：计算机程序设计、数字信号处理、人工智能概论、计算机视觉与模式识别、计算神经工程
Prerequisites：Computer Programming, Digital Signal Processing, Introduction to Artificial Intelligence, Computer Vision and Pattern Recognition, Computational Neural Engineering
学分：1
Credits：1

11.1.1 课程目的和基本内容（Course Objectives and Basic Content）

本课程是人工智能学院本科专业综合性实验必修课。

This course is a compulsory comprehensive experimental course for undergraduates in College of Artificial Intelligence.

本课程综合应用数字信号处理、机器学习、计算神经工程等相关知识，融合脑信号相关技术，要求学生了解脑电（EEG）、功能磁共振（fMRI）等实验方法和基本的范式设计思想，理解并能使用经典的脑信号处理方法。课程主要包括三大类内容共计 8 个实验：

(1) fMRI 大脑激活区分析、图像分类以及功能连接网络构建；
(2) 脑电视觉刺激呈现与采集、情绪识别及运动想象解码；
(3) 基于脉冲神经网络（SNN）的数字分类。

通过这些实验的动手实践，使学生熟悉脑信号的采集、预处理、特征提取、分类、分析与编解码等一般流程和重要的工具及方法。总之，本课程的实践可使学生对脑信号的认识更加深入，理解和运用脑信号的处理方法；培养学生综合运用所学知识解决实

际问题的能力,并激发他们探索大脑的兴趣。

This course integrates the knowledge of digital signal processing, machine learning, computational neural engineering, brain signal related technologies and so on. Students are required to master the scientific use of EEG and functional magnetic resonance experimental devices and basic paradigm design ideas, understand and use classical brain signal processing methods. The course mainly includes three categories of content, totaling 8 experiments:

(1) fMRI brain activation area analysis, image classification, and the construction of functional connectivity network;

(2) The presentation and collection of EEG signal of visual stimuli, emotion recognition and motor imagination decoding;

(3) Digital classification based on Spiking Neural Network (SNN).

Through these hands-on experimental practice, students will be familiar with the general processes and important tools and methods of brain signal acquisition, preprocessing, feature extraction, classification, analysis, encoding and decoding etc. Overall speaking, this course can help students deepen the understanding of brain signals, and understand and apply the signal processing methods of brain signal. Furthermore, it cultivates students' abilities to use the knowledge comprehensively in solving practical problems, and stimulates their interests in exploring the brain.

11.1.2　课程基本情况(Course Arrangements)

课程名称	脑信号处理 Brain Signal Processing								
开课时间	一年级		二年级		三年级		四年级		专业综合性实验
	秋	春	秋	春	秋	春	秋	春	
课程定位	本科生"专业综合性实验"课程群必修课						必修 (学分)		脑信号处理(1) 游戏 AI 设计与开发(1) 机器人导航技术(1) 创新设计思维(1)
学　分	1 学分						选修 (学分)		/
学　时	32 学时								
先修课程	计算机程序设计、数字信号处理、人工智能概论、计算机视觉与模式识别、计算神经工程								
后续课程									
考核方式	实验报告占 90%,考勤占 10%								
参考教材									
参考资料	1. Kim S P. Preprocessing of EEG[M]. Singapore:Springer Singapore. 2018. 2. Rajesh Singla. SSVEP-Based BCIs[M]. IntechOpen,2018.								
其他信息									

11.1.3 实验目的和基本要求(Experiment Objectives and Basic Requirements)

(1) 熟悉 fMRI 数据预处理,掌握其编解码的概念和过程,能利用支持向量机等方法进行数据分类;

(2) 熟悉一般线性模型及单被试统计分析过程,会利用 SPM 工具箱进行大脑激活区分析;

(3) 了解大脑网络和时间序列的概念,通过图像分析加深对 fMRI 数据的直观理解;

(4) 了解脑电视觉刺激实验程序的设计流程,熟悉脑电采集的实验过程和伪迹识别;

(5) 熟悉情感诱发的实验范式,掌握脑电功率谱密度分析和情绪脑电的分类识别方法;

(6) 熟悉运动想象实验范式和共空间模式特征提取方法,能够编程实现运动想象分类;

(7) 了解不同类型脉冲编码的特点,会使用脉冲神经网络进行图像分类。

11.1.4 实验内容及安排(Experiment Syllabus and Arrangements)

序号 Num.	实验内容 Experiment Content	课时 Class Hour	知 识 点 Key Points
1	基于一般线性模型的大脑激活区分析 Brain active area analysis based on general linear model	4	(1) 预处理:掌握基于 SPM 工具箱的 fMRI 预处理方法 (2) 模型设定:设置一阶统计分析参数 (3) 模型估计:根据给定的参数进行模型估计 (4) 统计推断:分析不同条件下各个体素的回归系数显著性,做出统计参数图 (1) Preprocessing: learn the fMRI preprocessing methods based on SPM toolbox (2) Model setting: set the first-order statistical analysis parameters (3) Model estimation: the model is estimated according to the given parameters (4) Statistical inference: analyze the significance of regression coefficient of each voxel under different conditions, and make a statistical parameter map

续表

序号 Num.	实验内容 Experiment Content	课时 Class Hour	知 识 点 Key Points
2	基于fMRI数据的图像分类 Image classification of fMRI data	4	（1）加载数据并进行数据预处理，分别提取输入数据和标签 （2）体素选择：通过MATLAB程序筛选出一个区域的体素，组成新的体素响应矩阵 （3）在MATLAB的分类器App中导入输入数据和标签，选择10折交叉验证，进行分类验证 （4）选择不同的分类器，比如SVM、LDA等，进行训练并检查交叉验证结果 （1）Load and preprocess data to extract the input and labels respectively （2）Voxel selection：Voxels in a region are selected through Matlab to form a new voxel response matrix （3）Import input data and labels into the classifier app of Matlab, and select 10-fold cross validation for classification validation （4）Select different classifiers, such as SVM, LDA, etc., conduct training and check thecross validation results
3	fMRI功能连接网络的构建 Construction of fMRI functional connection network	4	（1）数据重采样：自行选择方法完成感兴趣区域（ROI）与大脑模板维度的统一 （2）提取典型脑区的ROI时间序列，分别提取每个体素时间序列以及平均每个体素的时间序列 （3）典型脑区的功能网络构建及可视化 （4）基于给定模板的功能网络的构建及可视化 （1）Data resampling：freely select methods to unify the dimensions of regions of interest (ROI) and brain templates （2）Extract the ROI time series of typical brain regions, and respectively extract the time series of each voxel and average time series of each voxel （3）Construct and visualize functional network of typical brain regions （4）Construct and visualize the functional network based on the given template

续表

序号 Num.	实验内容 Experiment Content	课时 Class Hour	知 识 点 Key Points
4	脑电视觉刺激设计及呈现 Design and presentation of EEG visual stimuli	4	（1）用 Psychopy 软件编程实现左右手的运动想象刺激实验设计 （2）创建窗口，呈现文字刺激、按键等待、反应选择、循环、随机等内容 (1) Use the software Psychopy to realize experimental design of motor imagination for left and right hand movement (2) Create a window to display text stimulation, key waiting, response selection, circulation, randomization, etc
5	脑电的采集及伪迹 EEG signal acquisition and artifact	4	（1）了解实验前准备内容，学习佩戴脑电帽和注射导电膏的方法 （2）熟悉实验开始后的过程以及实验结束后的处理 （3）从数据中观察和了解不同类型的伪迹 (1) Understand the preparations before the experiment, and learn how to wear EEG caps and inject conductive paste (2) Know the process after the beginning of the experiment and the treatment after the end of the experiment (3) Observe different types of artifacts from the data

续表

序号 Num.	实验内容 Experiment Content	课时 Class Hour	知 识 点 Key Points
6	情绪脑电识别 Emotion recognition of EEG signal	4	（1）了解情绪诱发实验范式，从情绪脑电数据集中截取有用样本 （2）利用功率谱密度方法对情绪脑电信号进行特征提取，并对特征进行归一化 （3）使用线性判别分析方法对提取的特征进行分类 （4）利用交叉验证方法选择表现最好的模型 (1) Understand the experimental paradigm of emotion induction, and intercept useful samples from emotional EEG data set (2) Extract the features of emotional EEG signals by using the power spectral density method, and normalize the features (3) Classify the extracted features by using linear discriminant analysis (4) Select the best model using cross validation
7	运动想象解码 Motion imagination decoding	4	（1）了解运动想象实验范式，并从运动想象数据集中截取有用样本 （2）利用共空间模式对脑电信号进行特征提取 （3）使用支持向量机对提取的特征进行分类 （4）学习模型调优，使分类性能尽可能的高 (1) Understand the experimental paradigm of motion imagination, and intercept useful samples from the related data set (2) Extract the features of EEG signal by using the common space pattern (3) Classify the extracted features with support vector machine (4) Optimize the model to make the classification performance as high as possible

续表

序号 Num.	实验内容 Experiment Content	课时 Class Hour	知识点 Key Points
8	基于脉冲神经网络的数字分类 Digital classification based on spiking neural network	4	（1）应用不同类型的脉冲编码方法对图像进行编码 （2）使用 Python 构建脉冲神经网络，并通过所构建的网络对脉冲序列进行特征提取 （3）对比不同编码方式所获得脉冲序列的特点，实现简单图像的分类 (1) Encode images with different types of spiking coding methods (2) Use Python to construct the spiking neural network, and extract features from spiking sequence through the constructed network (3) Compare the characteristics of spiking sequences obtained by different encoding methods, and realized classification of simple images

大纲指导者：郑南宁教授（西安交通大学人工智能学院）

大纲制定者：陈霸东教授（西安交通大学人工智能学院）、张璇工程师（西安交通大学人工智能学院）

大纲审定：西安交通大学人工智能学院本科专业知识体系建设与课程设置第二版修订工作组

11.2 "游戏 AI 设计与开发"实验课大纲

课程名称：游戏 AI 设计与开发

Course：Game Design and Development with AI

先修课程：计算机程序设计、人工智能概论、机器学习

Prerequisites：Computer Programming，Introduction to Artificial Intelligence，Machine Learning

学分：1

Credits：1

11.2.1 课程目的和基本内容(Course Objectives and Basic Content)

本课程是人工智能学院本科专业综合性实验必修课。

This course is a compulsory comprehensive experimental course for undergraduates in College of Artificial Intelligence.

游戏技术涉及许多与现代游戏开发、设计和制作相关的核心技术领域。这些领域大多是由关键的人工智能技术驱动的,例如专家领域知识系统、搜索和最优化以及游戏中的计算智能。本课程的主要目标是理解、设计、实现和使用基本和新的 AI 技术,以在游戏中生成有效的智能行为。

本课程旨在向学生介绍基础和高级游戏人工智能主题的理论,并提供按照商业标准开发游戏 AI 算法的实践经验。其核心目标包括:使学生熟悉并能够理解基本和高级游戏人工智能技术以及提高学生按照商业标准制作开发游戏智能方法的能力。

本课程涵盖的主题包括:游戏开发背景下的 AI 方法、使用 AI 玩游戏、使用 AI 生成内容、前沿动态和前景展望等。

Game technology incorporates a number of core technical fields that are relevant for modern game development, design and production. Most of these areas are driven by key artificial intelligence techniques such as expert domain-knowledge systems, search and optimization, and computational intelligence in games. The primary goal of this course is the understanding, design, implementation and use of basic and nouvelle AI techniques for generating efficient intelligent behaviors in games.

The course aims to introduce students to the theory of basic and advanced game artificial intelligence topics and provide hands-on experience on the implementation of popular algorithms on commercial-standard games. The core aims of the unit are as follows: students get familiar with and are able to theorize upon basic and advanced game artificial intelligence techniques; students develop intelligent game agents for commercial-standard productions.

The topics covered in this course include: AI methods in games context, using AI to play games, using AI to model players, frontiers and outlook.

11.2.2 课程基本情况(Course Arrangements)

课程名称	游戏 AI 设计与开发 Game Design and Development with AI								
开课时间	一年级		二年级		三年级		四年级		
	秋	春	秋	春	秋	春	秋	春	**专业综合性实验**
课程定位	本科生"专业综合性实验"课程群必修课								必修 (学分)	脑信号处理(1)
									游戏 AI 设计与开发(1)
									机器人导航技术(1)
学 分	1 学分								创新设计思维(1)
									选修 (学分)	/
总 学 时	32 学时								
先修课程	计算机程序设计、人工智能概论、机器学习								
后续课程									
考核方式	实验成绩占 70%,实验报告占 20%,考勤占 10%								
参考教材	1. Georgios N. Y, Togelius J. Artificial Intelligence and Games[M]. Berlin: Springer, 2018. 2. Millington I. 游戏中的人工智能[M]. 张俊,译. 3 版. 北京: 清华大学出版社,2021. 3. Buckland M. 游戏人工智能编程案例精粹[M]. 罗岱,等译. 北京: 人民邮电出版社,2012.								
参考资料									
其他信息									

11.2.3 实验目的和基本要求(Experiment Objectives and Basic Requirements)

(1) 了解游戏 AI 的定义、发展历史、人工智能与游戏的联系;

(2) 了解并掌握游戏开发背景下的人工智能方法及典型工具的实践使用;

(3) 掌握游戏智能体建模方法,将 AI 方法应用于玩游戏,如基于规划方法、深度学习、强化学习、监督学习、混合方法;

(4) 选择并使用若干种基本方法进行游戏中智能体的设计与实现,包括单智能体、多智能体、对战、混战等;

(5) 掌握使用 AI 生成游戏内容,掌握应用于游戏内容生成的基于搜索的方法、基于求解器的方法、基于语法的方法、元胞自动机、噪声与分形与机器学习方法等;

(6) 了解游戏 AI 设计与开发的前沿进展。

11.2.4 实验内容及安排(Experiment Syllabus and Arrangements)

序号 Num.	实验内容 Experiment Content	课时 Class Hour	知 识 点 Key Points
1	概述 Overview	1	(1) 游戏 AI 的定义与发展历史 (2) 游戏开发与 AI 研究的联系 (3) 学术界与工业界的进展 (1) Definition and history of game AI (2) Benefits of AI for games (3) The academic and industrial development
2	游戏 AI 智能 体建模 AI agent modeling		(1) 建模方法介绍 (2) 有限状态机与行为树建模 (1) Introduction of Modeling (2) Modeling with Finite state machines,Behavior trees
3	游戏 AI 工具的 使用 Unity The use of Unity tool for AI game	3	(1) Unity 工具使用 (2) Unity 中状态机建模练习 (3) Unity 中行为树建模练习 (1) The use of Unity tool (2) State machine modeling exercise (3) Behavior tree modeling exercises

续表

序号 Num.	实验内容 Experiment Content	课时 Class Hour	知识点 Key Points
4	单智能体 AI 游戏设计 "吃豆人"及其他单机游戏智能体，如飞行鸟、贪吃蛇棋类游戏等等的 AI 游戏开发 The development of Ms Pac-Man, Flyer bird, Gluttonous Snake and other AI games	12	（1）游戏 AI 的交互接口 （2）基于 Unity 工具的设计 （3）基于动态规划的 AI 玩家 （4）基于深度强化学习的 AI 玩家 （5）基于进化算法的 AI 玩家 （6）基于符号方法的 AI 玩家 (1) Interface for AI players (2) Unity tool for AI games (3) Dynamic programming based AI player (4) Deep Q-learning based AI player (5) Evolutionary algorithms based AI player (6) Symbolic methods based AI player
5	高级多智能体 AI 游戏开发与对战设计与实践 Advanced intelligent multi-agent AI game development	12	（1）星际争霸等 AI 智能体开发 （2）荒野乱斗等 AI 智能体开发 （3）其他多智能体游戏智能体对战 (1) StarCraft based AI game (2) Brawl Stars AIgame (3) Other multi-agent AI game design
6	游戏中迷宫地图、风景生成及其他游戏相关主题设计 Maze/Scenery generation and other topic design for AI game	4	（1）基于图论的迷宫生成 （2）基于元胞自动机的迷宫生成 （3）基于生成式对抗网络的迷宫生成 （4）游戏设计中其他最新技术实验 (1) Graph theory based maze generation (2) Cellular automaton for maze generation (3) Generative adversarial nets for maze generation (4) Using latest AI game technology for game design

大纲指导者：郑南宁教授(西安交通大学人工智能学院)
大纲制定者：刘龙军副教授(西安交通大学人工智能学院)
大纲审定：西安交通大学人工智能学院本科专业知识体系建设与课程设置第二版修订工作组

11.3 "机器人导航技术"实验课大纲

课程名称：机器人导航技术
Course：Robot Navigation Technology
先修课程：现代控制工程、计算机视觉与模式识别
Prerequisites：Modern Control Engineering，Computer Vision and Pattern Recognition
学分：1
Credits：1

11.3.1 课程目的和基本内容(Course Objectives and Basic Content)

本课程是人工智能学院本科专业综合性实验必修课。

This course is a compulsory comprehensive experimental course for undergraduates in College of Artificial Intelligence.

本课程综合应用机器视觉技术、机器人控制技术和计算机视觉与模式识别的知识，在各课程已有实验的基础上，要求学生动手实践，完成一个基于视觉的低成本机器人的设计和实现，包括机器人的组装、对机器人的导航路线进行图像采集与拼接、机器人的路径规划、机器人的避障、机器人运动控制方法以及基于视觉和激光雷达的导航算法的实现。通过本课程，加深学生对机器人导航技术的全面理解，深入理解基于机器视觉技术的机器人导航技术的运行原理和实现机制，培养学生综合应用所学知识解决实际问题的能力。本课程的教学重点在于各课程之间知识点的衔接和综合应用。

This course integrates the knowledge of machine vision technology，robot control technology and computer vision and pattern recognition. On the basis of existing experiments in each course，students are required to do it. The design and implementation of a vision-based low-cost robot，including the assembly of the robot，the image acquisition and mosaic of the robot's navigation route, robot obstacle avoidance and robot motion control method, and the realization of the navigation algorithm based on the vision and Lidar, are completed. Through this course, we can deepen students' comprehensive understanding of robot navigation technology, deeply understand the operating principle and realization mechanism of robot navigation technology based on machine vision technology, and cultivate students' ability to solve practical problems by comprehensive application of the knowledge they have learned. The teaching emphasis of this course lies in the connection and comprehensive application of the knowledge points among the courses.

11.3.2　课程基本情况(Course Arrangements)

课程名称	机器人导航技术 Robot Navigation Technology							
开课时间	一年级		二年级		三年级		四年级		专业综合性实验
^	秋	春	秋	春	秋	春	秋	春	必修(学分): 脑信号处理(1); 游戏 AI 设计与开发(1); 机器人导航技术(1); 创新设计思维(1) 选修(学分): /
课程定位	本科生"专业综合性实验"课程群必修课								
学　分	1学分								
学　时	32学时								
先修课程	现代控制工程、计算机视觉与模式识别								
后续课程									
考核方式	实验成绩占70%,实验报告占20%,考勤占10%								
参考教材									
参考资料	1. Chatterjee A, Rakshit A, Singh N N. Vision Based Autonomous Robot Navigation [M]. Berlin: Springer, 2012. 2. 郑南宁. 计算机视觉与模式识别[M]. 北京: 国防工业出版社, 1998. 3. 朱大奇, 颜明重. 移动机器人路径规划技术综述[J]. 控制与决策, 2010, 25(7): 961-967. 4. 王春颖, 刘平, 秦洪政. 移动机器人的智能路径规划算法综述[J]. 传感器与微系统, 2018, 37(8): 5-8. 5. 王耀南, 梁桥康, 朱江, 等. 机器人环境感知与控制技术[M]. 北京: 化学工业出版社, 2019. 6. 陈兵旗. 机器视觉技术及应用实例详解[M]. 北京: 化学工业出版社, 2014.								
其他信息									

11.3.3 实验目的和基本要求(Experiment Objectives and Basic Requirements)

(1) 动手组装简易版机器人,了解用于感知外界环境信息的外设接口(内置的红外、增量编码器、激光雷达等传感器,摄像头的安装等),完成计算机与机器之间的数据传递;

(2) 进行图像采集以及处理(低通滤波、边缘检测、加粗并连接图像中的边缘、对边缘加粗的图像进行区域增长分割、将区域增长图像转化为地面区域等);

(3) 掌握障碍物检测与避障(掌握基于视觉和激光雷达的模糊导航系统以及基于红外传感器的模糊避障);

(4) 掌握路径规划技术(已知环境下静态环境路径规划、未知环境下静态环境路径规划、已知环境下动态环境路径规划、未知环境下动态环境路径规划);

(5) 掌握机器人的运动控制方法(基于运动学的机器人同时镇定和跟踪控制、基于动力学的机器人同时镇定和跟踪控制、基于动态非完整链式标准型的机器人神经网络自适应控制);

(6) 根据机器人接口的控制技术进行程序设计与优化,集中实现基于视觉的导航算法。

11.3.4 实验内容及安排(Experiment Syllabus and Arrangements)

序号 Num.	实验内容 Experiment Content	课时 Class Hour	知 识 点 Key Points
1	实验环境搭建 Experimental environment construction	2	(1) 机器人基本构造及组装 (2) 内置红外传感器的安装 (3) 内置增量编码器传感器的安装 (4) 激光雷达的安装 (5) 摄像头的安装 (1) Basic structure and assembly of the robot (2) Installation of built-in infrared sensor (3) Installation of built-in incremental encoder sensor (4) Lidar installation (5) Camera installation

续表

序号 Num.	实验内容 Experiment Content	课时 Class Hour	知 识 点 Key Points
2	图像处理(低通滤波、边缘检测、加粗并连接图像中的边缘、对边缘加粗的图像进行区域增长分割和将区域增长图像转化为地面区域) Image processing(low-pass filtering, edge detection, thickening and concatenating edges in the image, area-growth of images with bold edges, and conversion of area-growth images to ground areas)	6	(1) 算术均值滤波 (2) 坎尼边缘检测 (3) 膨胀与腐蚀 (4) 区域增长算法 (5) 图像坐标与机器人坐标之间的变换 (1) Arithmetic mean filtering (2) Canny edge detection (3) Expansion and corrosion (4) Regional growth algorithm (5) Transformation between image coordinates and robot coordinates
3	路径规划(已知环境下静态环境路径规划、未知环境下静态环境路径规划、已知环境下动态环境路径规划和未知环境下动态环境路径规划) Path planning(static environment path planning in known environments, static environment path planning in unknown environments, dynamic environment path planning in known environments, and dynamic environment path planning in unknown environments)	8	(1) 算术均值滤波 (2) 坎尼边缘检测 (3) 膨胀与腐蚀 (4) 区域增长算法 (5) 图像坐标与机器人坐标之间的变换 (6) 可视图法、泰森多边形法、栅格法构建环境 (7) 点对点路径规划：搜索路径最优，图搜索类算法(迪科斯特拉算法、A*算法、D*算法)，随机采样类算法(概率路标算法、快速随机数算法)、智能仿生算法(遗传算法、蚁群算法、粒子群算法) (8) 局部路径规划：人工势场法、模拟退火法、模糊逻辑法、神经网络法、动态窗口法、强化学习法以及基于行为的路径规划 (9) 遍历路径规划：随机遍历策略、沿边规划策略、漫步式探测路径规划

续表

序号 Num.	实验内容 Experiment Content	课时 Class Hour	知 识 点 Key Points
3	路径规划(已知环境下静态环境路径规划、未知环境下静态环境路径规划、已知环境下动态环境路径规划和未知环境下动态环境路径规划) Path planning(static environment path planning in known environments, static environment path planning in unknown environments, dynamic environment path planning in known environments, and dynamic environment path planning in unknown environments)		(1) Arithmetic mean filtering (2) Canny edge detection (3) Expansion and corrosion (4) Regional growth algorithm (5) Transformation between image coordinates and robot coordinates (6) Viewable method, Voronoi diagram, grid method build environment (7) Point-to-point path planning: search path optimization: graph search algorithm (Dijkstra algorithm, A* algorithm, D* algorithm), random sampling algorithm (probabilistic landmark algorithm, fast random number algorithm), intelligent bionic algorithm (genetic algorithm, ant colony algorithm, particle swarm algorithm) (8) Local path planning: artificial potential field method, simulated annealing method, fuzzy logic method, neural network method, dynamic window method, reinforcement learning method and behavior-based path planning (9) Traversal path planning: random traversal strategy, edge planning strategy, walk-through path planning
4	障碍物检测与避障(基于视觉或激光雷达的模糊导航系统和基于红外传感器的模糊避障) Obstacle detection and obstacle avoidance(vision or lidar based fuzzy navigation system and fuzzy obstacle avoidance based on infrared sensor)	8	(1) 图像去噪 (2) 亮度自动校正 (3) 视觉或激光雷达导航系统和避障时的模糊规则的表示 (4) 保存可能转向角绕行信息的算法 (1) Image denoising (2) Automatic brightness correction (3) Visual or lidar navigation system and representation of fuzzy rules in obstacle avoidance (4) Algorithm for saving possible steering around information

续表

序号 Num.	实验内容 Experiment Content	课时 Class Hour	知 识 点 Key Points
5	机器人运动控制方法(基于运动学的机器人同时镇定和跟踪控制、基于动力学的机器人同时镇定和跟踪控制和基于动态非完整链式标准型的机器人神经网络自适应控制) Robot motion control method (skin-based robot simultaneous stabilization and tracking control, dynamic-based robot simultaneous stabilization and tracking control, and robotic neural network adaptive control based on dynamic non-holonomic chain standard)	8	(1) 运动学控制律 (2) 动力学控制律 (3) 参数自适应律 (4) 反演控制方法 (5) 控制器设计 (6) 神经网络模型 (7) 混合控制律 (1) Kinematics control law (2) Dynamic control law (3) Parameter adaptive law (4) Inversion control method (5) Controller design (6) Neural network model (7) Mixed control law

大纲制定者：袁泽剑教授(西安交通大学人工智能学院)、王乐教授(西安交通大学人工智能学院)、左炜亮副教授(西安交通大学人工智能学院)

大纲审定：西安交通大学人工智能学院本科专业知识体系建设与课程设置第二版修订工作组

11.4 "创新设计思维"实验课大纲

课程名称：创新设计思维
Course：Innovation and Design Thinking
先修课程：计算机程序设计、电子技术与系统、现代控制工程、机器人导航技术
Prerequisites：Computer Programming, Electronic Technology and System, Modern Control Engineering, Robot Navigation Technology

学分：1
Credits：1

11.4.1 课程目的和基本内容（Course Objectives and Basic Content）

本课程是人工智能学院本科专业综合性实验必修课。
This course is a compulsory comprehensive experimental course for undergraduates in College of Artificial Intelligence.

创新设计思维是一种以人为本、问题解决为导向的方法，一般用于解决复杂问题和工程创新。本课程综合运用计算机程序设计、电子技术与系统、现代控制工程、机器人导航技术、高级自动驾驶技术与系统等多门课程中获得的知识和能力，依托智能移动机器人教具平台，通过对一个具体智能系统完成从设计到搭建的全过程，锻炼和培养学生的创新设计思维能力。创新设计的过程一般包括"共情""定义""构想""原型""验证"五个阶段，本实验内容将上述五个阶段在一个自主智能系统的设计过程中具象化。通过本实验，加深学生对自主智能系统特别是无人驾驶技术的全面了解，掌握智能移动机器人的基本原理和算法功能，培养学生的动手实践能力，特别是解决复杂问题时综合应用所学知识进行创新设计的能力。课程的教学重点在于各课程知识点的衔接、团队合作、创造力的培养以及动手搭建、调试真实物理平台的工程实践能力。

Innovation and design thinking is a human-oriented problem-solving method, which is used to solve complex problems in engineering innovation. This course comprehensively utilizes the knowledge and ability acquired from courses including computer programming, electronic technology and systems, modern control engineering, robot navigation technology, autonomous driving platform etc. Based on the intelligent mobile robot teaching platform, this course requires students to complete the whole process from design to construction of this system, so as to cultivate students' innovation and design thinking abilities. The process of innovative design generally includes five phases: "empathy"-"definition"-"ideate"-"prototype"-"verification". This course concertize them in a real design process of an autonomous intelligent system. Through this practice, students can deepen their comprehensive understanding of autonomous intelligent systems, especially autonomous driving technology, master the basic principles and algorithms of intelligent mobile robots, and cultivate their hands-on skills, especially the ability to comprehensively apply the knowledge learned to solve complex problems in innovative design. The teaching focus includes the connection of knowledge points from various courses, team collaboration, cultivation of creativity, and the engineering practical ability to build and debug real physical platforms.

11.4.2 课程基本情况(Course Arrangements)

课程名称	创新设计思维 Innovation and Design Thinking								
开课时间	一年级		二年级		三年级		四年级		专业综合性实验
^	秋	春	秋	春	秋	春	秋	春	必修 (学分): 脑信号处理(1) / 游戏 AI 设计与开发(1) / 机器人导航技术(1) / 创新设计思维(1) 选修(学分): /
课程定位	本科生"专业综合性实验"课程群必修课								
学 分	1 学分								
学 时	32 学时								
先修课程	计算机程序设计、电子技术与系统、现代控制工程、机器人导航技术								
后续课程									
考核方式	平时成绩占 20%,实验成绩占 80%								
参考教材									
参考资料	1. 李力,王飞跃.智能汽车:先进传感与控制[M].北京:机械工业出版社,2016. 2. 胡春旭.ROS 机器人开发实践[M].北京:机械工业出版社,2018.								
其他信息									

11.4.3 实验目的和基本要求(Experiment Objectives and Basic Requirements)

(1) 了解和学习无人驾驶技术的意义,确定开发场景及需求;

(2) 学习无人驾驶技术相关各功能模块的原理,完成具体场景到智能车计算模型的问题定义;

(3) 利用头脑风暴、思维导图、工具软件等进行智能小车的软硬件设计;

(4) 充分利用各类工具完成各个功能模块的系统集成，形成完整的软硬件一体化智能小车原型；

(5) 进行原型测试，验证解决方案是否有效，对原型进行迭代和改进。

11.4.4　实验内容及安排（Experiment Syllabus and Arrangements）

序号 Num.	实验内容 Experiment Content	课时 Class Hour	知　识　点 Key Points
1	共情阶段 Phase of empathize	4	（1）了解发展无人驾驶的意义 （2）了解国内外无人驾驶技术的发展历程 （3）学习智能车基本技术与算法原理，了解该技术的发展前沿、难点及挑战 （4）选择一个具体的场景，如静态障碍物规避、自主泊车、交叉路口避让行驶等，明确该场景下的智能车系统设计开发需求 (1) Understanding the significance of developing autonomous driving (2) Understand the domestic and international development history of autonomous driving technology (3) Learn the basic technology and algorithm principles of intelligent vehicles, and understand the forefront, difficulties, and challenges of this technology (4) Select a specific scenario, such as static obstacle avoidance, autonomous parking, avoidance on intersection etc., clarify the design requirements of the intelligent vehicle system in that scenario

续表

序号 Num.	实验内容 Experiment Content	课时 Class Hour	知 识 点 Key Points
2	定义问题 Definition of problem	4	(1) 无人驾驶相关传感器的原理：激光雷达、RGB-D 相机、惯性测量单元等 (2) 阿克曼轮、差速轮、麦克纳姆轮等车辆底盘的动力学原理 (3) WiFi、蓝牙、UWB 等室内 AGV 常用无线通信方式的特点 (4) 完成所选择具体场景到智能车计算模型的问题定义，将行驶场景需求具体化为适当传感器与感知、通信、路径规划模块的软硬件组合 (1) Principle of unmanned driving related sensors: laser radar, RGB-D camera, IMU etc. (2) Dynamics principle of various chassis: Ackerman wheel, differential wheel, and Mecanum wheel etc. (3) Characteristics of common indoor AGV communication ways: WiFi, Bluetooth, UWB etc. (4) Complete the problem definition from the selected scenario to the computational model of intelligent vehicle, and concretize it into appropriate software and hardware combination of sensors, perception, communication, and path planning modules
3	构想阶段 Form ideas	8	(1) 根据算力需求选择合适的嵌入式计算平台 (2) 根据所选择的传感器，完成供电及数据传输方案设计 (3) 选型适当的车辆底盘 (4) 设计基于 ROS 环境的软件架构 (5) 运用 3D 设计软件，进行车辆外观设计及结构设计 (1) Choose appropriate embedded computing platform according to the requirements of computational power (2) Complete the designs of power supply and data transmission for the selected sensors (3) Select appropriate type of vehicle chassis (4) Design the software architecture based on ROS environment (5) Complete the appearance design and structural design for vehicle using 3D design software

续表

序号 Num.	实验内容 Experiment Content	课时 Class Hour	知识点 Key Points
4	制作原型 Produce the prototype	8	(1) 使用金工工具完成整车车架的结构安装 (2) 完成计算单元、传感单元、执行单元等的安装，使用电工工具完成各单元模块的电路调试 (3) 使用 Python 及 C++编程实现预期的软件功能 (4) 使用 3D 打印设备完成车辆外壳的制作 (1) Complete the installation of entire vehicle frame using metalworking tools (2) Complete the installation of the computing units, sensing units, execution units etc., and debug the circuit of each unit using electrical tools (3) Implement the expected software functions using Python and C++ programming (4) Produce the vehicle shell using 3D printing equipment
5	验证阶段 Phase of verification	8	(1) 验证车辆底盘及控制模块进行直线、弧线、S弯循迹行驶的稳定性及精确性 (2) 验证建图、定位、规划等基本自主导航功能 (3) 验证多传感器融合定位与感知、多智能体网联、"端-边"结合的深度学习模型部署等高级自动驾驶功能 (4) 根据上述验证效果，在"构想"-"原型"-"验证"三个阶段进行反复的迭代和调整，直至找到最合适的解决方案 (1) Verify the stability and accuracy of the vehicle chassis and control module on straight, curved, and S-shape tracking tasks (2) Verify basic autonomous navigation functions such as mapping, positioning, planning etc. (3) Verify the advanced autonomous driving abilities such as positioning and perception via multi-sensor fusion, multi-agent networking, deep learning via "end-edge" computation etc. (4) Based on the above verification effects, iterate and adjust repeatedly in the three phases of "ideate", "prototype", and "verification" until the most suitable solution is found

大纲指导者：郑南宁教授（西安交通大学人工智能学院）

大纲制定者：刘剑毅副教授（西安交通大学人工智能学院）

大纲审定：西安交通大学人工智能学院本科专业知识体系建设与课程设置第二版修订工作组

后 记

当前,人工智能正在以惊人的速度向纵深和更高级的方向发展,人类社会的几乎所有领域对人工智能技术都有着越来越迫切的需求。因此,掌握了人工智能的基本理论和方法,就能使我们以熟悉的方法与工具进入到生疏的研究与应用领域,发现新的知识和自然规律。

虽然我们结合自身的实践,完成了人工智能专业知识体系构建和课程设置,但还需要在实践中进一步完善。随着人工智能的发展,其知识体系与课程内容、人才培养方式与模式也应与时俱进。特附上主编于2019年3月在《西安交通大学学科前瞻三十年》上发表的文章《深化学科交叉,发展人工智能》,抛砖引玉,以飨读者。

附:

深化学科交叉　发展人工智能

2019年3月7日

人工智能是人类历史上最重要的一个演变。过去40亿年当中,所有的生命完全按照有机化学的规则演化,但人工智能的出现和发展使这一规则发生了变化,即生命可以在某种程度上根据计算机智能设计,人类社会将迎来以有机化学规律演化的生命和无机智慧性的生命形式并存的时代。当前,人工智能已成为引领新一轮科技革命和产业变革的战略性技术。以此为契机的人工智能及相关技术的发展和应用对于整个人类的生活、社会、经济和政治都正在产生重大而深远的革命性影响,人工智能已成为国家综合实力与发展的核心竞争力的重要体现。

一、人工智能是新一轮科技革命和产业革命的引擎

人工智能是以机器为载体,模拟、延伸和扩展人类或其他生物的智能,使机器能胜任一些通常需要人类智能才能完成的复杂工作。人工智能的萌芽可以追溯到2300多年前亚里士多德提出的逻辑三段论和形而上学的思想。逻辑打开了人工智能的可能性,亚里士多德提出的三段论使逻辑走向形式化的发展,后人在此基础上不断完善和发展,使逻辑学取得了极大的进步。无论未来人工智能发展到何种水平,逻辑学这门基础科学在其中的重要作用都无法忽视。1956年在美国达特茅斯学院举行了为期两个月的关于"如何用机器模拟人的智能"的夏季研讨会,第一次正式采用"人工智能"(Artificial Intelligence,AI)术语,标志着人工智能正式成为一门新兴的交叉学科。人工智能具有多学科综合、高度复杂的特征,渗透力和支撑性强等特点。

近年来，布局发展人工智能已经成为世界许多国家的共识与行动。以习近平同志为核心的党中央高度重视人工智能发展。习近平总书记多次就人工智能做出重要批示，指出人工智能技术的发展将深刻改变人类社会生活、改变世界，要求抓住机遇，在这一高技术领域抢占先机，加快部署和实施。习近平总书记特别强调"人工智能是新一轮科技革命和产业变革的重要驱动力量，加快发展新一代人工智能是事关我国能否抓住新一轮科技革命和产业变革机遇的战略问题"；2017年7月，国务院正式发布《新一代人工智能发展规划》，将我国人工智能技术与产业的发展上升为国家重大发展战略。《新一代人工智能发展规划》要求"牢牢把握人工智能发展的重大历史机遇，带动国家竞争力整体跃升和跨越式发展"，提出要"开展跨学科探索性研究"，并强调"完善人工智能领域学科布局，设立人工智能专业，推动人工智能领域一级学科建设"。2018年4月，为贯彻落实国家《新一代人工智能发展规划》，教育部印发了《高等学校人工智能创新行动计划》，强调了"优化高校人工智能领域科技创新体系，完善人工智能领域人才培养体系"的重点任务。

同时，美英法加日和欧盟等主要发达国家和经济体也都相继制定了人工智能的重大发展战略，不断加大对人工智能发展的国家引导力度。此外，为应对解决计算普及和人工智能崛起带来的全球机遇和挑战，世界人工智能教育和科研的佼佼者——美国麻省理工学院（MIT）于2018年10月宣布投资10亿美元加强人工智能与其他相关学科的交叉融合和发展，实施60多年来最重大的所有学科结构的变革，以计算和人工智能重塑MIT。

二、人工智能面临的三大挑战

2016年，围棋软件"阿尔法围棋"战胜围棋世界冠军李世石，让人们惊叹人工智能发展取得的成就。这是否意味着机器即将获得类人智能呢？现在得出这样的结论还为时过早。

发展新一代人工智能将面临以下三大挑战：

1. 让机器在没有人类教师的帮助下学习。人类的很多学习是隐性学习，即根据以前学到的知识进行逻辑推理以掌握新的知识。然而，目前的计算机并没有这种能力。迄今为止最成功的机器学习方式被称为"监督式学习"，需要人类在很大程度上参与机器的学习过程。要达到人类水平的智能，机器需要具备在没有人类过多监督和指令的情况下进行学习的能力，或在少量样本的基础上完成学习，即机器无须在每次输入新数据或者测试算法时都从头开始学习。

2. 让机器像人类一样感知和理解世界。触觉、视觉和听觉是动物物种生存所必需的能力，感知能力是智能的重要组成部分。在对自然界的感知和理解方面，人类无疑是所有生物中的佼佼者。如果能让机器像人类一样感知和理解世界，就能解决人工智能研究长期面临的规划和推理方面的问题。虽然我们已经拥有非常出色的数据收集和算法研发能力,利用机器对收集的数据进行推理已不是开发先进人工智能的障碍,但这种推理能力建立在数据的基础上,也就是说机器与感知真实世界仍有相当大的差距。如果能让机器像人类那样进一步感知真实世界,它们的表现也许会更出色。要达到人类水平的智能,机器需要具备对自然界的丰富表征和理解能力,这是一个大问题。尽管围棋很复杂,让计算机在棋盘上识别最有利的落子位置也很难,但与精确地表征自然界相比,描述围棋对弈的状态显然要简单得多,两者之间的差距还要几十年甚至更长时间才能弥合。

3. 使机器具有自我意识、情感以及反思自身处境与行为的能力。这是实现类人智能最艰难的挑战。具有自我意识以及反思自身处境与行为的能力是人类区别于其他生物最重要、最根本的一点。另外,人类的大脑皮层能力是有限的,如果将智能机器设备与人类大脑相连接,不仅会增强人类的能力,而且会使机器产生灵感。让机器具有自我意识、情感和反思能力,无论对科学和哲学来说都是一个引人入胜的探索领域。

三、西安交通大学模式识别与智能系统学科的发展

虽然在我国现有的学科体系中,尚未设立人工智能的一级学科,但在 20 世纪 80 年代,在控制科学与工程一级学科内就已设置了"模式识别与智能系统"二级学科,在当时所有自然科学门类中这是唯一与人工智能相关的学科。

1986 年春天,我从国外留学归来不久,在我国模式识别领域著名学者、西迁教师宣国荣教授的带领下,我们在自动控制专业计算机控制教研室的基础上组建了西安交通大学在人工智能领域第一个专职科研机构——"人工智能与机器人研究所"(简称人机所)。当时学校在论证研究所设置的会议上,建议的名称为"机器人研究所",我们坚持在研究所名称中加上"人工智能",并给出英文名称为"Institute of Artificial Intelligence and Robotics"(简称 IAIR)。从今天的人工智能发展来看,当时的学术判断和坚持是具有前瞻性的。在我们研究所成立至今的 33 年历程中,世界范围内人工智能的发展曾经历过寒冬,但我们始终坚持人工智能,特别是计算机视觉与模式识别的应用基础理论研究,并与国家重大需求相结合,没有放弃当初建所时的学术目标和追求,培养了一大批优秀人才,取得了一批丰硕的科研成果,为西安交通大学在人工智能领域奠定了坚实的基础。今天,人工智能与机器人研究所已成为在国内外学术界乃至工业界具有重要影响的研究机构,它是"模式识别与智能系统"国家重点二级学科、"视觉信息处理与应用国家工程实验室"及"高等学校学科创新引智基地"等国家级科研平台的支撑单位。特别是近十余年来,我们围绕人工智能前沿基础理论及其在国家航天重大工程、无人驾驶、医学图像处理、视觉大数据智能化处理及其芯片等领域的人才培养和科学研究,取得了一系列在国内外具有重要影响力的突出成就,建立了一支能力突出、结

构合理的具有一流水平的教学和科研团队,形成了独特的育人文化和制度,培养了国际人工智能领域45岁以下顶尖科学家、前微软亚洲研究院首席研究员、现任旷视科技首席科学家孙剑博士为代表的一批学术界和产业界的领军人才。

人工智能不仅是科技发展竞争的焦点,更是大学发展和学科建设的新机遇。在当前的人工智能浪潮中,人工智能技术在高等教育、人才培养和各个学科的应用与发展也必将重塑国内外一流大学的格局和地位。

四、人工智能与机器人研究所下一个三十年的学术目标

人工智能与机器人研究所下一个三十年的发展将立足国家发展全局,聚焦人工智能重大科学前沿问题和应用基础理论瓶颈,重视面向国家重大需求的研究和应用,加强多学科的深度交叉融合,并重点围绕如何设计更加健壮的人工智能、人机协同的混合增强智能,以及人工智能技术的核心芯片与新型计算架构开展系统性的研究,并在新的发展时期进一步做好西迁精神和人机所团队文化的传承,做强做大西安交通大学人工智能学科,为我国人工智能科技水平跻身世界前列,为加快建设创新型国家和世界科技强国做出更大的贡献。

未来三十年的主要学术目标:

1. 设计更加健壮的人工智能。尽管当前深度神经网络在诸多领域获得了成功的应用,但其泛化能力差、过度依赖训练数据、缺乏推理和对因果关系的表达能力等缺陷也被广为诟病。经典人工智能的形式化方法不可能为所有对象建立模型,不可能枚举出一个行为的所有隐性结果,"未知的未知"问题对构建稳健的人工智能系统提出挑战。谷歌流感预测的失败证实数据并非越大越好,一个鲁棒的人工智能系统必须在一个非完备的世界模型下正常运行。而人类大脑不是通过一个统一的未分化的神经网络实现单一的全局优化原理来学习,而是具有独特且相互作用的子系统支持认知功能,如记忆、注意、语言和认知控制。研究大脑网络的聚合和敛散性可以洞察大脑的认知机理,类脑神经计算的潜力在于能够将直觉与经验和以数据为基础的演绎归纳相结合,从而能够在不完整的世界描述中产生正确的行为。因此,需要从脑认知机理和神经科学获得启发,发展新的人工智能计算模型与架构,让机器具备对物理世界最基本的感知与反应,使机器具有"常识"推理的能力,它能快速思考、推理和学习,能够像人一样凭直觉了解真实世界,从而实现更加健壮的人工智能系统。

2. 实现人机协同的混合增强智能。人类智能与机器智能的协同在人工智能发展中是贯穿始终的,任何智能程度的机器都无法完全取代人类,将人的作用或认知模型引入人工智能系统,形成混合增强智能形态,是人工智能可行的、重要的成长模式。人工智能具有标准化、重复性和逻辑性的特点,擅长处理离散任务,而不是自身发现或打破规则;人类智能则具有创造性、复杂性和动态性的特点,两者优势高度互补。人在回路的混合增强智能是新一代人工智能的典型特征,通过人机交互、人机协作逐步提高机器的自主学习和自适应能力,并逐步发展到人机融合。脑机协作的人机智能共生是耦合程度最高的混合增强智能方式,采用脑机交互有望实现人与机器在神经信息连接基础上的智能融合增强。该领域取得新突破的关键在于脑功能建模、脑机接口以及全脑模拟等方面的探索。云机器人可能是人机混合增强智能研究转换为应用最快的领域之一,通过云计算强大的运算和存储能力,给机器人提供一个更智能的"大脑",构成"1+1>2"的人机协同混合智能系统。

3. 探索新型计算架构及其核心芯片。人工智能的发展需要突破硬件平台和处理器设计架构等基础设施建设的掣肘,人工智能技术的核心芯片已经成为国内外产业界高度关注的创新领域。随着摩尔定律的失效,通过减小工艺尺寸改善硬件计算效能遇到了瓶颈。灵活的可重构计算架构已引起学术界和工业界的广泛关注,被认为是能够同时达到高灵活性和高能效的计算架构设计技术;神经形态计算研究力图在基本架构上模仿人脑的工作原理,使用神经元和突触的方式替代冯·诺依曼架构体系,使芯片能够进行异步、并行、低速和分布式处理信息数据,并具备自主感知、识别和学习的能力。因此,实现计算、存储和通信高效协作的混合计算架构在新一代人工智能发展战略中起着核心的平台支撑作用。我们将继续深入研究内存与计算融合的新型存储设备、神经网络功能连接的实现机制、认知计算框架等基础科研问题,并积极探索结合冯·诺依曼计算架构和生物智能计算特征的混合计算架构和新型人工智能芯片设计技术。

五、人工智能的基本方法和哲学思考

人工智能的多学科交叉属性需要我们把来自不同学科的具有创新思维的科学家、工程师聚集在一起,对新一代人工智能的基本科学问题及实现进行深入研究,要准确地把握问题的所在,并能给出合适的方法和数学工具,这样才能为未来的研究铺平道路。

同时,我们要清楚地认识到,一些人工智能发展的重大问题,在现时很难纳入已有的或成熟的理论框架之中,因此一些新的研究方向是不确定的,但一个重要的基本途径是:从脑认知和神经科学寻找发展新一代人工智能的灵感,推动人工智能的学科交叉研究已成为必然的趋势。

在推动新一代人工智能发展的过程中,还需要有科学的哲学思考。在每一个看似极其复杂、而难以用已有方法解决的人工智能重大问题的背后,总是存在一种简化的基本原理,找到这种基本原理,就能使我们深刻理解问题的本质及其产生的规律。例如对于人工智能领域一大类具有不确定的复杂性问题,往往具有约束条件和先验知识,其机理并非都是杂乱无章的,揭示这类复杂性的机理,实现机器理解的计算模型,就可以找到不确定问题求解的方法。

由于人工智能模糊了物理现实、数据和个人的界限,衍生出复杂的伦理、法律和安全问题。随着人工智能的逐渐普及,如何应对人工智能所带来的深刻的社会问题已成为全球性的问题。人类社会需要审慎管理人工智能来应对这一转变。在这一方面,人文社会学科领域和哲学学科将大有作为。

六、深化人工智能应用,助力新一代人工智能发展

人工智能已给人们的生产、生活方式带来革命性变化,未来的世界科技强国也一定是人工智能强国,中国要成为世界科技创新强国,发展人工智能已成为这一伟大事业的重要基础。当前,我们要充分利用和发挥互联网大国的优势,把我国数据和用户的优势资源转换为人工智能技术发展的优势,深化人工智能技术的推广应用,做强做大人工智能产业。

人工智能是人类最伟大的梦想之一,将是未来三十年对人类发展影响最大的技术革命。"前事不忘,后事之师",人工智能成为一门独立学科已走过六十三年的历程,也经历了两次高潮和低谷,20世纪人工智能领域在实现其"宏伟目标"上的完全失败,曾导致人工智能研究进入"冬天"。

在当前人工智能发展新一轮的热潮中,我们要保持清醒的认识,进一步加强信息科学、认知科学、脑科学、神经科学、数学、心理学、人文社科与哲学等学科的深度交叉融合,踏踏实实地开展人工智能的基础研究,避免不切实际的预言和承诺,而使研究"落入一张日益浮夸的网"中;另一方面,我们必须重视人工智能面向重大应用工程的研究和市场的创新开拓,但同时要避免在产品研发和市场推广中的"低水平、同质化"现象。

西安交通大学人工智能与机器人研究所不会满足于过去的辉煌,未来我们将继续促进与其他学科的深度交叉融合,推动人工智能技术新的应用,催生新的学科生长点,助力中国新一代人工智能事业的发展。

(郑南宁,中国工程院院士,西安交通大学人工智能与机器人研究所教授)